COMMONALITIES IN ART, SCIENCE AND RELIGION

To Caroline Isard

Commonalities in Art, Science and Religion

An Evolutionary Perspective

WALTER ISARD
Emeritus Professor of Economics
Cornell University
and
Adjunct Professor of Regional Science in Economics
University of Pennsylvania

Avebury
Aldershot • Brookfield USA • Hong Kong • Singapore • Sydney

Published by
Avebury
Ashgate Publishing Limited
Gower House
Croft Road
Aldershot
Hants GU11 3HR
England

Ashgate Publishing Company
Old Post Road
Brookfield
Vermont 05036
USA

British Library Cataloguing in Publication Data

Isard, Walter
Commonalities in art, science and religion : an evolutionary perspective
1. Learning and scholarship 2. Science and the arts 3. Arts and religion 4. Religion and science
I. Title
001

ISBN 1 85972 475 2

Library of Congress Catalog Card Number: 96-86353

Printed in Great Britain by Ipswich Book Co. Ltd., Ipswich, Suffolk.

Contents

Figures and tables

Acknowledgements

In writing this book I am deeply indebted to Helena Wood for her patience and excellent secretarial and art work in the preparation of materials. In Chapter 7A I have drawn upon joint research with José Lobo and very heavily upon that with Panagis Liossatos. Once again I profited greatly from the assistance of the user-friendly Reference Department of Cornell University under Caroline Spicer, its head. Very helpful comments and suggestions were at times provided by Elizabeth Delehanty and Professors Lloyd Rodwin and Peter McClelland.

Needless to say, I am most indebted to my wife, Caroline, who now for over a half century has provided me with constant support, encouragement and assistance in writing books.

Permission to reproduce figure 2.5 on page 37, figure 9.1 on page 156 and the table on page 213 in *Cosmogenesis: The Growth of Order in the Universe* by David Layzer has been granted by Oxford University Press; to reproduce the figure on page 17 in *Chaos* by James Gleick (1987) has been granted by Michael V. Carlisle, Agent; to reproduce figures 3, 4, 5 and 6 on pages 74, 75, 76, and 77, respectively, in 'A Two Dimensional Mapping with a Strange Attractor' by M. Hénon in *Communications in Mathematical Physics*, Vol. 50 has been granted by Springer-Verlag GmbH and Co. KG and by the author; to reproduce figures 14 and 17 on pages 275 and 278 in 'A Rayleigh-Bénard Experiment' in *Nonlinear Phenomena at Phase Transitions and Instabilities* by T. Riste (ed.) has been granted by Plenum Press and the authors; to reproduce figure 6.4 in this book has been granted by John Hubbard; to reproduce figure on page 41 in *Chaotic and Fractal Dynamics* by F.C. Moon (1992) has been granted by John Wiley and Sons, Inc.; to reproduce figures 1, 2, and 3 in *Exploring Complexity* by G. Nicolis and I. Prigogine (1989) has been granted by W. H. Freeman and Company; to reproduce figure 16 in *Order Out of Chaos* by Ilya Prigogine and Isabelle Stengers has been granted by Bantam Doubleday Dell Publishing Group; to reproduce figures 6.2, page 125 and figure 7.14, page 187 in *Noise-Induced Transitions* by W. Horsthemke and R. Lefever (1984) has been granted by Springer-Verlag New York, Inc. and by the authors.

Preface

This book is a collection of ideas intermixed with scientific findings, reasonable hypotheses and theories, both logical and somewhat unconventional, and finally impressions. They all relate to commonalities of art, science and religion, and in part to their analytic synthesis and implications for unity of thought. In this book I treat art as an area that covers painting, sculpture, architecture and music. I take science to cover all the physical, biological and social sciences. I view religion broadly; however, because of the limits to the amount of information and knowledge that a single mind can handle, I have confined myself to the widespread Western religions with only scant reference to others. Also, for the same reason I do not cover literature and related humanities.

When I speak of commonalities, I have in mind such phenomena as symmetry breaking, and hierarchic construction and disintegration, which seem to make their appearance everywhere I look—this the reader will note as he/she wades through this book.

When I speak of implications for the unity of thought in art, science and religion, I think back to my teenage fascination with Einstein's effort to develop a unified field theory and the subsequent work on a Grand Unified Theory by particle physicists. By analogy, this effort led my mind to ask the question: Can the principles, relationships and processes in different fields of study be synthesized or brought together jointly to influence or explain a phenomenon? Specifically, what I have in mind when I talk of the unity of thought in art, science and religion is the development of a greater, hopefully much greater understanding and recognition of the interdependence and interconnections of these areas of study *centered around commonalities*—an understanding and recognition that allows a mind to comprehend, absorb and fuse a wider range of knowledge and analytical procedures.

Realistically, an effort at such unity must fall far short of that goal. Because of the current inadequacy of the state of knowledge in all fields—the many unknown variables and relationships that are yet to be identified and the many we will undoubtedly fail to identify—there is no hope for a complete unity, and certainly not in the near and even far future given the rapid pace of technological and cultural (socio-politico-economic) development. All one can hope for is a partial unity. However, significant commonalities can be identified which can lead to the enhancement of a number of areas of study simply from transfer and incorporation of useful concepts, principles and tested relationships found in others. Herein is where one of the contributions, if any, of this book must lie.

In the presentation of materials for this work several approaches could have been followed. Perhaps the most logical would have been to consider evolution over cosmological time. Here one could characterize this process as comprising three phases. Phase I would be the purely physical phase, starting with the widely accepted view of a Big Bang and followed by the appearance of all forms of matter and energy governed by the laws of physics. Phase II would see the emergence of the biological system within the physical system, starting with the appearance of the first replicator and covering all forms of life thereafter and involving feedbacks between the physical and biological systems and constraints that each system imposes on the other. Phase III would begin with the development of a brain in living organisms, in particular with the appearance of reflection and consciousness, and which for purposes of this book, may be viewed as a first step in the emergence of a human social system.

Our world today is the outgrowth of the interrelations (inclusive of feedbacks) among the physical, biological and social systems with each providing constraints on the operation of the others. If we had followed the above approach we would have started with most of the materials covered in chapters 5 (Physical and Cosmological Processes, and Hierarchical Structure), and 6 (Order Within Seeming Disorder, Chaos and System Transition) and then moved on to Chapter 3 (Biological Evolution Theory Up to the Present Time), and finally onto chapters treating elements of the social system, and the interconnection of the physical, biological and social systems.

Initially, I had planned to pursue this three phase approach. However, I came to do otherwise. I realized that if I were to follow this approach, I would be discussing culture and religion in the last part of a fairly long manuscript. Knowing my own shortcomings, in particular that my interest in the manuscript might dwindle after having spent a number of years in its development, I realized that I probably would not give appropriate time and attention to culture and religion, especially since my training has been in the sciences. To avoid this potential shortcoming I have chosen to begin with a brief chapter concerned primarily with the deep religious thinking of a notable scientific figure, then to move on to evolution from a biologist's standpoint, and proceed thereafter in ways that seem to be most effective in connecting the various materials to be presented.

Also, there is undoubtedly much disciplinary bias in the concepts and other materials I develop—a bias that reflects my training as an economist whose research has moved away from historical into much mathematical analysis. I fall back upon the use of concepts from economics, regional science, physics and mathematics more often than those from other disciplines—especially since I am more at ease in writing about the former and too often unaware of the strengths and relevance of the latter and their ways of thinking. This bias clearly appears in my inclination to think in terms of concepts and materials relevant for the operation of comprehensive and integrated global models.

I trust this book makes some contribution to our stock of knowledge—to our understanding of the interconnections between the physical, biological and social systems and the implications of these ties for social issues confronting society, in particular the co-play of science and religion. On this issue and others, natural scientists have not treated, as thoroughly as I believe they should, the social, economic, political and related factors revealed by modern developments in the social sciences and their rigorous analyses. Nor have great religious figures who also have acquired sound and extensive knowledge of the natural and biological sciences, as we explicitly note in Chapter 2 with regard to Teilhard's thinking. It is by introducing more adequately and at a deeper level the play of social forces with others, as revealed in current rigorous social science analysis, that I trust this book constitutes a step forward, though at best endowing us with a stock of knowledge still inadequate for a full understanding of the great social issues.

In this book the synthesis of basic concepts and ideas does not take place until the later chapters. Since I have not followed a logical approach and since much technical material and unfamiliar terminology will be encountered prior to the synthesis, I recommend that most readers not attempt to read carefully the chapters in the order presented. To do so might lead them to question where the book is going and to put aside the book after reading one hundred pages or so. I therefore make the following recommendations for readers with different backgrounds.

For each scholar, the overview in Chapter 1 should be read and perhaps a skimming through of Chapter 2 on Teilhard's thinking wherein the terminology may be unfamiliar and statements too broad. Then one of several paths may be taken. Those with background in the *physical sciences and engineering* may proceed to Chapters 5 and 6 (skipping parts of these chapters which are uninteresting or on which they are fully informed), and then to Chapter 3 reading it in a like manner (possibly omitting section 3.7). From there on, they may go to: (a) Chapter 10, reading sections 10.1-10.3 and the early part of section 10.4, and, if so inclined, continue to finish the chapter; or (b) section 9.9 (and if particularly interested in the history of religious development, to the earlier parts of Chapter 9); or (c) to section 8.9 (and, if particularly interested in the evolution of painting, to the earlier parts of Chapter 8). Always, they should go back to parts of earlier chapters which discuss concepts with which they are less familiar and on which they need further knowledge, for example the diverse types of economies discussed in section 7.5 and the deterministic and linguistic modes discussed in section 4.2. Then they may desire to finish the chapters of which they have already read parts, especially 4 and 7. If they have the mathematical competence and interest, they may delve into Chapter 7A. Finally, they may selectively read other chapters and their parts and end up with the concluding Chapter 11.

After skimming Chapter 2 those interested in *history and cultural development* may read Chapter 3 (possibly omitting section 3.7), and sections 5.1 to 5.3. They may then proceed to Chapter 10, delving into section 10.4 after lightly scanning sections 10.2 and 10.3. As concepts are encountered in the reading of section 10.4 which are unfamiliar, they may refer back to relevant sections of the preceding chapters, for example: (a) sections 6.2 and 6.3 when they wish to know about bifurcation, period doubling, strange attractor, fractal structure, far-from-equilibrium states and other concepts of chaos and Prigogine-type theory; (b) sections 4.2 and 4.3 when they desire to know more about deterministic and linguistic modes and brain theory; and (c) section 7.5 when they are inquisitive about concepts like participation potential and the roles of specialization and scale, localization and other economies. Finally, they should read Chapter 9 and desirably Chapter 8, concluding with Chapter 11.

Scholars and intellectuals interested in *theology* and the *evolution of religion* and related matters may do more than just skimming Chapter 2, read Chapters 3 and 5, peruse section 6.3 on Prigogine jumps and far-from-equilibrium transitions, and then plunge into Chapter 9. As each reads section 9.9, he/she should start digesting selectively contents of Chapters 4, 6, 7, and 8. Chapters 10 and 11 should be read in their entirety at the end or earlier.

Following a skim through Chapter 2, those interested in *art,* and especially *painting,* should read Chapters 3 and 5 (including the Appendix to Chapter 5), peruse section 6.3 on Prigogine jumps and far-from-equilibrium analysis, and then plunge into Chapter 8. As concepts are encountered which are unfamiliar, especially in section 8.9, the reader may refer back to relevant sections of the preceding chapters, for example, he/she may need to: (a) go through sections 4.2 and 4.3 (to grasp the significance of deterministic and linguistic modes), (b) review the Appendix to Chapter 5 to understand better continuity in sensation build-up, (c) expose him/herself to the concepts associated with chaos theory (as discussed in section 6.2) to gain deeper insight into modern painting, and (d) plough through Chapter 7 to understand better hierarchical structuring in painting and the interaction of the painter with his diverse audiences and their feedback. Finally, those interested in art should read Chapter 10 for materials on the arts outside the area of painting, at least section 9.9 of Chapter 9 which discusses the coplay of religion and art, and the concluding Chapter 11.

The *biologist,* especially, the *evolutionary biologist,* may skim through Chapter 3 (with whose contents he/she is knowledgeable) after doing so with Chapter 2. Next, he/she may proceed to.Chapters 5 and 6. Then after Chapter 6 he/she may pursue the same path as recommended for the physical scientist, perhaps giving more attention to the contents of Chapter 7.

After skimming through Chapter 2 those interested in *social science, social philosophy* and related areas may first read Chapter 3 or 5, and then the other. Next they may proceed to Chapter 10, lightly scanning sections 10.2 and 10.3, and then delving into section 10.4. Once they have moved along in section 10.4 they may go back to earlier chapters to become informed on relatively unfamiliar concepts, for example to section 6.3 on Prigogine jumps and far-from-equilibrium transitions, to section 7.5 on participation potential analysis and diverse economies, to section 6.2 on concepts associated with chaos theory, and to section 4.2 on dynamic and linguistic modes.

1 Introduction and overview

This book is concerned with: (1) commonalities among disciplines and professional studies; and (2) the existing and potential interconnections between these disciplines and studies and their implications for unity of thought and analysis. To achieve a satisfactory depth in this investigation we limit its scope primarily to art, science and religion. Also, we shall examine these commonalities and interconnections as they relate to the emergence and demise of empires and civilizations.

For reasons set forth in the Preface, we choose to start with a brief presentation in Chapter 2 of concepts useful for considering the role of religion and its interconnections with art and science. (However, as suggested in the Preface, many readers might best skim through this chapter as well as choose one of the paths outlined in the Preface for selectively absorbing the contents of this book.) In Chapter 2 there is set forth the evolutionary perspective of Teilhard de Chardin, who was an outstanding religious figure and at the same time a notable paleontologist. His perspective incorporates scientific findings within a space-time path of personal, interpersonal, moral and ethical development aimed to carry an increasingly organized mankind ultimately to an ideal social welfare state.

Chapter 3 proceeds to present findings and thinking of biologists on evolution which also enrich and bring up-to-date the evolutionary thought of Teilhard. There the accepted (standard) doctrine pertaining to natural selection, gene complexes, altruism and other concepts are discussed as well as the views expressed in the debates on the unit of selection, speciation, the relevance of sociobiology and other issues. In particular we play up the concept of meme (a unit of cultural influence) and hierarchical structure in biological organization.

Chapter 4 proceeds to develop concepts related to the meme, and in particular to the phenomena of reflection and consciousness (so much emphasized by Teilhard), and the functioning of the brain (so extensively addressed today by a number of different fields). Of particular significance is the thinking of Pattee, a general systems theorist who distinguishes between the deterministic mode (which covers the rate-dependent processes that govern so much of the physical world) and the linguistic mode, which could be better designated the cognitive-linguistic-spiritual mode (which covers rate-independent processes as in reflection, consciousness, decision making, use of language and in general those in which mental activity is involved). Appendices to this chapter treat: computational experimentation; cellular automata; concurrent, distributed network models; and adaptive fuzzy systems models that can both furnish insights into the brain's operation and add to our understanding of emergent behavior and creativity.

Chapter 5 begins our study of the physical system. (As remarked in the Preface, a more logical approach in considering evolution over cosmological time, which is fundamental to the discussion on commonalities, would have placed the material of this chapter at the beginning of this book.) The chapter starts our effort to expose commonalities such as symmetry breaking, hierarchic construction and stratified stability in physical, biological and social systems. Beginning with the Big Bang we move to discussion of the formation of diverse particles and the appearance of the basic four forces of physics, then to the emergence of chemical elements and their structural features, and finally to theory regarding the formation of stars, galaxies and their clusters. The Appendix to Chapter 5 presents some very general conceptual materials of the physicist and philosopher Bohm who views reality as undivided wholeness, as an implicate order in which all is interdependent and which becomes unfolded into the explicate order of which we are aware. Bohm's view of consciousness as an unending process of movement and his other notions provide insights into the interplay of the processes in the deterministic and cognitive-linguistic-spiritual modes and enriches the background against which the analysis in later chapters takes place.

The identification of commonalities is further advanced in Chapter 6. The concepts and hypotheses associated with the recent development of chaos theory are examined, such as order within seeming disorder, strange attractors, fractals and turbulence. Also examined are equilibrium, transitions to far-from-equilibrium states, irreversibility, the joint operation of chance and determinism, and other basics of the Prigogine school of thought, particularly as they can help us understand the space-time processes and dynamics among the interconnected physical, biological and social systems.

One set of commonalities that plays a major role in all three systems—physical, biological and social—concerns hierarchies, their construction and disintegration. Chapter 7 presents an extensive discussion of these processes. First, the various types of and perspectives on hierarchies of different disciplines are covered. Emphasis is given to the interactions among and within hierarchies of behaving units, whether replicators or not, with attention to inputs to and feedback from the environment. To help provide a coherent framework for various relevant causal chains emanating from the diverse disciplines, a Pattee heterarchical model is proposed. Further, a participation potential model is set forth to capture more effectively the above interactions. With this model and along with a presentation of our present knowledge of economic and social forces (advantages and disadvantages) leading to hierarchical structure, an attempt is made to complement operational models from the physical sciences with additional explanatory elements and interpretation for further understanding of the dynamics of the social system.

An extensive Appendix to Chapter 7, designated Chapter 7A, is presented to demonstrate the scientific rigor of many of the relationships and processes on space-time development discussed in Chapter 7 as well as make explicit the limits which our current scientific knowledge confronts. It is intended for those having substantial mathematical knowledge.

The next set of chapters deals with the history of selective aspects of the social system drawing upon the analytic background provided by the preceding chapters. Chapter 8, the first of this set, treats the evolutionary development of painting. This development is taken to be at least somewhat representative of the way the arts in general have been affected by the play of forces over the historical record.

We begin by sketching the stages (experiences) of painting from caveman art to that of today—its upsurges and periods of little if any major creativity. We then indicate how the concepts, processes and analytical frameworks of the preceding chapters facilitate our understanding of painting's experience. We do so with regard to the particle physicist's symmetry breaking and field theory, Bronowski's stratified stability and irreversibility, Prigogine's transitions to far-from-equilibrium states and autocatalysis, Dawkins' memes, coadapted gene complexes and the biologist's coevolution, Bohm's doctrine of unfoldment of the implicate order and consciousness as movement, Teilhard's psychic energy and increasing hominisation, Pattee's principle of optimum loss of detail and the heterarchical model as a set of causal chains, chaos theory's strange attractors and determinism, and others such as hierarchical construction and disintegration, participation potential, coadapted meme complexes, self-similarity at different scales, fractal structure, and so forth.

The next chapter, Chapter 9, treats in a like manner the evolution of Mediterranean and Western religions—in particular Judaism, Christianity and Islam. We sketch the historical record from prehistoric man to current ecumenical movements, noting again development stages with emphasis on upsurges and declines. As with painting, we end up indicating how the concepts, processes and analytical frameworks of Chapters 2 to 7 provide invaluable insights into forces governing religion's historical experience, and understanding of many of its specific aspects and their emergence.

Chapter 10 proceeds with this approach in examining the ups and downs of each of the major civilizations and empires in the West—from the end of the Pleistocene Era to the close of the 16th century when the amazingly powerful Roman Catholic empire of the West was effectively challenged by firmly entrenched Protestant movements. We do so, again in terms of the concepts, processes and analytical frameworks covered in Chapters 2 to 7, as partially listed above in the remark on Chapter 8. Also, we develop complementary materials on the evolution of architecture, sculpture and music and pay particular attention to forces of decline.

Chapter 11 brings to a head the thinking and main themes of this book. In providing an overview on commonalities, it discloses a great deal of the unity of thought and analysis in art, science and religion and enhances our understanding of the development of each of these areas and of the ups and downs in empires and civilizations. At the same time, this chapter points up the need for science and religion to have *complementary* roles in an effective unitary society—in essence for the operation of the respective deterministic and cognitive-linguistic-spiritual modes, where the latter embraces creative and emergent decision making (policy formulation) and moral and ethical considerations. We close by examining the implications of our study of commonalities for possible states of the global society in the future (21st century and beyond) and advance an exceedingly highly speculative set of steps for its 'improvement'.

2 The Teilhard scientific-religious framework

2.1 Introduction

As discussed in the Preface we start with a brief chapter that provides a framework for the study of evolution in which religion plays a significant role. We do so in order to insure that religious forces are given proper weight and not brought in at the end of a long book, possibly with inadequate attention. We thus eschew what would appear to some readers the more logical approach that would begin with the physical world and laws governing its behavior subsequent to the Big Bang.

In this chapter we present in succinct form some of the basic thinking of Teilhard de Chardin in his classic work, *The Phenomenon of Man* (1968). The breadth, depth and comprehensive nature of Teilhard's thinking in both the areas of science and religion qualifies it for the presentation of a religious framework useful for the purposes of this book. Other frameworks, of course, could have been followed.

2.2 Basic elements of Teilhard's thinking

Teilhard's conception of evolution leading to the emergence of humankind and his path to an ideal society designated Omega is one of amazing vision. He attempts to unify the realm of science with that of religion, the knowable material and physical world with that of the mind and spirit. Equally, if not more important, his framework is a dynamic one, embracing the process of development on the basis of paleontologic, biologic, and other physical evidence coupled with Darwinian-type analysis and intellectual thought and theology—a process aligned with the emergence of thinking, consciousness and technologic know-how. His rich, evolutionary framework allows him to project speculatively the future directions of economic-socio-political development, a development that is completely fused with his understanding of and deep emotional attachment to religion, and in particular, Christianity. In the latter regard, he goes well beyond the concept of progressive psycho-social evolution developed by Huxley (1968). The result is Teilhard's projection, at an undefined future time, of the ultimate attainment of Omega—a 'Heaven on Earth,' a perfect society *sans* conflict.

Teilhard has organized his *Phenomenon of Man* in four books titled: (1) Before Life Came, (2) Life, (3) Thought, and (4) Survival. As Huxley (1968, p.15) has stated, Teilhard's concept of complexification involves

> the genesis of increasingly elaborate organisation . . . as manifested in the passage from subatomic units to atoms, from atoms to inorganic and later to organic molecules, thence to the first subcellular living units of self-replicating assemblages of molecules, and then to cells, to multicellular individuals, to cephalized metazoa with brains, to primitive man, and now to civilized societies.

The recent work of Dawkins (1989a, 1989b) and other evolutionary theorists, to be summarized in the next chapter, brings up-to-date critical lines of thought relating to biological evolution; and hence we will not discuss Teilhard's studies that cover the period up to the emergence of primitive man. Rather, we will emphasize the points that need to be made explicit in order for us to comprehend Teilhard's humanistic dynamics.

Start with Teilhard's notion of 'consciousness,' a concept that he takes to embrace all forms of psychism, from interior perception at the most rudimentary level imaginable to the human phenomenon of reflective thought. Associated with psychism is a type of energy, namely, psychic energy internal to man, which is distinct from the energy associated with matter which physicists study. Teilhard would say that a major jump in the magnitude of psychic energy was realized when the first replicator came on the scene, or one might claim that psychic energy first appeared at that time. And, following Teilhard's thought, we may say that another major jump in the magnitude of psychic energy occurred when the replicator by a series of mutations evolved into a replicating cell, and still another when cellular combinations were achieved.

In surveying the evolution of diverse species up to the emergence of man, Teilhard claims that with man, consciousness and psychic energy reaches its epitome and at that point there evolves reflection. Teilhard (1968, p. 183) defines reflection as

> the power acquired by a consciousness to turn in upon itself, to take possession of itself *as of an object* endowed with its own particular consistence and value: no longer merely to know oneself; no longer merely to know, but to know that one knows.

With reflection, another world is born, and accordingly:

> Abstraction, logic, reasoned choice and inventions, mathematics, art, calculation of space and time, anxieties and dreams of love—all these activities of *inner life* are nothing other than the effervescence of the newly-formed centre as it explodes onto itself.

In man cephalization has been realized, which as expressed by Huxley (1968, p. 21) involves 'the differentiation of a head as the dominant guiding region of the body, forwardly directed, and containing the main sense-organs providing information about the outer world and also the main organ of coordination or brain'.

In the leap from instinct to thought and reflection as embodied in man, the beginnings of the process of hominisation take place, the consequence of the progressive phyletic spiritualisation in human civilisation of the forces in the animal world—a process which allows man to realize more and more of his potentialities. Thus the era of biogenesis is promoted to an era of psychogenesis. With still further hominisation, namely development of the mind and its higher functioning, there will emerge the era of noogenesis, an era wherein the advanced mind will continue to grow and develop and experience still more involution and interiorization.

Viewed from another standpoint, the realization of psychogenesis has been, and the anticipated noogenesis will be, associated with the *personalization* that Teilhard links to the process of cephalization. The fully developed individual will be not only a more highly individualized individual but one who has crossed the threshold of self-consciousness and has attained conscious integration with the outer world of man and nature. He will be personalized. The set of interthinking personalized individual behaving units will then comprise and be unified within a noosystem, wherein the noosphere has replaced or better yet has been superimposed upon the biosphere.

Teilhard believes that: (1) increase in population numbers; (2) major advances in technology and communication which facilitate the diffusion of knowledge and ideas among humans; and (3) the consequent easing of migration will foster the development of man as a single interbreeding species, precluding the fragmentation (splintering) and like centrifugal tendencies characteristic of so many species. The results of differentiation within mankind will become organized into a unified pattern. Man's growth and spatial diffusion and the consequent cultural spread, all within a bounded, geographically varied earth, will lead along with the process of involution to the increase in man's energetic tension and to a convergent integration. This integration with its increasingly complex and extensive units of organization will involve extensive cultural melding, yet allow room for diversity.

At this point, we must introduce Teilhard's thesis regarding the harmonious interplay of science and spirituality, of objective analysis and the personification of the individual that leads to religious ferment—and thus to the unification of the cult of progress and the passion for the glory of God. As man probes more and more deeply into his 'within' (his soul) with ever greater intensity, as his interdependence allows him to stretch out to reach more and more of the restricted geographic space of the earth, he becomes a creature exuding greater and greater love and embracive of more and more cooperative behavior. The process of Christogenesis grows stronger and stronger. (To Teilhard, of all religions that have existed and continue to exist, the Christianity of the West with its civilization is the only one having ultimate survival capability—to which view we do not need to subscribe, and do not, in the discussion which follows). Mankind comes to approach more and more closely that perfect state of Omega.

Thus we see how Teilhard develops a philosophy (or generalized theory) that reconciles Christian theology with the scientific theory of evolution, pointing up the immense potential for mankind in the future. His approach, he claims, is phenomenological, based on human history, human values and biology and provides a deducible direction (but not a deterministic space-time path) for the realization of a state of ultrahominisation, of hyperpersonal organization, of high psychosocial energy leading to a noetic system of high tension and extensive love, and ultimately and infallibly spiritual perfection and Omega.

2.3 Questions to be asked, problems to be addressed

In their doctrines and analyses many theologians, like Teilhard, maintain the possibility of, if not optimistically forecast, the realization of an Omega or some equivalent perfect (ideal) world. This is so whether or not they relate to Western religions, and whether to Catholicism at one extreme, or to Quakerism at the other.

However, there remain basic questions to be asked, and problems to be addressed. Recognizing that we live in a world subject to the constraints set by physical laws, for example gravity, and physical phenomena like climate and weather, and in general a scarcity of resources no matter what the state of technology, how do we proceed from our current state to a significantly better or an ideal one? In specifying relevant directions, one must also recognize the limits imposed by our biological structure and the operation of natural selection which despite recent advances in genetic engineering introduces competition among behaving units. Moreover, after having randomly given some organisms the ability to learn, natural selection sees that by and large only those who do learn survive. Further, if some do not pass on their knowledge to their offspring they will tend to be eliminated by those who do. But to pass on knowledge, ways must be developed. Demonstration is only one way. Speech, language development, and in time formal education are other ways. Such of course necessitates the development of institutions.

Simultaneously, ways and institutions are required to facilitate cooperation, not only to undertake tasks that one organism alone cannot perform—like moving a boulder or killing a mammoth—but also to counter the inroads of a competing species or other group, to confront environmental hazards and so forth. It becomes necessary to define not only the particular task of each behaving unit, but also how to share the kill, or other gains. Thus institutions and organizations must and have developed, simple at first, but in time complex and hierarchical. What are the principles that ought to govern such development, particularly in any future approach to an Omega? In the writings of Teilhard and other theologians, specific questions of this sort are given little attention. (Of course, in the case of Teilhard, one could not have expected him to be aware of the advanced thinking in the social sciences of today—on the relationships of man to man and of man to man's groups.) If some form of world federation or government is implied by Teilhard and others, what are the criteria that would define such an organization for a significantly better world? To insure an equitable distribution of income, guarantee civil rights and full development of each individual, how must one go about establishing the required set of economic and social institutions? What should be the hierarchical structures of these institutions for the billions of people in an advanced technological society to foster the greater love among individuals that is envisaged—love that would be generated in the daily, weekly, monthly . . . interactions among individuals? Moreover, how do we go from where we are to there? In essence, what principles should govern the space-time path to Omega?

To these questions and problems we turn in succeeding chapters. The next chapter looks for understanding and insights to be gained from evolutionary theory and knowledge of biological processes.

References

Teilhard de Chardin, P. (1968), *The Phenomenon of Man*, New York: Harper.

Huxley, J. (1968), 'Introduction' in Teilhard de Chardin, P., *The Phenomenon of Man*, New York: Harper, pp.11-30.

Wildiers, N. M. (1968), *An Introduction to Teilhard de Chardin*, New York: Harper.

3 Biological evolution theory up to the present time

3.1 Introduction

In Chapter 2 we presented Teilhard's vision, extending from the time when the earth was a 'soup' to an almost infinite (if not infinite) point in time when the state of Omega will be reached. We now examine the thinking on evolution up to the present time. In section 3.2 we present the basic framework drawing heavily on the writings of Dawkins. Sections 3.3, 3.4 and 3.5 discuss, respectively, the nature and role of altruism, the concept of extended phenotype, and memes and the propensity for cooperation, again drawing heavily upon Dawkins' works. Section 3.6 deals with hierarchical structure and biological organization, as Layzer views it; and in section 3.7 we examine sociobiology, speciation, the individual unit and other matters relating to human organization. Section 3.8 presents a summary.

3.2 The basic framework: the selfish gene

We start off with a focus on the selfish gene notion. At some time in the distant past, some three to four thousand million years ago, we may posit the existence of a primeval soup of organic matter of diverse types. By and large these types were stable, the unstable ones by definition having only a transitory existence. Then, at some point, a remarkable replicator appeared, a molecule capable of creating copies of itself, of reproducing. As a consequence, this type of molecule must have spread rapidly throughout the soup, absorbing smaller molecules which may have been required in its construction.

No matter how faithfully the replicating molecules copied themselves, at times, however infrequent, mistakes in copying, in effect mutations, occurred, and we may imagine that some of the error copies were stable and still capable of replicating themselves. Hence, different types of replicating molecules may have come to exist in the soup.

Before the soup became a limited resource, the relative prevalence of the different types of replicators would have been affected by several factors. One would be their longevity. Some types would have been more stable than others and subject to less decomposition and deterioration, and their longevity would also have depended on the longevity of the aggregates or bodies of replicators (survival machines) of which they became a part or within which they were embedded. A second factor would have been their fecundity, or the speed at which they made copies. A third would have been their copy-fidelity, the extent to which they copied themselves accurately and without error. Errors in copying by and large produce a replicator that is unable to survive, or to prevail in a struggle for scarce resources, although at times an error copy may turn out to be a more efficient organism, even much more efficient, than the replicator producing it. This latter phenomenon will be further discussed.

At first, the soup may have seemed to be an inexhaustible resource, an inexhaustible supply of materials for the construction of new copies of replicating molecules. But in time this soup must have become a scarce resource, limiting the extent of the replicating taking place. At that point, competition among the replicators for such resources must have arisen.[1]

The replicators themselves need not have been and probably were not aware of the existence of competition. Nonetheless it was present. Of two (or more) processes at work converting resources into copies of the replicator associated with each, clearly the more efficient process(es) would have prevailed given limited resources, especially if the replicators were contained in survival vehicles that have limited life.

Enter the gene. The gene is a replicator. It is part of a chromosome, which is a long chain of genes. In each cell of the human body there are 46 chromosomes, constituting 23 pairs, one chromosome of each pair issuing from each parent. Modern genes are made of DNA molecules. A DNA is a set of instructions on how to make a body—instructions written in the A (for adenine), T (for thymine), G (for guanine) and C (for cystonine) alphabet of the nucleotides, each nucleotide being a small molecule in the long chain that comprises the DNA.

As a replicator, the gene requires raw materials (resources) as inputs for its reproduction. As already mentioned, these raw materials in the primeval soup could have been smaller molecules, or parts of disintegrating larger molecules, or parts of relatively unstable, short-lived replicators, or parts of replicators that had decomposed as a result of chemical reactions induced by the presence of other replicators. Moreover, some of these other replicators may, as a result of mutation, have developed means to resist chemical breakdown, such as protective walls of protein around themselves, thereby forming a predecessor of the cell. Thus, they would have come to have increased longevity, and *ceteris paribus* would have increased in numbers, relatively speaking. They would have unconsciously become more successful in the struggle for scarce resources.

Still further, a replicator within its protective walls, which may be viewed as a very simple version of a survival machine, may have come to develop into a bigger, and more elaborate and effective survival machine. In time it may have bumped into, or otherwise interacted with another replicator, and joined it to form a larger organism as a still more effective survival machine. And so on, with larger and larger aggregations of replicators emerging as still larger and more effective survival machines. In discovering unawaringly how to break up rival replicators and their survival machines, chemically or otherwise, a combination of replicators embodied in a survival machine may have become the first protocarnivore.

It is unnecessary to spell out further the process of cumulative development of replicator aggregates, cellular structures, and complex survival machines, all of course becoming possible as a result of mutations, rare or not. The rationale for such cumulative development lies in the play of various forces that lead to efficiency. One is the advantage of the specialization of functions that becomes possible. As aggregations of replicators grow in size and survival machines become larger, soon on each replicator or subgroup of replicators within the aggregate can specialize on a particular function and perform that function more effectively than an unspecialized replicator. One subgroup may specialize in detecting potential protocarnivores, or replicators with chemical disintegrating power—another in attacking prey, and so forth. Further, up to a point, there could be economies of scale; larger aggregations might be more effective against attack, or in gobbling up smaller aggregates, or might permit still finer and greater advantages from specialization (economies of scale and other agglomeration economies), especially when functional specialization is centrally coordinated.

Ultimately, we see the emergence of plant and animal species and homo sapiens. We see the advent of the amazingly well integrated physical structures of complex functionally interconnected organs that characterize today's advanced animals and humans in societies within the world system—societies that are coordinated, but too often engaged in physical violence and war.

With this general depiction of organism development in mind, we need to be more specific about its key aspects. First let us establish the gene's selfish mode of operation justifying the designation *selfish gene*. At no time is the DNA molecule conscious of, or involved in planning this development. What happens is that selection among genes takes place. The gene goes on replicating automatically if not constrained by its environment. It does so even though it consumes resources that other replicators or organisms need for survival. It does so even if by mutation it comes to emit a chemical that decomposes another replicator. It does so even if it happens to join other genes to form a survival vehicle where one of the genes or combination of genes operates to cause the survival machine to be carnivorous. The gene is selfish in the sense that it maintains or increases its longevity, fecundity and copy-fidelity, continues to make copies of itself and forms combinations with other genes so that the survival vehicle in which it is embedded prospers relative to other survival vehicles and the genes contained in the latter. It is simply not aware of the natural selection that is going on; and in effect, in continuing to replicate it is selfish.

What is the gene's mode of operation? Biologists speak of *mitosis*, the process of normal cell division where a cell divides into two cells, genetically alike, each containing the same 46 chromosomes. However, there is also *meiosis*, the process which takes place in the testes and ovaries where a cell with the full set of 46 chromosomes (23 from each of the parents) divides to form sex cells (sperms or eggs), each having a single set of 23 chromosomes. This single set can be constructed in a multitude of ways (via crossing over, replacement and elimination), any given one of the 23 chromosomes being formed by pieces, tiny or large, from a pair of chromosomes, one from each parent. The new chromosome in an offspring corresponds in location and function to the pair from which it is derived. When a sperm with its 23 chromosomes fertilizes an egg with its 23 chromosomes, a new cell of 46 chromosomes forms which contains all the necessary instructions for the development of a tremendously complicated and finely honed organism, such as a human being. It is important to note that the 23 chromosomes in the sperm come from the 46 chromosomes of the male's grandparents (each grandparent having contributed 23 chromosomes) and the 23 chromosomes of the egg come from the 46 in the sex cells of the female's grandparents. Hence, the new born individual obtains its 46 chromosomes from the grandparents. None of these chromosomes were devised or manufactured by the parents. Thus the parents can in no way influence the evolution of the organism directly. They can do so only indirectly via gene mutation (as by X-ray), or the gene engineering that imposes constraints on and controls the behavior of the survival machines in which genes reside—matters which we will discuss later. In effect, the process of evolution is bottlenecked. From conception on, genes are in the cell and in the cells later formed by replication; these cells remain unchanged (except from mutation and now, modern engineering) throughout their life. Sperm and egg emerge directly from the unique male cell (derived from grandparents) and female cell (derived from grandparents), respectively. The life cycle of all multi-celled animals and most plants starts and ends with the infinitesimally small bottleneck of a fertilized egg. Should a parent lose a leg, be a brilliant pianist, or develop a path-breaking general theory, such will have no direct effect on the evolutionary process in which he/she serves as a survival machine. It can only affect his/her existence as a survival machine, and thus the perpetuation of the genes he/she carries. In this sense, the gene in the words of Dawkins (1989a, p.44) may be viewed as 'the nearest thing we have to a fundamental independent agent of evolution'. Survival machines such as humans are in effect short-lived gene complexes. They do not have the long life of the average gene by virtue of its ability to replicate, generation after generation, despite the possibility of elimination in the cross-over process during meiosis.

Other important aspects of the development of the organism need to be spelled out. Given that: (1) the first replicator and others resulting from the mutations came to replace the non-replicating organisms in the primeval soup—in effect to form a gene pool; (2) that sex and cross-over has kept this pool well stirred up with genes literally dispersed in all directions; (3) that replicators at times joined up or cooperated and formed survival vehicles which over time became more and more intricate, involving increasing interdependence and coordination among genes, to produce such amazing finely-honed organs as the eye, ear, heart and such an elaborate communication network as the nervous system—given all this, still more must have transpired. For a gene to be embodied in a survival machine it must have been around at the right time and place so that its attributes could effectively complement those of other genes that happened to be there in the gene pool and whose attributes were also required for the possible new survival machine. That is, a condition of mutual compatibility among genes and/or aggregates of genes in the same vicinity or environment had to exist in order for a survival machine to form.[2]

Bear in mind that once gene complexes developed some form of consciousness, not all machines constructed by genes turned out to be survival machines. To survive, an organism must care about survival and struggle to achieve it. For suppose this were not the case. Then, even if at the time of formation the organism was a superior machine, subsequently other machines of equal or greater capability with the desire to survive would have been constructed and would have beat out the former in the struggle for scarce resources. Hence, survival machines that have persisted over time must have cared about survival and must have entered into the competition for scarce resources; accordingly the genes and gene combinations embodied in them would have survived, gone on to replicate and, via the bottlenecked process of a sperm fertilizing an egg, continued to produce another and another of its own kind of survival machine.

3.3 The nature and role of altruism

We have already noted that gene selfishness can lead to individual selfishness. But in reality we also confront individual altruists—an altruist is an entity which behaves to increase the welfare of another entity while decreasing its own. Dawkins, however, claims that individual altruism also stems from gene selfishness. The suicidal behavior of kamikaze bees seems to be inconsistent with ensuring the survival of its genes through reproduction. But if the kamikaze bee is sterile and cannot bear offspring of its own, self-sacrifice perpetuates its genes through increasing the chance of survival of its relatives. Likewise, for other sterile workers among social insects. In fact, in cases where sterile soldiers and normal reproductive adults are genetically identical, as in aphids, and where therefore the environment determines which individuals are soldiers and which reproductive adults, what better way for a soldier to preserve his genes than to save the lives of the reproductive adults.

According to Dawkins, other so-called altruistic behavior, such as bird calling or gazelle stotting can easily be interpreted as individual selfishness. For example, stotting by a gazelle is conspicuous behavior, but it can tell a predator that the high jumping of the gazelle attests to his health and fitness and that a predator preferring an easy catch should look elsewhere, perhaps for an old and/or unhealthy gazelle.

One might consider as altruistic the mutual cooperation among members of different species, but as already indicated in the discussion of advantages of specialization and increased size, such mutual cooperation is easily interpreted as mutual selfishness, and may be said to derive from gene selfishness.

3.4 The concept of the extended phenotype

As already noted, the gene, as a replicator, does not sense the world, does not consciously pursue an objective; it makes machines that do all of these things. The gene and the individual organism (survival machine) in which it is embedded are not competitors; they perform complementary roles. Natural selection smiles on some genes rather than others not because of characteristics that the gene may possess but because of the capability of the survival machine that contains it.

Normally, when one considers the effects of a gene one thinks of effects on just the individual organism, the phenotype, which is the joint product of the genes and their environment during the process of development of the individual organism—the process of ontogeny. One thinks of the color of the eye, the size of the ear, the beat of the heart. But phenotype effects can and do extend beyond the individual organism. They can and do extend to the outside world. Dawkins discusses, for example, how the genes of a certain kind of parasite fluke (the flatworm) are able to influence snail physiology, causing the snail to produce a thicker more protective shell than it otherwise would. This thicker shell makes the replication of the fluke-gene more successful. However, this is at a cost to the snail itself for it uses up resources that otherwise would be spent on producing more offspring, and thus has a negative impact on the passing on of the snail's genes. In this way, the phenotype effects of the fluke's genes extend beyond its own body to the outside world. Dawkins presents other examples of parasite-host relationships where phenotype effects take place beyond the body. Note that a host, say the snail, might over time have developed features that would effectively resist manipulation by a parasite, but this might have involved excessive 'overhead' costs—costs in developing the necessary resistance function.[3]

The doctrine of the extended phenotype embodies phenotype effects of a gene that also extend over distance. For instance, a case of effects 'at a distance', of a long reach of the gene, is that of the parasite-host relationship where a bird (a reed-warbler, for example) seeing the red gape of a cuckoo nestling is diverted from its path and proceeds to drop into the cuckoo's mouth food it is carrying for its own young. Here a cuckoo gene has the phenotype effect of producing the color and shape of the young cuckoo's gape and the extended 'at-a-distance' effect (via chemical or other reaction) of inducing another type of bird to behave as foster parent and provide food for the young cuckoo. Many other phenotype 'at-a-distance' effects can be found. To quote Dawkins (1989a, pp.253,265-6).

> . . . in more modest ways nature teems with animals and plants that manipulate others of the same or different species. In all cases in which natural selection has favored genes for manipulation, it is legitimate to speak of those same genes as having (extended phenotypic) effects on the body of the manipulated organism. It doesn't matter in which body a gene physically sits. The target of its manipulation may be the same body or a different one. Natural selection favours those genes that manipulate the world to ensure their own propagation. This leads to what I have called the Central Theorem of the Extended Phenotype: *An animal's behavior tends to maximize the survival of the genes 'for' that behavior, whether or not those genes happen to be in the body of the particular animal performing it. . .*
>
> With only a little imagination we can see the gene as sitting at the centre of a radiating web of extended phenotypic power. And an object in the world is the centre of a converging web of influences from many genes sitting in many organisms. The long reach of the gene knows no obvious boundaries. The whole world is criss-crossed with causal arrows joining genes to phenotypic effects, far and near.

3.5 Memes and the propensity for cooperation

Up to this point we have largely focused on the selfish gene, its act of replication, its construction in cooperation with other genes of survival machines, and the relatively short-run orientation of these machines, often in combination with other machines in integrated and coordinated super-machines and organisms. We now consider long-run behavior, generally viewed as wholly cooperative but often serving the long-run interests of organisms that are involved and hence selfish from the long-run viewpoint.

Dawkins (1989b, p.290) has coined the term *meme*. He defines it as "a unit of cultural inheritance, hypothesized as analogous to the particulate gene, and as naturally selected by virtue of its 'phenotypic' consequences on its own survival and replication in the cultural environment." As he puts it, by making brains Darwinian evolution produced the 'soup' in which the first memes arose.[4] Brains have the ability to imitate and once the brain stumbled upon (produced) a good idea (a meme)—the use of sound for warning purposes, for example, or the practice of leverage, or knowledge on how to make a fire, or how to construct and use a wheel—the idea spreads rapidly among other brain-endowed machines. Imitation, the leaping of an idea from brain to brain, is analogous to replication, and via imitation and selection good ideas (memes) survive in the meme pool. Poor ideas (memes) are short-lived and eliminated (for example, the 'ether' theory of physics). Longevity, fecundity (diffusion effect, popularity) and copying fidelity are characteristics affecting the survivability of a meme. A fertile idea plays the role of a parasite that invades the brain and converts the brain into a vehicle for the meme's propagation.

Memes, like genes, are subject to mutation. The development of language provides a classic illustration—for instance, the spelling of a word. Changes in bird songs and fashions are others. The number is almost infinite. Moreover, just as coadapted gene-complexes evolve, so do coadapted meme-complexes in which memes are selected for their mutual compatibility—as the set of rituals of a religion, or as elements of advanced technologies, cultural patterns and ultimately whole culture themselves.

Thus, with memes as replicators, and with the elaborate, coordinated and integrated coadapted meme complexes that come to characterize a highly advanced cultural system, the short-run selfish interests of survival machines evolving from selfish genes can be combatted by long-run interests of human societies. When we look beyond the immediate short-run we can devise ways to cope with conflicts, if not fully resolve them, to the benefit of all. We can counteract the selfishness inherent in gene replication and restrict the games that humans and groups of humans play to those which are variable-sum and that guarantee improvement for all parties concerned.

3.6 Hierarchical structure in biological organization

At this point, it is important to recognize the hierarchical structure in biological organization. Layzer (1990) has examined this question at considerable depth. He views evolution as basically hierarchic construction. Table 3.1 depicts Layzer's hypothesis on the sequential development of different levels of biological organization. It begins with the molecular building blocks, the twenty amino acids, the four nucleotides (A, G, T, C), sugars and fatty acids. The same four nucleotides are assembled in the

DNA macromolecules of all known forms of life, from bacteria to advanced plants, to humans; and the DNA direct the assembly of the same twenty acids in the formation of protein macromolecules, using the same genetic code. From the macromolecules listed in Table 3.1, evolution later constructed macromolecular complexes (ribosomes, enzyme complexes, membranes), and still later bacterial cells, and cellular organs (organelles). In time, there appeared higher cells, then tissues, and in order organs, organ systems and higher organisms. Finally, communities emerge which we shall discuss in later chapters. Systems on each level are aggregates of systems on the prior level. Macromolecules are aggregates of molecular building blocks . . . organs are aggregates of tissues, etc.—aggregation being a basic process in hierarchical construction. Layzer (1990, p.212) discusses and illustrates how

> gene duplication followed by mutations can have a variety of outcomes. Gene duplication can give rise to a single gene coding for a polypeptide chain twice as long as the original one, as in the example of ferredoxin; to genes coding for individual chains in a protein consisting of several chains, as in the evolution of hemoglobin; to genes coding for enzymes that fulfill complementary, specialized functions, such as the digestive enzymes chymotrypsin, elastase, and trypsin; and to genes coding for enzymes that fulfill biologically unrelated functions, such as trypsin and thrombin. In every case, new structures and functions arise through combined and simultaneous processes of *differentiation* and *integration*. . .
>
> Evolutionary changes in a given structure or function often require, or open the way for, coordinated changes in related structures and functions. Consider, for example, the evolution of the three small bones of the mammalian middle ear, which transmit sound waves from the eardrum to the oval window. The middle ear of reptiles has only one bone. The two additional bones in the middle ear of mammals evolved from two bones that in reptiles form a hinge between the upper and lower jaws. The fossil record shows that these two bones became redundant when reptiles developed a new jaw articulation. Their subsequent integration into the mammalian three-bone structure made possible, and was driven by, improvements in the hearing of early mammals. These evolutionary changes in turn stimulated a variety of structural and behavioral changes that allowed mammals to exploit the increased flow of auditory information about their environment.

Further Layzer shows how through blind genetic variation and natural selection genetic mechanisms evolve that focus and regulate genetic variation itself, making possible the macroevolution of biological organization. Layzer considers such organization, in which functional hierarchy is characteristic, as resulting from a process of hierarchic construction, in which, to repeat, differentiation and integration of existing functional units yield new functional units.

Table 3.1 Levels of biological organization

Levels	Examples
Molecular building blocks	Amino acids
	Nucleotides
	Sugars
	Fatty acids
Macromolecules	Proteins
	DNA, RNA
	Starch
	Lipids
Macromolecular complexes	Ribosomes
	Enzyme complexes
	Membranes
Bacterial cells, cellular organs (organelles)	Cell nuclei
	Mitochondria
	Chloroplasts
Higher cells	
Tissues	Xylem, phloem
	Blood
	Smooth muscle
	Bone
Organs	Stem
	Taproot
	Kidney
	Heart
Organ systems	Root system
	Circulatory system
	Digestive system
	Nervous system
Higher organisms	
Communities	

Source: *Layzer (1990), p. 213.*

3.7 Sociobiology, speciation, and the individual and species as units

The previous sections were primarily based on the writings and views of Dawkins and on Layzer's thought on biological organization. When we move on to consider questions of human organization and its hierarchical structure there are strong views of biological-oriented evolution theorists who differ sharply on the interpretation of the empirical evidence. It is appropriate to present their views here. In later chapters we will look at human organization as perceived by a number of different social and natural scientists, philosophers and theologians.

One of the major differences that has led to heated discussion among scholars of evolution centers around claims for and accusations of reductionism—recently intensified with the development of the new field of sociobiology. Paraphrasing E.O. Wilson (1975, 1978), the leading figure in this field, one may define sociobiology as the systematic study of the biological basis of social behavior in every kind of organism, presently focussing on animal societies but clearly applicable to man. It draws upon knowledge from biology, psychology, anthropology, ethology (the natural history of animal behavior), genetics, ecology and other disciplines. It combines and fuses concepts, hypotheses and empirical findings from such disciplines in ways that provide deeper insights and understanding of human behavior and society. While sociobiology places heavy emphasis on the comparison of societies of different kinds of animals and of man, this is to draw analogies. Such comparison is designed to devise and put to test theories about the underlying hereditary basis of social behavior. Sociobiologists investigate how the diverse species adapt to the special opportunities and exigencies in their environment to gain deeper insights into principles that human societies have and can employ.

Basically, the analyses of Wilson in his writings on sociobiology and Dawkins as summarized above are similar; there are minor differences in definitions and uses of concepts. Both have been accused of reductionism. For instance, Lewontin, Rose and Kamin (1984, p. 5) consider that philosophically, the sociobiology

> view of human nature is very old; it goes back to the emergence of bourgeois society in the seventeenth century and to Hobbes's view of human existence as a 'bellum omnium contra omnes', a war of all against all, leading to a state of human relations manifesting competitiveness, mutual fear, and the desire for glory. For Hobbes, it followed that the purpose of social organization was merely to regulate these inevitable features of the human condition. And Hobbes's view of the human condition derived from his understanding of human biology; it was biological inevitability that made humans what they were.

According to Lewontin et al. (1984) sociobiologists are reductionists who try to explain the properties of complex societies in terms of the units of which those societies are composed. Like other reductionists who argue that the properties of a protein molecule are uniquely determined and predicted in terms of the properties of the electrons, protons, etc. constituting its atoms, they also argue that the properties of human society are no more than the sums of the individual behaviors and tendencies of the individual humans constituting that society. For instance, societies engage in warfare because their members are warlike. Formally speaking, Lewontin et al. state, sociobiologists claim that the compositional units of a whole are ontologically prior to the whole—that is, the units and their properties exist before the whole, and causation runs from the units to the whole.

More specifically, Lewontin et al. aver that sociobiologists consider an individual's actions to be the inevitable consequences of the biochemical properties of the cells that make up the individual, such characteristics in turn being uniquely determined by the individual's genes. Hence, the behavior of a human society can be causally traced back via each individual to the sets of genes that are present in that society. All human nature is fixed by those genes.

It is such notions of biological determinism that Lewontin et al. reject; they insist upon an integrated development of relationships between the biological set and the social set of forces where neither set can be claimed to be dominant.

Another distinguished critic, Gould (1978), maintains that taking humans as animals does not imply that specific human behavior and social organization are in any way directly determined by the genes of humans. He asserts that if certain behaviors are invariably found among humans and their closest primate relatives, they may simply be analogous and may have no direct genetic specification in humans. Moreover, altruism and reciprocal altruism while common among man and his closest primates can only be a similarity of result based on biological potential and not stem from biological determinism. Still more, Gould claims that the adaptive behavior of humans must be viewed in the light of the marked increase in their brain size which makes possible *nonprogrammed* learning. It permits tremendous flexibility in considering actions in a given situation, without predisposition to any. For man/woman, with his/her vast range of potential behavior, the idea of biological determinism, with specific genes for specific behavioral traits flies against reality.

The response of Dawkins (1989a, p.332) to his critics is that they misunderstand his writings. One can maintain that genes exert a statistical influence on human behaviour while simultaneously admitting that nonbiological factors can modify, override or reverse this statistical influence—this statistical influence having derived via natural selection. For example, he states, human sexual desire has evolved by natural selection, and genes certainly influence sexual desire. Yet humans have no trouble curbing their sexual desires when socially dictated.

> What is dualist about that? Obviously nothing. And no more is it dualist for me to advocate rebelling 'against the tyranny of the selfish replicators'. We, that is our brains, are separate and independent enough from our genes to rebel against them. . . and we do so in a small way every time we use contraception. There is no reason why we should not rebel in a large way, too.

Wilson (1978) replies in a more philosophical manner. He points to the fact that the brain is a machine of ten billion nerve cells and that the mind can somehow be explained as the summed activity of a finite number of chemical and electrical reactions. There are a multitude of other biological type boundaries that limit the human prospect. Biologically we are restricted. Darwinian natural selection, genetic chance and environmental necessity made the human species and not God. (We note here this seeming inconsistency with Teilhard type of thinking, which we shall discuss in later chapters.) Wilson (1989a, pp. 1-3) states:

> Deity can still be sought in the origin of the ultimate units of matter, in quarks and electron shells. . .but not in the origin of species. However much we embellish that stark conclusion with metaphor and imagery, it remains the philosophical legacy of the last century of scientific research.
>
> No way appears around this admittedly unappealing proposition. It is the essential first hypothesis for any serious consideration of the human condition. Without it the humanities and social sciences are the limited descriptors of surface phenomena, like astronomy without physics, biology without chemistry, and mathematics without algebra. With it human nature can be laid open as an object of fully empirical research, biology can be put to the service of liberal education, and our self-conception can be enormously and truthfully enriched.
>
> But to the extent that the new naturalism is true, its pursuit seems certain to generate two great spiritual dilemmas. The first is that no species, ours included, possesses a purpose beyond the imperative created by its genetic history. . . The reflective person knows that his life is in some incomprehensible manner guided through a biological ontogeny, a more or less fixed order of life stages. He senses that with all the drive, wit, love, pride, anger, hope, and anxiety that characterize the species he will in the end be sure only of helping to perpetuate the same cycle. . .

Wilson then speaks of two dilemmas. The first is that the human species has no goal that is external to its biological nature. It has no particular place to go. While educators and other leaders talk about fulfillment and realization of individual potential as primary aims beyond the achievement of a stable ecosystem, a well-fed population and global security, he asks: what is potential, and then what ends are realized by potential?

In discussing the second dilemma, Wilson (1978, p.5,6-7) states

> innate censors and motivators exist in the brain that deeply and unconsciously affect our ethical premises; from these roots, morality evolved as instinct. If that perception is correct, science may soon be in a position to investigate the very origin and meaning of human values, from which all ethical pronouncements and much of political practice flow. . .
>
> Which of the censors and motivators should be obeyed and which ones might better be curtailed or sublimated? These guides are the very core of our humanity. They and not the belief in spiritual apartness distinguish us from electronic computers. At some time in the future we will have to decide how human we wish to remain—in this ultimate, biological sense—because we must consciously choose among the alternative emotional guides we have inherited. To chart our destiny means that we must shift from automatic control based on biological processes to precise steering based on biological knowledge . . .
>
> Above all, for our own physical well-being if nothing else, ethical philosophy must not be left in the hands of the merely wise. Although human progress can be achieved by intuition and force of will, only hard-won empirical knowledge of our biological nature will allow us to make optimum choices among the competing criteria of progress.

At this point in the debate, consider the compromise view in Durham's coevolutionary theory of human biology and culture. Durham (1978, p. 425) states, that 'what was presumably genetically selected for in our ancestors was an increasing ability to modify phenotypes through learning and experience'. Such permitted the development of culture, whose attributes arise from internal, individual-level selection on the basis of their capability to enhance the survival and reproductive prospects of their carriers. Thus without any direct genetic basis, human phenotypes could alter, extend and transmit culture, and individual inclusive fitness differentials could result. In the words of Durham (1978, p.444), cultural selection

> functionally complements natural selection by retaining in time those cultural variants whose net effect best enhances the inclusive fitnesses of individuals. Where cultural selection operates in this way, human phenotypes would then evolve *subject to both biological and cultural influences* in the direction of character states that maximize inclusive

> fitness under prevailing environmental conditions. . .
>
> Put different, the operation of cultural selection—what some would call only a 'proximate' mechanism—may at times replace and preclude the operation of natural selection—which then cannot be considered the 'ultimate' mechanism. In short, this theory can explain the biocultural evolution of human attributes without presuming a genetic basis or predisposition for all adaptive forms.

In what follows in this book, we are inclined to adopt Durham's type of theory—yet with a predisposition to recognize that, at least up to the current era of biological human engineering, the biological ontogeny, a more or less fixed order of life stages, involves constraints (whose control over time the human has and will be able increasingly to relax) to which human, cultural and social development must conform in their full, flexibly chosen time paths.[5]

Before leaving evolutionary biology and its related fields, it is fruitful to discuss another point of heated debate. This concerns the question of unit of selection—the individual, the species, the multi-space community. In introducing his concept of variational evolution Darwin was concerned with the individual organism, its variation and selection. His analysis was at the level of the deme (population) wherein the gene pools of the successful individuals yield the next generation. But as Mayr (1942, 1963) and later Layzer (1990) have forcefully pointed out, variational evolution takes place at a second hierarchical level, namely the species. Mayr (1989) criticizes Teilhard's work as in the tradition of transformation theories which place emphasis on steady progression and directed movement toward an ultimate goal (aspired perfection)—a belief in cosmic teleology—without proper recognition of the full scope of variational evolution. Mayr holds that in a highly opportunistic manner many new species steadily emerge—particularly at peripheral locations (usually isolated). Most are doomed to rapid extinction. The founder populations of the small number that do survive and diffuse, provide the basis for the major and most rapid evolutionary change.

Recently, evolutionary doctrine has been invaded by a second concept which also emphasizes hierarchical structure, namely that of punctuated equilibrium or saltation. As formulated by Eldridge and Gould (1972), evolutionary change by and large occurs in rapid bursts through speciation (formation of new species) over geologically short periods of time, and that there is relative stasis (conditions of very little change) in the new (and old) species after the punctuational burst. While the notions of sudden burst and prolonged stasis thereafter seem to find wide acceptance, there still remains the question of species selection. Mayr (1989) maintains that species selection occurs via a selection process imposed on the individuals of a species. He argues that a species turns out to be superior because its individuals are better adapted and can better exploit the resources of the environment. The individuals in a superior species thus have greater expectation of survival and successful reproduction than those of an inferior species. Consequently, in competition for a given niche or resource base, individuals of the superior species will win out and be the ones to survive. The inferior species will become extinct.

Gould (1989, p.62) and others, however, would view species sorting (a term to be preferred to species selection) as possible not only through selection upon organisms as described by Mayr but also through 'selection upon emergent and irreducible characteristics of species treated as entities,' where the species itself is a source of variation.[6] Such then yields a hierarchical theory of selection with transfers and interactions up and down and across levels.

The recent type of hierarchical thinking of evolutionary biologists—which they consider as linking macroevolution (at the species level) with microevolution (at the gene pool level)—can of course be extended conceptually upwards to recognize multispecie communities, agglomerates of multispecie communities, and so forth (wherein the memes of Dawkins operate), and downward in explicitly regarding sub-specie communities of decreasing complexity lying below the level of the species.

This discussion then raises the question of what is the appropriate unit for evolutionary doctrine.[7] Mayr insists that it is the individual organism. Gould would maintain that the specie itself is a unit, too. Others take different positions. A most general view that is possible is suggested by hierarchical theory where at each level of an hierarchy a unit can exist having a certain amount of independence, yet also related, and thereby constrained in its behavior, to the community of units both above and below the community of which it is a member. Here one might fruitfully refer to the study of termites by Lewis Thomas (1974).

Inhabiting the Australian termite's digestive tract is the parasite myxotricha paradoxa, a single-celled creature which supplies the enzymes that break down cellulose of the wood finely chewed by the termite. Edible carbohydrate results and also nondegradable lignin which is excreted by the termite as tiny pellets and used in building the termite's nest. The tiny myxotricha turns out to consist of whole populations of even tinier creatures symbiotic with each other, yet retaining autonomous individuality —for example spirochetes regularly attached to the surface of the motile myxotricha which they propel. Thomas (1974, pp. 28-30) goes on further to state:

> The whole animal, or ecosystem, stuck for the time being halfway along in evolution, appears to be a model for the development of cells like our own. . . There is an underlying force that drives together the several creatures comprising myxotricha, and then drives the assemblage into union with the termite. If we could understand this tendency, we would catch a glimpse of the process that brought single separate cells together for the construction of metazoans, culminating

> in the invention of roses, dolphins, and, of course, ourselves. It might turn out that the same tendency underlies the joining of organisms into communities, communities into ecosystems, and ecosystems into the biosphere. If this is, in fact, the drift of things, the way of the world, we may come to view immune reactions, genes for the chemical marking of self, and perhaps all reflexive responses of aggression and defense as secondary developments in evolution, necessary for the regulation and modulation of symbiosis, not designed to break into the process, only to keep it from getting out of hand.
>
> If it is in the nature of living things to pool resources, to fuse when possible, we would have a new way of accounting for the progressive enrichment and complexity of form in living things.

The existence of hierarchies in nature, and also in society raises the question: are there principles or laws which apply to all levels of a given hierarchy, and to all the diverse types of hierarchies? To this question we will turn in a later chapter, as well as to the associated question of the applicability and utility of a general systems theory aimed at constructing theoretical models and identifying general principles applicable to biological, social, and symbolic systems.

Finally, as one possible implication of biological evolutionary theory, note the thinking that views the world as a single functioning organism. One manifestation of this thinking is the envisaged noosystem of the spiritually-oriented paleontologist Teilhard. Another is the notion of Gaia, an hypothesis set forth by Lovelock (1988). To paraphrase him, Gaia is a total planetary being. It is more than the biosphere, defined as that part of the Earth where living things normally exist. It is more than the biota, the collection of all individual living organisms. It is more than the biosphere and biota when taken together. It has continuity with the past back to the origins of life, and is projected into the future with the existence of life as long as life persists. As Lovelock (1988, p. 9) states, the

> Gaia hypothesis, when introduced . . . supposed that the atmosphere, the oceans, the climate, and the crust of the Earth are regulated at a state comfortable for life because of the behavior of living organisms. Specifically, the Gaia hypothesis said that the temperature, oxidation state, acidity, and certain aspects of the rocks and waters are at any time kept constant, and that this homeostasis is maintained by active feedback processes operated automatically and unconsciously by the biota. The conditions are only constant in the short term and evolve in synchrony with the changing needs of the biota as it evolves. Life and its environment are so closely coupled that evolution concerns Gaia, not the organisms or the environment taken separately.

While evidence is accumulating on feedback loops among processes covered by the Gaia hypothesis, still adequate support for it has yet to be built up. As the critic Doolittle (1981, pp.61-62) has asked: 'Can we not assume that there is an immense number of planets on which life independently arose and then through some global catastrophe was extinguished before the evolution of self-awareness?' What is to prove that the existence of the world that we know is not a random development, purely accidental, extremely fragile and not the operation of an efficient or optimal global cybernetic system?

3.8 Summary

To summarize, begin with the thinking presented in Dawkins' work, which up to the emergence of primate man, we take to be fundamental and consistent by and large with Teilhard's thinking and that of evolutionary scientists.

Starting with the primeval soup, there emerged, by chance, a replicator. In the process of replication and diffusion throughout the soup, mutation occurred, leading to a diversity of replicating units. In time, some of these units happened to join up with others, and in certain cases forming more efficient survival vehicles for assuring the replication of the genes that became embodied. The more efficient vehicles were those which were more successful in finding the resources required for gene replication—resources which had become or were fast becoming scarce. With time more and more genes became hooked up into larger and larger survival vehicles, each gene joining an existing mutually compatible gene or combination of genes in an environment that was favorable. The more efficient vehicles, often manifesting altruistic behavior, were basically selfish; they outcompeted the less efficient, and the latter disappeared. In time, the finely coordinated, beautifully integrated set of organs of advanced animals and humans evolved—survival machines. Though of limited life-span, these machines lasted long enough to perpetuate the bottlenecked process (meiosis) whereby the genes in the cells of a male and a female are mixed to produce the replicating genes contained in the next generation of machines. While the genes singly and jointly determined the sequences of protein manufacture that ultimately led to the full development of the phenotype (the bodily manifestation of the genes), mutations of genes occurred leading to changes in the resulting phenotype.

Some of the effects of the genes on the phenotype, as on the color of eyes and the shape of the nose, and in general on the organs of the body in which they reside, are clear. Their effects on attributes and behavior outside the body are less clear, but nonetheless are there. The latter can be 'at a distance', as in the case of the cuckoo and foster parent bird—and even at a very great distance. Hence, the notion of an extended phenotype of a gene or set of genes within a cell which can come to have a radiating effect upon the outside world.

When the genes in a new mutually compatible combination stumbled upon the brain and thus a new machine that turned out to be at least as efficient as others in obtaining the resources for survival, there emerged the setting for the spread of an idea, the meme. Today, highly successful memes and coadapted mutually compatible memes are all around us. Via these memes humans are able to combat the short-run selfishness of the genes embodied in myopic survival machines and nurture cooperative relationships among humans and groups consistent with long-run selfish interest, if not altruism.

The evolutionary process extending from the first replicator to the communities of plants, insects, animals and humans of today reflects the development of a very complex hierarchical structure, as discussed by Layzer. Involved was the basic process of aggregation—molecular building blocks forming macromolecules, macromolecules forming macromolecular complexes, etc. Not only did specific genes perform specific functions, but also there came to be present mutator genes that functioned to foster variability and creativity. Thus there continuously emerged hierarchical biological and human structures, organizations and communities with new functions based upon differentiation and integration of existing functional units.

The views of Dawkins and Layzer are consistent with and bring the detail of current thought into Teilhard's framework; and the emergence of memes, as discussed by Dawkins, can at some state, if not at the start, be viewed as consistent with Teilhard's notion of reflection, an advanced stage of consciousness discussed in Chapter 2. Dawkin's thinking on memes is also consistent with that of Wilson and other sociobiologists. However, there arises the issue of biological reductionism: to what extent do the operations of the selfish gene (even when on the surface they are altruistic) condition the social behavior of man and his institutions (groups and organizations) as Wilson and his fellow sociobiologists would maintain, or to what extent can social behavior be independent of the functioning of the selfish genes, and in fact fully control such behavior. This issue can be hotly debated and one is inclined to adopt the coevolutionary theory of Durham, namely the joint operation upon behavior of both human biology and culture.

Another issue concerns the unit of analysis. While Mayr accepts the hypothesis of punctuated equilibrium and related notions on speciation, he considers the individual organism as the fundamental unit—the key survival machine. For while a specie may not survive for one reason or another, this can happen only if each member of that specie fails to function as a survival machine. Yet, when for example, a species and its institutions fail to control the spread of a fatal disease such as AIDS, the fundamental decision-making unit for analysis is the species. The species itself, Gould claims, could be the basic unit for analysis and understanding.

We shall not attempt to settle this issue. However, it does bring up the nature of hierarchical organization in the physical world—the linkage of microevolution and macroevolution—and the importance of feedback effects, up and down an hierarchy. It raises the basic question: are there subsystems of species and behaving units which are so interconnected that the subsystem itself becomes the effective unit of analysis? Or, is in reality all elements of our earth-bound world interdependent as in the Gaia hypothesis? Does this imply that effective analysis is possible only when we consider our earth-bound world as one living unit? Or given our sensual and cognitive limitations—which are many—can we conduct analysis on only subsystems of much smaller magnitude?

Whatever the answers to these questions, it becomes clear that the Teilhard framework with its involution and interiorization process, expanded and modified by the selfish gene hypothesis, punctuated equilibrium, and by concepts of hierarchical structure and both simple and complicated (direct and indirect) feedback loops, is a workable one, especially when the Christogenesis process is flexibly interpreted as welfare-oriented institutional and cultural strivings with spiritual-religious development as a core aspect.

Notes

1 If the soup had become a scarce resource before a second type of replicator had evolved through mutation, and in particular a much more efficient replicator, competition would then have arisen at the appearance of this second type.

2 However, recall that by mutation a new type of gene may form and because of some superior attribute succeed in spreading rapidly in the gene pool, thereby destabilizing the pool to some if not a significant extent for a period of time. During this period of instability, conditions may be more favorable for a given stable gene to hook up with others, even the new one, to form more efficient and productive survival machines.

3 However, a gene of a parasite may have the same interest as the gene of the host; it may for example be interested in the host's survival and propagation, perhaps because the egg or sperm of the host permits transfer of the parasite's gene to a new host. In sharing a common destiny, the genes of the parasite and host may eventually merge with the host's tissue and the parasite may become unrecognizable as such. Such merging has been taking place all during the evolution of human beings, who are in a real sense the result of ancient parasite mergers.

4 A major new mutation in a gene, or in the functioning of a combination of coadapted genes, may open up a vast area for new development. That is, such a mutation, like the brain, can be viewed as having been generated by a new gene or combination of genes that favors creativity—that is as a selective factor in favor of evolvability.

5 This depiction has similarities to the characteristic structure of a mathematical programming model wherein the objective may be to maximize some welfare function or the extent to which a goal is achieved subject to constraints, the constraints being gradually relaxed over time as knowledge increases.
6 For some recent relevant discussion see Bocchi (1991).
7 See the writings of Huxley, Buss, Margulis and Sagan in Barlow (1991).

References

Barlow, C. (1991), *From Gaia to Selfish Genes,* Cambridge, MA: MIT Press.

Bocchi, G. (1991), 'Biological Evolution: The Changing Image' in Laszlo, Ervin (ed.), *The New Evolutionary Paradigm*, New York: Gordon and Breach.

Dawkins, R. (1989a), *The Selfish Gene*, New York: Oxford University Press.

Dawkins, R. (1989b), *The Extended Phenotype*, New York: Oxford University Press.

Doolittle, W. F. (1981), 'Is Nature Really Motherly?', *Coevolution Quarterly*, No. 29, Spring, pp. 58-62.

Durham, W. H. (1978), 'Toward a Coevolutionary Theory of Human Biology and Culture' in Caplan, A.L. (ed.), *The Sociology Debate*, New York: Harper and Row, pp. 428-48.

Eldridge, N. and Gould, S.J. (1972), 'Punctuated Equilibria: An Alternative to Phyletic Gradualism', in Schopf, T.J.M. (ed.), *Models in Paleobiology*, San Francisco: Freeman, Cooper, pp. 305-32.

Gould, S. J. (1978), 'Biological Potential vs Biological Determinism' in Caplin, A. L. (ed.), *The Sociology Debate*, New York: Harper and Row, pp. 343-51.

Gould, S. J. (1989), 'Punctuated Equilibrium in Fact and Theory' in Somit, A. and Peterson, S. A. (eds.), *The Dynamics of Evolution*, Ithaca, N.Y.: Cornell University Press, pp. 54-84.

Layzer, D. (1990), *Cosmogenesis: the Growth of Order in the Universe*, New York: Oxford University Press.

Lewontin, R.C., Rose, S. and Kamin, L. (1984), *Not in Our Genes*, New York: Random House.

Lovełock, J. (1988), *The Ages of Gaia*, New York: Oxford University Press.

Mayr, E. (1942), *Systematics and the Origin of Species*, New York: Columbia University Press.

Mayr, E. (1963), *Animal Species and Evolution*, Cambridge, MA: Harvard University Press.

Mayr, E. (1989), 'Speciational Evolution or Punctuated Equilibria,' in Somit, A. and Peterson, S.T. (eds.), *The Dynamics of Evolution*, Ithaca, N.Y.: Cornell University Press, pp. 21-53.
Thomas, L. (1974), *The Lives of a Cell*, New York: The Viking Press.
Wilson, E. O. (1975), *Sociobiology: The New Synthesis*, Cambridge, MA.: Harvard University Press.
Wilson, E. O. (1978), *On Human Nature*, Cambridge, MA: Harvard University Press.

4 Linguistic and dynamic modes, rate-dependent and rate-independent processes, and systems theory

4.1 Introduction

Before moving on to insights from the physical sciences, we wish to dig more deeply and formally into the evolutionary process. We start with the work of the systems-oriented theorist Pattee. We shall not hesitate to use extensively his statements, expressions and terminology. To give his ideas more specificity we will relate them to particular concepts, analytical methods and theories of social science—and in particular to economics, regional science and the emerging interdisciplinary field of peace science where primary concern is with conflict management and cooperation basic to the achievement of Teilhard's Omega and to the notion of long-run interests and social welfare discussed by evolution theorists.

4.2 The dynamic and linguistic modes

In earlier articles discussing classical physics, Pattee (1977, 1978) distinguishes two modes of description, *dynamic* associated with *rate-dependent processes*, and *linguistic* associated with *rate-independent processes*. According to Pattee, the laws of classical physics involve rate-dependent processes. To illustrate a familiar rate-dependent process, take Newton's second law of motion. It states that the time rate of change of the velocity (directed speed), which in calculus is written as dv/dt, and which is defined as acceleration a, is directly proportional to the force F and inversely proportional to the mass m of the body. That is:

$$a = F/m \text{ or } F = ma \tag{4.1}$$

Neglecting air resistance, the force F acting on a freely falling body of mass m is its weight acting down and it produces familiar downward acceleration equal to the acceleration of gravity.

To be specific on this point in a social science context, consider a simple classical rate-dependent equation (often called a law of motion) for the production system of an economy. It states that during a given period of time the increase or decrease in its capital stock K is equal to its production output Y minus its consumption C minus its exports E plus its imports M. Thus the time rate of change of capital stock, which in calculus is written as dK/dt, and which we may define as an acceleration a, is given by:[1]

$$a = dK/dt = Y - C - E + M.$$

Pattee takes the classical laws of physics and the laws of other sciences to the extent that they can be viewed as strictly rate-dependent processes to be inexorable and universal and to refer to continuous and reversible time. They do not need a reference zero point. Anybody's calendar is acceptable. These laws define one type of structure.

In contrast to the deterministic mode and its rate-dependent processes stands the linguistic mode and its rate-independent processes. (At times, we shall use the more broad and directly meaningful term, cognitive - linguistic - spiritual mode, to replace the linguistic mode.)

To make clear the distinction between these two modes, Pattee points out that the universal and inexorable laws of physics, such as 'like charges repel one another' or the physicist's law of motion, do not require a description for their operation. They need no outside observer to enforce them. However, for these laws to operate in the real living world, they require a starting point (initial conditions) or an end point (terminal conditions) which serve as constraints and which come from the linguistic mode. For example, to be able to calculate the force (impact) F for a free falling body from a building when it hits the ground, we need to know (have measured) the height from which it starts its fall (an initial condition) and then apply (after adjustment for air resistance and other factors) acceleration due to gravity which is 32.2 feet per second per second, a number derived from experiments. Such initial conditions can be established only by measurement, a process which is entirely independent of the formal laws. Measurement is a logically disjoint primitive act; measurements are not reducible to or derivable from laws, and laws are not reducible or derivable from measurement. Yet measurement is an indispensable act in relating the world to these laws.

Measurement, as embodied in the linguistic mode, and other processes associated with it, are complex activities especially when we go outside of classical physics into quantum physics, biology and social systems. To help point this out, as well as to indicate the relation of these processes to the complementary rate-dependent processes in the production system and in general, we construct Figure 4.1. In this figure, drawn from the field of peace science, S represents the system here depicted for only two regions, nations or other spatial units which we designate J and L. (The analysis that follows also holds for a system of many more spatial units, A,B,C,. . .etc.) This figure contains boxes which refer to the information subsystem (M) and cognitive decision-making subsystems (C^J and C^L). If J refers to a nation and if we are focussing on the international system and its conflicts, then the C^J box may be taken to refer to national decision making and foreign policy formulation within nation J. This would be dependent upon key sociological, psychological, economic and political variables and elements such as attitudes (hostility, competitiveness, etc.), perceptions and misperceptions, stress-coping strategies and decision-making pathologies, scripts, schema and stories, perspectives, expectations, aspirations, interpretive packages (group think), market performance, industrial structure, public opinion and many other elements. Each nation's cognitive system is interconnected with those of others, as indicated by the arrows going to and from the I box, providing inputs to and receiving feedback from it. (Likewise for the C^L box.) The I box, another cognitive decision-making subsystem, may be taken to refer to the international arena wherein key variables and elements are arms control policies, tariff and other trade policies, international tension, world monetary stability, exchange rate fluctuation, World Bank policy, ocean resource development, global environmental regulations, United Nations peace-keeping and other operations, and the like.

The information subsystem M in Figure 4.1 can be taken to cover all knowledge—historical events, measurement of outputs and interpretations, all resulting data sets and records, all theories, all hypotheses, validated or not, all know-how and methods for analysis and measurement, and so forth—all feeding, though greatly filtered, into the C boxes.

The C^J, C^L, I and M boxes and all measurement and activities that take place within them then constitute the linguistic mode.

Note that in Figure 4.1 an arrow proceeds from the C^J and C^L boxes to the P^J and P^L boxes, respectively. Each arrow represents the information, orders, commands, decisions from the linguistic mode that provide the initial conditions and other constraints on the operation of the production subsystem (the dynamic mode) of a nation. These conditions and other constraints indicate the available resources, what to produce, what technology to use, what to set aside for government use, household consumption, etc. Note that the production subsystems P^J and P^L are interconnected to each other (as indicated by arrows) via exports to and imports from each other which are taken to flow through a world market, the R box, at which at least exchange rates and thus world prices are established.

Let us elaborate more on Pattee's thinking. A social system must constantly evaluate its performance (output) for if it fails to do so, it may become inefficient and unable to meet the competition of other systems (species). In short, survival and development requires constant evaluation to insure satisfactory, if not optimal, performance. But to evaluate means that we must measure a perceived 'objective' structure, for measurement provides information needed for evaluation. In turn measurement requires rules for measurement. Further, if part of the evaluation requires description of what is going on, we need a language, a symbol system and some syntactical rules for combining symbols into statements which have meaning and can be interpreted. That is, we need a mechanism for writing descriptions, later for reading stored information, and still later for executing the statements embodied in orders and commands. While the laws of motion governing the dynamics of the system depicted in boxes P^J, P^L and R may be fairly complex, they are not nearly as complex as the constraints involved in a syntax and in the mechanisms employed in writing, reading and executing. These constraints together with symbols are basic elements of Pattee's *linguistic mode*—of the M, C^J, C^L and I boxes. And while several efforts have been made to incorporate measurement and evaluation as time-dependent processes in dynamical systems models (see the presentations and discussions in Chapter 7A), they have failed to come close, let alone replicate, the reality of measurement and control.

It is in the M, C^J, C^L and I boxes which contain non-dynamic rate-independent processes where measurement and other information is written down (recorded) for subsequent reading. Here is where the brain performs and operates and where individuals and groups cognitively interact to interpret measurements and other information and to design policies and construct orders and commands that are passed on and operate as constraints on the workings of an organism, organization or social system.[2]

To repeat, the linguistic mode is *not* governed by laws of motion. It is *not* part of the dynamic works of the system. It corresponds to an entirely different mode of description and explanation. For example, measurement of GNP is not a rate-dependent process, nor is the computation of the constant π of geometry, nor the conduct of a census, nor the interpretation of statements and formulation of policy. The linguistic mode relates to processes and operations in *ordinal* time. What is unique (and irreversible) is the order in which events are recorded, statements made, demands articulated, decisions taken, policies formulated, etc. The extent of

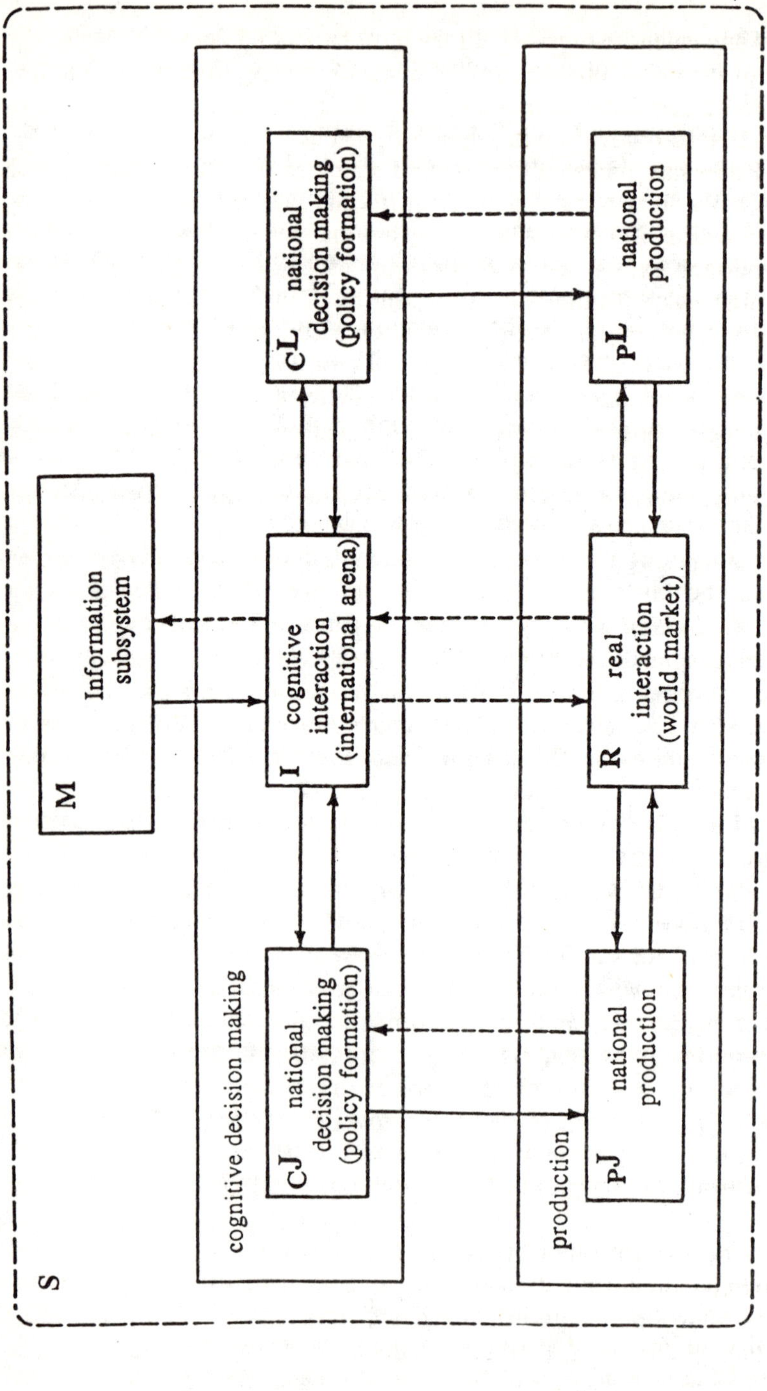

Figure 4.1 A system's structure
Source: *Adapted from Isard, 1988, p. 12.*

time—say any calendar time—that may have transpired does not change the sequence of events. Moreover, in the linguistic mode, time is irreversible. Records and measurements are taken, decisions made, policies instituted; they cannot be fully reversed in a real-life system.[3]

In fact, Pattee has indicated that systems which are fully governed by laws of motion have no way by which to insure that they do not degenerate or lose their function. In contrast, he points out that genetic description, which to an important extent social description processes parallel, is a self-programming system that *does* allow survival and evolution in spite of errors within the system and even within the description of the system. We still do not understand how the genetic description has come about, and how the genes (enzymes) go about reading and executing such descriptions; nor in the social system are we able to explain very much how policy or control constraints are set and fed back to the P^J, P^L and R boxes.

Pattee compares the two modes (linguistic and dynamic), covered by his principle of complementarity for a unitary behaving system, with diverse dualities: subject and object, image and event, knower and known, genotype and phenotype, soft program and hardware, and policy and mechanistic behavior. Each element in each of these dualities is incompatible with the other. Each requires a separate mode of description. Each can provide only one part or one approach to explanation of a situation. Both are needed for full explanation. Likewise, with his two modes which cannot be unified into one analytic framework. Yet both are required for the full understanding of a system.

Pattee's framework provides additional insight into the reductionist debate. As already stated, the reductionist approach to understanding living systems would reduce all processes covered by the linguistic mode (e.g. measurement, writing, reading, and interpreting) to natural, inexorable, universal laws of motion. But, to reiterate, before these laws can take a real form, the deterministic equations which express these laws must have the linguistic mode provide at the minimum prespecified initial (or terminal) conditions and parameters. This is so for the simplest deterministic models of physics as well as economics. These initial conditions and parameters are what physicists sometimes call timely and local experiences. As Pattee notes, they are not predictable nor completely regular. Rather they are irregular, but at the same time not completely irregular. Such conditions must be provided as inputs into the dynamic mode by the linguistic mode of a meaningful and efficient behaving system.

But the linguistic mode must do more. Physicists concerned with quantum mechanics and quantum field theory, and certainly all social scientists concerned with empirical testing of hypotheses, know that the process of measuring what is, was, or will be at some reference point of time, cannot escape subjectivity. In quantum theory, according to Pattee and many others, there is no precise conceptual distinction between *laws* and *measurement*, or between the *objective structure* being measured and the *subjective agency* of measurement. The agency must choose what possible measurements are to be carried out (what information is to be collected) and how. Both implicitly and explicitly this involves inputs from the observer's senses, that is *cognition* and *volition*.[4] All this in turn affects the determination of the set of alternatives (e.g. the set of relevant parameters, or the set of possible values of a relevant parameter) which are to be reduced in number by observation. Thus, as Pattee notes, in quantum physics it is recognized that the system that is being discussed in fact includes the measuring agent and its arrangement with the object. The measuring rod is not assumed to have vanishing small or no physical interaction with the system, as in the formal classical physical models.

In biological and social systems it is much more widely recognized that information is about the object *plus* the measuring device as a system. This partly reflects the fact that both cognition and volition and the role of information are so important in governing the behavior of the biological or social system. For example, take the Census, a stock of information on the social system extensively used to evaluate the performance of the system as well as to provide 'data' for derivation of the initial conditions and parameters. What is reported in the Census is what is perceived and judged in various ways by government officials, political leaders, scholars and others to be important. This in turn has conditioned the kinds of laws-of-motion models that economists, regional scientists, peace scientists and others can now construct and use to 'explain' the system.

In biological systems, the role of cognition and volition in measurement—namely what to measure and how—is less obvious. Yet it is there. In the genetic DNA, there exists stored information. Again, as Pattee remarks, the code and construction rules that it contains reflect the specific types of measurements that are being performed by organisms of species that have survived the process of natural selection.[5] The reading of the rules to form linear discrete sequences of amino acids may be said to reflect cognitive and volitional (interpretive) processes in the cells of the organisms of successful species. On the other hand, the folding of the linear sequence into a three-dimensional catalytic machine whose only function is to control rates of specific reactions is clearly governed by a law of motion.

In parallel fashion, the claims, demands, prescriptions, or suggestions that each participant makes in a multiparticipant decision-making situation in a social system correspond to a discrete rate-independent string of statements. When the surrounding environment is just right and when these statements have been properly ordered (in terms of a constitution, laws and other accepted procedures), they may be said in many instances to fold spontaneously into an n-dimensional catalytic machine, the policy statement, whose function is to control specific rates of production, consumption, investment, socialization, education, and other processes.

According to Pattee the above points thus invalidates the 'reductionist' approach. He denotes its invalidity in another way. Suppose the observer and measurement were subject to laws of motion. Then they could be included in an extended set of laws of motion pertaining to a larger system, one that would include the observer and measurement; and the behavior of the observer and his measurement would be regular and deterministic. But for this larger extended system, it would still be necessary to establish at the minimum the relevant initial conditions, and so measurement would be required. Such measurement, it could be argued, might be subject to laws of motion, and thus imply a still more extended system. In turn, the latter would require initial conditions and measurement, etc., etc., leading to infinite regress, a highly questionable result, at least operationally.[6]

Thus, Pattee (1977, 1988, 1992) holds that the observer's behavior and his measurement must be considered to be nonregular.[7] Hence, too, the formulation of policy and control inputs cannot be considered to be deterministic and unique, fully derivable once initial conditions and parameters are set. Rather, to paraphrase Pattee once again, the observer with his measurement devices must be considered to be associated with *selection from among alternatives*. However, this conclusion does not mean that the behavior of the observer and measurement can be completely irregular. A measurement to be valid must be locally repeatable. That is, it must conform to a set of rather fixed rules evolved by an organism or society. Otherwise there would be generated magnitudes which would not be biologically consistent with the survival of an organism, or sociologically acceptable by and large in a community.

Selection among alternatives, the setting of a constraint, and the choice of one of many ways to read and interpret information all testify to the non-deterministic nature of the linguistic mode. In the social system, this is clearly appreciated when one studies the diverse religions, cultures, and languages that have evolved throughout the world. As Pattee has indicated, measurements of outputs, when used, couple information about a rate-dependent system to a rate-independent system, while control inputs couple a rate-independent system to a rate-dependent one.

Before concluding this section, note again that it is a universal property of language (and hence all description) that the structure of symbol vehicles or signs (e.g. the letters of the alphabet, nucleotides, words, codons, etc.) are related to their referent or their effect by arbitrary rules. These rules are not derived from, or reducible to the laws of nature. Often they can best be described as frozen, historical accidents. They can be *accidents* because they came about not as a result of laws but, like some habits and customs, as random experiences which were found to 'work' and be useful. They can be *historical* because the random experience occurred only once, and *frozen* because they are embedded in an organism, organization or social system as a coherent, stable, hereditary constraint. While these rules can be arbitrary and rate-independent, note that they require structure (whose elements may be rate-dependent) to execute them.

In a language system, the syntax is a generally accepted coherent set of rules employed for descriptive purposes. Typically, this set is local, and thus we find in a real world many language systems—among nations and very frequently among ethnic groups in the regions of a nation.

Yet under sufficient pressure, an alphabet, the way to spell a word, the process by which words are put together to form statements, and the meanings attached to a statement can change—most often gradually, but on occasion drastically, as when an entirely new language is adopted by a community. Such changes and the associated informational control processes in general are guided by complex values and not by simple logic. Some scholars may consider them as aesthetics, especially when they result from debate and discussion and from the use of principles to reach compromise social actions, policies and explanations which are ultimately found acceptable and reasonable.[8]

4.3 Evolution, emergence and brain theory

In later articles Pattee (1992a, 1992b, 1993) has extended and made much more specific his thinking on evolution (emergence), on physical, computational and network models, and on the operation and function of a brain in man. In doing so, he also deepens his thinking on measurement.

As already stated, the laws of classical physics are rate-dependent, inexorable, universal. Each law must hold in all places, at all times, and for all states of the observer during measurement. The laws are completely independent of the observer's personal experiences, which experiences enter into measurement that provide initial conditions and constraints. These conditions and constraints are the pre-established observables of the underlying theories, and say nothing about new, useful observables that emerge from creative activity. Further, the classical laws assume that the mathematical formulations which express theories can yield computable output.

In quantum physics, as already indicated, the specific choice of a measuring rod by the observer affects the observed structure that is identified; that is the observer's input affects the output. Thus quantum measurements yield alternative objective structures. But, as with classical physics, nothing is said about *non-observables that emerge* and relate to novel structures and behaviors.

Like physical models, computational ones have little, if anything, to offer for describing evolution and creative thought. This claim made by Pattee can be maintained despite the many virtues of the programmable computer. As cellular automation adherents have observed (see Appendix A to this chapter), a computer program can express and apply an algorithm (procedure) for converting an input into an output specified by any given scientific law. Further, it can move into areas too complicated not only for the brain to comprehend but also to be attacked with the use of traditional mathematical methods. This is so, when these methods require excessive computations to reach an output. Moreover, when mathematicians, because of limited resources, explore such complicated areas with the use of less appropriate mathematical functions, the computer working with the proper functions can come up with more accurate results. Further the computer can perform highly desirable simulations of processes in which there are innumerable units in action (molecules moving, individuals behaving) which mathematicians have been unable to handle and where averaging and the use of a differential equation pertaining to the behavior of the average are clearly inappropriate—although such has often been done in the past by physical and social scientists in order to have tractable models. Equally important, computer programming allows the investigator directly to interact with it; it can spout out results speedily when an investigator wants to explore the implications of diverse policies or changes in various aspects of a model—the values of its parameters, the magnitudes of its constraints, and the very functions the model embodies. For further discussion see Appendix A to this chapter on Computational Experimentation and Cellular Automata.

But after all is said and done, computers cannot simulate evolution and the process of creative thought. As Pattee (1992b, pp. 4-5) concludes:

> Computers can represent complex behavior that in some ways is beyond the capacity of the brain, and therefore the significance of computer models is often difficult to evaluate. A new appreciation of nonlinear dynamical behavior (strange attractors, chaos, fractals, cellular automata, concurrent networks, etc.) resulted from intelligent observers discovering new patterns or observables in the output displays of computers. These discoveries have caused us to reconsider our views of determinism and random behavior as well as our choice of significant observables in complex systems. As a result of these computation-based discoveries, we are more willing to consider the possibility that much detailed physical and biological behavior is in

> principle unpredictably complex, and therefore emergent in practice, if not in principle.
>
> However, no matter how complex computer patterns may be, computers cannot *autonomously* extend their symbolic domain. All programmable computations require sharp separation of the formal rules from the physics of the devices which execute them. Hence the clear distinction of software and hardware. Moreover, the software inputs, outputs, and all intermediate steps must be in well-defined symbolic forms, *all* of which must be treated as simple observables... In other words, the programmed operations do not distinguish observables and non-observables.

In essence, programmed operations cannot and do not deal with or reveal non-observables and model the evolution of creative ideas, although they can reveal patterns too complex for the senses and mind to perceive. This motivates Pattee to examine the functioning of the brain with regard to emergence. In the billion of years prior to the development of physical laws and science and computational models, there was after all evolution, emergence and the coming of a brain that produced creative ideas. The brain did not consciously employ laws of motion, nor computational models utilizing formal systems of symbols, either alone or together. Yet it permitted a further development of the measurement and control processes that underly the functioning of nervous systems—systems which in turn evolved as elaborations of the early essential functioning of enzymes. [An enzyme functions to establish a strictly local measurement-control activity coupling the recognition of the substrate (the substance to be acted upon) to the specific control that the enzyme exercises as a catalytic agent.] Currently, the brain provides the coherent dynamics of: (1) measurement-control constraints, (2) complex pattern recognition, and (3) action—a set of functions well beyond the scope of physics and the steps of a system of a fixed, discrete set of simple symbols that requires a program.

Pattee goes on to consider the virtues of concurrent, distributed network models. One virtue of these models is their ability to 'learn' new classifications. These result from the many-to-one mappings that they produce. Thus a statistical stationary state, designatable as a class, is attainable from many input patterns, each viewed as an initial condition. At times a new set of classes (observables) can be identified at the statistical level without employing a physical or biological law—an outcome that can be taken as learning, and contributory to the basis for emergent behavior. See the discussion in Appendix B of this chapter.

It is with regard to this last feature that Pattee views concurrent, distributed network models and the closely associated neural network and fuzzy system models (e.g. see Hopfield, 1982, 1984 and Kosko, 1992) as resembling more closely the functioning of the brain. (See Appendix C for discussion of adaptive fuzzy system models.) They are adaptable measuring devices. They have the potential to learn, to measure any number of complex patterns, and provide insight as to how the brain confronts uncertainty. In this respect they seem to furnish more natural models of measurement and cognition than computational and physical models. Yet in their ability to exhibit gradual adaptive behavior better than the latter models, they are still far from being able to model the measurement-control and complex pattern recognition activities of the brain.

In bringing his thinking on emergence and brain theories to a close, Pattee forcefully states that the brain is not just a physical symbol system equivalent to a computer. It is not just a dynamical system upon which nonlinear constraints are suitably imposed. It is not a concurrent distributed network which has limited abilities at learning. Rather, it is an organ that carries on the continuous and closely-related intentional activities of control and measurement with at times construction of new concepts and devices. It is a generalized measuring device that uses in a complementary pragmatic manner law-based and rule-based models (from physics and computer programming, respectively) as disjoint modes. However, how the brain chooses to use these modes, how it (as an organ of an observer) chooses to predict and measure, how it learns, discovers previous unobservables, and carries on creative activity are all unanswered questions—the major challenge for brain theory.

4.4 Some concluding remarks

Just as Chapter 3 added considerable detail on the evolutionary processes from the standpoint of natural selection acting on genes and species, and on genetic influence through hierarchic construction, and presented materials that enrich the noetic framework conceived by Teilhard, this chapter has added another set of important details and relationships based on growth in our stock of knowledge over the last decades.

While the increasing interdependence and interrelation of man is an observable when we consider his economic, social and political activities, we still must acknowledge that socio-economic-political development is not a deterministic space-time path. In this regard, the analysis of Pattee on the complementary roles of the dynamic and linguistic modes, of the law-based and rule-based models, is highly significant. So too is his view of the brain as a generalized measuring device (giving rise to Dawkins' memes and their spread) which accounts for his notion of emergence and the evolution of creative ideas, and which is consistent with Huxley's notion of progressive psycho-social evolution, and Teilhard's conception and growth of a noosystem. Nor does Pattee's thinking deny the Gaia hypothesis, or Dawkins' selfish gene, Gould's punctuated equilibrium and Mayr's speciation hypotheses. (It does, however, question Wilson's sociobiologic theses). There remains, as has always been the case in man's consciousness and reflective thinking, unanswered questions—as there are likely always to be as man's stock of knowledge increases as well as his capacities that generate at every point of time new unobservables to be identified and relationships to be understood.

Notes

1 In a more satisfactory way, we may state for a given region J that its production output (Y^J) depends on (is a function f^J of) capital input (K^J) and labor input (L^J), that is

$$Y^J = f^J(K^J, L^J).$$

The goods available for investment in J (I^J) is equal to its production output (Y^J) less what the region consumes (C^J) less what it exports (E^J) plus what it imports (M^J), that is

$$I^J = Y^J - C^J - E^J + M^J.$$

The increase in its capital stock (ΔK^J) equals its investment (I^J) where we ignore the need to replace capital that is worn out, that is

$$\Delta K^J = I^J$$

Or when the unit of time Δt is conceived as very small (infinitesimal) in order to exploit the basic principles of calculus, ΔK^J is written as dK^J/dt so that at any given point of time t the equation becomes

$$dK^J/dt = I^J$$

When we replace I^J in the second equation above by dK^J/dt, we obtain

$$dK^J/dt = Y^J - C^J - E^J + M^J,$$

a rate-determined process, where the dK^J/dt corresponds to acceleration of the economic system.

For the system of many regions, the equations are conservative, that is over all regions, total imports are equal to total exports

2 For example, in economics the dynamic input-output technique models rate-dependent processes governed by constraints. The bill of goods (final demand) represents one set of constraints reflecting many policy decisions, e.g. export and import policy and magnitude of government programs. The production coefficients represent another. The household consumption coefficients and capital coefficients represent still others. The last three sets of constraints come from the information subsystem M. The first set may involve policy constraints from the C^J, C^L and I subsystems.

Pattee (1978, p. 198) refers to social dynamics as what happens once constraints from the linguistic mode and the environment external to the system are imposed on interactions of society (the P^J, P^L and R boxes). The social dynamics correspond to the football play that takes place once the center, receiving his cue and given the weather, hikes the ball to the quarterback. A social policy or goal or plan is what some group or behaving unit decides to implement as a constraint. It corresponds to the outcome of the football huddle that is the result of a cognitive process which may take any amount of time (within the maximum time allowed) and which is rate-independent.

3 In well-specified dynamic systems, we can set terminal conditions instead of initial ones, and then derive the initial from these terminal. We can in effect turn time backwards, which we cannot do in real life and with linguistic processes in general. Processes in the linguistic mode do not need a unique reference point in time, nor a unique unit of time. Since the life-cycle of any behaving unit involves linguistic processes, the life of a behaving unit cannot be reversed.

4 Even when experts agree on what is to be measured and what constitutes perfect measurement (e.g. a perfect lens), they may sharply disagree on a measure of lack of fidelity or distortion.

5 Recall that natural selection operates directly on the phenotype and determines which phenotypes survive. By doing so it indirectly determines the genotype and actual DNA that are living at any given point of time.

6 This point can be specifically made with respect to Figure 4.1. Suppose the information box M, which covers description is a dynamic model of the system's workings (P and R) and of its cognitive processes (C and I), assuming that these processes could be described by laws of motion. Then to predict or forecast the behavior within the P, R, C and I boxes,

the hypothesized model M in some sense must run faster than the models in the P, R, C and I boxes, Now if the system is to adapt or improve its model M, it must do so on the basis of the total system behavior (M+C+I+P+R). This implies an improved dynamic model M' of (M+C+I+P+R) that must run faster than M. But the M' model then requires an M" model which in turn would require an M"' model—leading to infinite regress. Hence to avoid infinite regress, time must be left out of the rules and symbols of M, and M must therefore be a nondynamic descriptive model.

7 See also Rosen (1974).

8 The notion of aesthetics as the ultimate criterion in the choice of social policy suggests that there is no such thing as a social welfare function but simply rules and procedures that are feasible, generally acceptable and hence in general aesthetically pleasing.

References

Hopfield, J.J. (1982), 'Neural Networks and Physical Systems with Emergent Collective Computational Abilities,' *Proceedings of the National Academy of Science*, Vol. 79, pp. 2554-58.

Hopfield, J.J. (1984), 'Neurons with Graded Response Have Collective Computation Properties Like Those of Two-state Neurons,' *Proceedings of the National Academy of Science*, Vol. 81, pp. 3088-92.

Isard, W. and Smith, C. (1982), *Conflict Analysis and Practical Conflict Management Procedures*, Cambridge, MA.: Ballinger (reprinted, Department of City and Regional Planning, Cornell University).

Isard, W. (1988), *Arms Races, Arms Policy and Conflict Analysis:* New York: Cambridge University Press.

Kosko, B. (1992), *Neural Networks and Fuzzy Systems: A Dynamic Systems Approach to Machine Intelligence*, Englewood Cliffs, N.J.: Prentice-Hall.

Pattee, H.H. (1977), 'Dynamic and Linguistic Modes of Complex Systems,' *International Journal of General Systems*, Vol. 3, pp. 259-66.

Pattee, H.H. (1978), 'The Complementarity Principle in Biological and Social Structures,' *Journal of Social and Biological Structures*, Vol. 1, pp. 191-200.

Pattee, H.H (1992a), 'The Measurement Problem in Physics, Computation, and Brain Theories,' in Carvallo, M.E. (ed.), *Nature, Cognition and System*, Winschoten, Netherlands: Klewer Academic Publishers, Vol. 2, pp. 179-92.

Pattee, H.H. (1992b), 'Emergence in Physical, Computational, and Network Models,' in Patel, M.J. and Harvey, I. (eds.) *Complex Adaptive Systems and Evolution of Information.*

Pattee, H.H. (1993), 'The Limitations of Formal Models of Measurement, Control and Cognition,' in Conrad, M. (ed.), *Applied Mathematics and Computation*, Vol. 56: pp. 111-30.

Rosen, R. (1974), 'Planning, Management, Policies, Strategies: Four Fuzzy Concepts,' *International Journal of General Systems*, Vol. I, No. 4, pp. 245-52.

Rosen, R. (1991), *Life Itself*, New York: Columbia University Press.

Smolensky, P. (1988), 'On the Proper Treatment of Connectionism,' *Behavior and Brain Science*, Vol. 11, pp.1-23, commentary, pp.23-74, and continuing commentary, (1990), *ibid.*, Vol. 13, pp.399-412.

Appendix A: Computational experimentation and cellular automata

Pattee has pointed out the virtues of computer simulations for obtaining patterns and results which are beyond the capacity of the brain to perceive and derive, with or without physical laws and mathematics. He also noted their limitations in creative problem solving and projecting emergence phenomena. Other scholars have also examined these matters, and we now wish to report upon their writings, particularly on cellular automata. In doing so we draw heavily upon the work of Wolfram (1984, 1986), and paraphrase him extensively.

Consider a cellular automaton. It is a simple type of system with the following characteristics.

1. It has a discrete number of identical mathematical cells as components.
2. These cells can be arranged in a regular manner along a single dimension (as a sequence of equally spaced points along a line), or in two dimensions (as a set of squares or hexagons), or in any number of dimensions (as a regular lattice).
3. Each cell has a value at each point of time, and this value is updated in a sequence of discrete time steps, each commonly designated as a 'tick'.
4. The possible values in each cell are the same; they are finite in number, often just 0 or 1.
5. The value in each cell is updated simultaneously according to a fixed deterministic rule.
6. The rule for determining this value for a given cell depends only on the values in neighboring cells within some finite distance (for example, the values in a first tier of neighbors, or in the first two tiers, etc . . .) at (i) the current point of time, or (ii) the current and previous point of time, or (iii) the current and some fixed number of previous points of time.

As Wolfram (1984, pp. 194-6) states:

> cellular automata can serve as explicit models for a wide variety of physical processes. Suppose ice is represented on a two-dimensional hexagonal grid by cells with the value 1 and water vapor is represented by cells with the value 0. A cellular-automaton rule can then be used to simulate the successive stages in the freezing of the snowflake. The rule states that once a cell is frozen it does not thaw. Cells exposed at the edge of the growing pattern freeze unless they have so many ice neighbors that they cannot dissipate enough heat to freeze. Snowflakes grown in a computer experiment from a single frozen cell according to this rule show intricate treelike patterns,

> which bear a close resemblance to real snowflakes. A set of differential equations can also describe the growth of snowflakes, but the much simpler model given by the cellular automaton seems to preserve the essence of the process by which complex patterns are created. Similar models appear to work for biological systems: intricate patterns of growth and pigmentation may be accounted for by the simple algorithms that generate cellular automata.

While we are not particularly interested in the specific stages of formation of a snowflake or a pigmentation pattern we are interested in the processes involved and all other processes such as those associated with the DNA that can be depicted or viewed as the operation of a computer. Scientific laws, as already noted, give algorithms (procedures) for converting input into output, for taking one body of information and converting it into another. In effect the algorithms determine how systems behave. A computer program can express and apply an algorithm. It can modify the numbers and symbols that represent objects and mathematical structures in the way specified by a given scientific law.

Thus a 2-dimensional, zero-one, cellular automaton, which is a *universal computer*—that is a computer which can solve any computable problem (can simulate any process that can be precisely described and perform any calculation that is performable)—can depict the operation of scientific laws and deduce their consequences.

Carrying through a computer program can also be viewed as performing an experiment. The objects in a computer experiment may be real ones, and the program may be designed to replicate a real input-output process —real in the sense of what an investigator perceives. However, the experiment need not be bound by the laws of nature. The computer program can generate output from input based on any law having a consistent form. It therefore can investigate processes of an hypothetical world. For example, it could examine spatial interactions and development among an hypothetical set of abstract regions (as in Isard and Liossatos 1979, chaps. 7-9) or multination trade based on a social scientist's adaptation of a gravity model once appropriate parameters and constraints are specified (Isard, 1960). It can thereby extend and foster experimentation.

Moreover, the computer can move into areas too complicated for the use of traditional mathematical methods. When such methods fail to assist in tackling a problem, a substitute set of much less appropriate mathematical functions and constructs is often developed for use—as much for their mathematical simplicity and tractability as for their capacity to model the salient features of a phenomenon. In economics, for example, trade relationships among nations or regions are frequently depicted by a constant elasticity of substitution function. This function yields a simple ratio with which to project substitution among sources of a commodity when relative prices change—a function which with a computer program can help often to lead to better projection of trade but at best can only be regarded as far-fetched for representing the real underlying behavior of trading partners.

Another place in the social sciences where conventional means fail is in depicting the time-path of variables such as investment, labor force and pollution (for example in the Isard-Liossatos 1979 regional growth model), especially when coupled with spatial diffusion processes, such as that of capital stock. Thus, given their capability at treating the more complicated mathematical functions that are identified as appropriate, computers can help provide deeper understanding of trade growth, spatial diffusion and many other processes.

Further, in many situations conventional mathematical means can be used, but require an excessive amount of computation. This is so in an effort to describe the motions of an electron in a complicated mathematical field, or the production of any single industry in a linear system that interrelates say 500 industries. Here computer use is practical. It is also preferable for use to avoid undesirable simplifications which the investigator with limited research resources would be required to undertake, such as combining 500 industries into 20 aggregate industrial sectors in order to use the conventional preexisting means.

The use of a computer has further attractions. One is that such use permits the investigator to interact directly with it. Since computers can spout out results rapidly the investigator can find it practical to consider numerous changes in various aspects of a model of phenomenon or system operation—changes, for example, in the reaction parameters, constraints and so forth. For the policy analyst, this means that he/she can examine implications of many different hypothetical structures of say an economy and many different policy scenarios. This possibility is, for example, a feature basic for justifying the construction of integrated world region models and integrated multiregion models of economists and regional scientists (see Isard and Smith, 1983). Such models have grown out of an increasing recognition by these scholars that in a real world no policy can or does operate in a vacuum. A policy is frequently interconnected with at least one other policy. An environmental policy is interconnected with energy policy, transportation policy, employment policy, industrial development policy and fiscal policy, among others. Hence, an investigator needs to examine the implications of several combinations of interconnected policies, each being a policy scenario that can be inputted into a model—and then after examining these implications to experiment with changes in the combinations—ultimately to identify those combinations which are likely to be satisfactory, desirable or among the best.

The computer can also perform highly desirable simulations of processes in which there are innumerable units in action (molecules moving, individuals behaving) and where averaging and the use of a differential equation pertaining to the behavior of the average turns out to be an inappropriate representation. This is so for the self-avoiding random walk and space-time processes where distributed differential equation sets are relevant but cannot be depicted by the behavior of a simple average (as in the Master equation approach)—except under very special conditions (Isard and Liossatos, 1979, chap.10). In general, one must resort to a computer experiment to obtain the evolution of spatial patterns, and at any point of time the possible fan of future distributions.

As already noted, the 0-1 two-dimensional cellular automaton is a universal computer capable of treating all problems that are computationally reducible. But it can also provide simulations when problems are *irreducible*, that is when an outcome can be found effectively only by simulating explicitly the evolution of a process, as is the case with models of fluid turbulence and other processes that are subject to deterministic chaos. It can generate valuable information under different assumptions on forthcoming outcomes (evolution) otherwise unobtainable and on problems that are theoretically *undecidable*—that is when no finite simulation involving a fixed set of steps can be guaranteed to yield the ultimate behavior of the system being investigated.

References

Fredkin, E. (1988), Interview in Robert White, 'Did the Universe Just Happen,' *Atlantic Monthly*, April, pp.29-44.

Isard, W. and Liossatos, P. (1979), *Spatial Dynamics and Optimal Space-Time Development,* Amsterdam: North Holland.

Isard, W. and Smith, C. (1983), 'Linked Integrated Multiregional Models at the International Level,' *Papers*, Regional Science Association, Vol. 51, pp.3-19.

Isard, W. (1990), 'Progress in Global Modeling for World Policy on Arms Control and Environmental Management,' *Conflict Management and Peace Science,* Vol. 11, No. 1, pp.57-94.

Layzer, D. (1990), *Cosmogenesis: The New Synthesis,* Cambridge, MA.: Harvard University Press.

Toffoli, T. and Margolus, N. (1987), *Cellular Automata Machines*, Cambridge, MA.: MIT Press.

Wolfram, S. (1984), 'Computer Software in Science and Mathematics,' *Scientific American,* Vol. 251, September, pp.188-203.

Wolfram, S. (1986), *Theory and Applications of Cellular Automata*, Singapore: World Scientific Publishing Co.

Appendix B: Concurrent, distributed network/connectionist models

Smolensky (1988, 1990) presents one of the best discussions of connectionist models. (However, also see Hanson and Burr, 1990). He defines the *symbolic paradigm* (the paradigm using symbols) as that involved in the development of Artificial Intelligence (AI) and like computer programs to serve as models of psychological performance. (See Newell and Simon, 1976; Newell, 1990.) He then defines the *subsymbolic paradigm* (the paradigm using subsymbols) as that followed by *connectionist* models. His connectionist models relate to large networks of simple parallel computing elements. The level of analysis of these models is lower than that in the AI symbolic models (suggesting still another level in an analytic hierarchy). At the same time, however, they are more closely related to the level of AI models than to that of neural models of the brain's functioning. He views the elements of connectionist models as fine-grained and suggests that they can be said to correspond to *constituents* of the symbols used in the symbolic paradigm. He thus designates these elements as *subsymbols.*

Each element i of a connectionist model carries a numerical *activation value* computed by some simple numerical formula from the value of each neighboring element j in its network. The influence I_{ji} of j on i is the activation value of unit j times the strength of the connection from j to i; thus the influence can be *excitatory* when *j's* activation value is positive, and *inhibitory* when this value is negative. The connection strengths are usually regarded as encoding the system's knowledge.

As Smolensky (1988, p.1) states:

> in a typical connectionist model, input to the system is provided by imposing activation values on the *input units* of the network; these numerical values represent some encoding, or *representation*, of the input. The activation on the input units propagates along the connections until some set of activation values emerges on the *output units*; these activation values encode the output the system has computed from the input. In between the input and output units there may be other units, often called *hidden units*, that participate in representing neither the input nor the output.

At a given point of time the numerical activition values of the input and hidden units, the processors, form a large state vector. The equation governing the changes in this vector, reflecting the interactions of the processes (thus the influence I_{ji} on one another), is designated the *activation evolution equation*, being a differential or appropriate difference equation. It may, for example, represent a quasilinear dynamical system, where each unit computes its value by using a nonlinear equation of the weighted sum of inputs (the influences I_{ji}) from other units. To yield (establish) learning which is associated with change in connection strengths among units there is another differential (or difference) equation, *the connection evolution equation.*

From the above description, it is clear that along with the semantic difference between the symbolic and subsymbolic paradigms, there is a *syntactic difference*. Subsymbol operations are not carried on, as symbol operations are, by a single discrete manipulation. Rather they involve a large number of finer-grained (numerical) operations. As a consequence, concurrent, distributed networks of a connectionist approach can learn new classifications. Classifications result from a many-to-one mapping. For example, in a physical system a class must correspond to a statistical stationary state attainable from many initial conditions. So, too, with networks; a number of input patterns can come to be identified with a particular class—and thus at times a new set of classes (new observables) can be identified at the statistical level without the intervention of a physical or biological law. Thereby learning results and the basis for emergent behavior comes to exist.

References

Hanson, S.J. and Burr, P.J. (1990), 'What Connectionist Models Learn: Learning and Representation in Connectionist Models,' *Behavioral and Brain Sciences*, Vol. 13:3, pp.471-510.

Newell, A. and Simon, H. (1976), 'Computer Science as an Empirical Inquiry,' *Comm. of the Association for Computing Machinery*, Vol. 19, pp.113-26.

Newell, A. (1980), 'Physical Symbol Systems,' *Cognitive Science*, Vol. 4, pp.135-83.

Smolensky, P. (1988), 'On the Proper Treatment of Connectionism,' *Behavioral and Brain Sciences*, Vol. 11, pp.1-74 and continuing commentary (1990), *ibid.*, Vol. 13, pp. 399-412.

Appendix C: Adaptive fuzzy systems models

Still another major extension of the role of computer modelling and experimentation is associated with recent developments on the combined employment of an artificial neural network model and a fuzzy systems analysis, sometimes designated an *adaptive fuzzy system model*. Such can provide additional insights on how the human mind might confront uncertainty, imprecision and noise in its perceptual activity, concept formation and common-sense reasoning—phenomena that the symbolism and symbolic manipulation of Artificial Intelligence and like cognitive models cannot address.

As Kosko (1992a, p. 13) states:

> neural networks and fuzzy systems estimate input-output functions. . . Unlike traditional statistical estimators, they estimate a function without a mathematical model of how outputs depend on inputs. They are *model-free* estimates.

According to Kosko they 'learn from experience' with sample data, being either numerical or verbal statements. In this respect, together they behave as intelligent systems, usually adaptively estimating continuous functions without specifying mathematically an input-output relation while at the same time emitting appropriate problem-solving responses to problem stimuli. The functions are continuous in the sense that small input changes produce small output changes.

As with connectionist models, artificial neural networks consist of numerous, simple processing units or 'neurons' with many 'synaptic' connections between neurons. These networks can be trained to store, recognize and associatively retrieve patterns or database entries. Their capability, particularly at distributed encoding, enables them to complete partial patterns, to filter out noise from messy patterns—and in general to estimate changing continuous functions as additional sampling comes from experience. They behave as adaptive function estimators.

Fuzzy systems, unlike neural, are abstract and do not claim biological or psychological plausibility. They store banks of common-sense rules but reason with parallel associative inference. To infer an output, conclusion or answer (which can be a fuzzy or multivalued set) from an input (which also may be a fuzzy or multivalued set), a fuzzy system fires each fuzzy rule in parallel, but to different degree.[1]

In a combined operation, a neural system may take in representative numerical samples (including training data) and then generate parameter estimators to be embedded directly in the fuzzy rules of an overall fuzzy architecture[2]—thereby forming an adaptive fuzzy system. Or they can be combined in the reverse order and operated over several runs. To illustrate the latter, consider the task of Sawargi et al. (1992) of understanding geological images transmitted from remotely-sensing satellites—in particular to develop as best as possible a map representing the distribution of linear features on the surface of the earth. Employing a class of geological hypotheses called 'wrench tectonics,' the researchers apply fuzzy clustering methods to the satellite images that serve as inputs to derive quantified 'belief' values (estimates of values) for each of eighteen labelled elements [in particular for six geological features, each one of which can be further classified by each of three subfeatures (representing three types of distinct uneveness labels.)] The eighteen labelled elements are taken as eighteen nodes ('neurons') in an artificial neuron network. The weights of connections between nodes are learned from sampled information (the training data) and serve as constraints; and the model is run starting with the quantified belief value as the activation value for each node. Subsequently with feedback to the fuzzy clustering stage and rerunning of the neural model, the above combined fuzzy analysis and neural network operation yields an image understanding of the satellite map representing an intellectual, focused man-machine communication.

Kosko views AI models and like symbolically-oriented cognitive models as rule systems developed on the basis of logic and traditional scientific investigation. They take in an initial well-defined input and may reach a response directly in one step by the use of a well-defined existing rule, or indirectly by the use of a chain of such rules. No intuition or judgment enters into the process. In contrast, the combination of an artificial neuron network and fuzzy system analysis can take in fuzzy data and perceptual information (e.g. pattern recognitions which are undefinable) as input (often as falling within some category of input data). By the employment of fuzzy relationships and weights (based upon intuition and judgment) applied to data and relationships, the analysis develops a principle (a possible forerunner of a new rule) on which to reach a response (often an existing response category[3]) and on occasion a new response[4] (response category). Thus, in the latter case, an artificial neuron model-fuzzy system analytic combination can yield a valuable creative response. While its creative nature falls far short of what the brain does as described by Pattee, it does point up a potential for the use of computational methods involved in both artificial neuron models and fuzzy system analysis for advancing our knowledge of how the brain actually operates, and in turn our understanding of human behavior.

Notes

1 To be more specific, we may consider the unknown function

$$y = f(x) \quad 4.C.1$$

where stimulus input x generates the response output y. Around input x there may be other stimuli x' where the closer x' is to x, the more it resembles x. Together these stimuli may form a ball B_x. In like manner there may be y' close to y that form a ball B_y. Hence, loosely speaking we may imagine that our function f relates the two balls

$$B_y = f(B_x) \quad 4.C.2$$

in the sense that balls are so constituted that for every y' in B_y we can find some x' input in B_x. Actually, take an input space X of some dimension and an output space Y of the same or other dimension so that we can specify a reasonable (meaningful) input-output product space $X\text{x}Y$. In effect the fuzzy system estimates a joint probability density $p(x,y)$ where the solution points $(x, f(x))$ reside in high-probability regions of the product space $X\text{x}Y$. That is, it identifies for any given x in this high probability region the probability of each of the diverse outputs y in this region, each output being generated by a particular FAM (fuzzy associative memory) rule; and looking backwards, for any given y in this region the system identifies the probability of each x in this region being the specific x in the function $y = f(x)$.

2 Sometimes, fuzzy rules may be given by experts.

3 As implied by the text discussion, these input and output categories are not clearcut, and any new sample input-output data are assignable to these categories in a nearest-neighbor sense, that is to a lesser or greater degree. Any prototype analysis that is employable here does not have the clearcut character of that employed in AI-type models where categories are defined to cover elements that satisfy precisely a list of well-defined properties.

4 Typically, novel or creative behavior represents a new response that involves a small change from a previous response to the same input (stimuli) or to a small change in the input. When a drastically changed input (stimuli) that had not been experienced previously requires a creative response, often impractical inappropriate ones are designed (constructed) before an appropriate one, if ever, is hit upon, perhaps

unexpectedly. The latter, if found, might correspond to Pattee's emergence phenomena.

References

Hopfield, J.J. (1982), 'Neural Networks and Physical Systems with Emergent Collective Computational Abilities,' *Proceedings of the National Academy of Science*, Vol. 79, pp. 2554-58.

Hopfield, J.J. (1984), 'Neurons and Graded Response Have Collective Computation Properties Like Those of Two-state Neurons,' *Proceedings of the National Academy of Science,* May, Vol. 81, pp. 3088-92.

Kosko, B. (1992), *Neural Networks and Fuzzy Systems: A Dynamical Systems Approach to Machine Intelligence*, Englewood Cliffs, N.J.: Prentice-Hall.

Sawaraji, T., et al. (1992), 'Integrative Cognitive Architecture for Image Understanding Using Fuzzy Clustering and Structured Neural Network,' *IEEE International Conference on Fuzzy Systems*, Piscataway, N.J.: IEEE Service Center, pp. 21-9.

Zadeh, L. A. (1987), *Fuzzy Sets and Applications,* New York: Wiley.

5 Physical, chemical and cosmological processes, and hierarchical structure

5.1 Introduction

In previous chapters, we have had occasion to speak about space-time, interaction, hierarchical structure and a number of other concepts. We now examine these concepts from the standpoint of physics, chemistry, and cosmology. From such examination, we hope to obtain more precise or extended definitions and views of these concepts, especially that of hierarchical structure. In this way we hope to deepen the thinking of Teilhard, Dawkins, Mayr, Wilson, Pattee, Smolensky, Bohm, Wolfram, Kosko and their colleagues in evolutionary and brain theory, and in computer experimentation. Perhaps our statements pertaining to concepts may not be as rigorous as some scientists might insist upon; nevertheless we hope they will prove adequate, given the objectives of this book. We should also bear in mind that the fields of physics and cosmology are perhaps more than ever in a state of flux, and what is considered sound or best doctrine today should not be taken as sacrosanct. From experience and the extensive experimentation being carried on, major change in accepted theory and doctrine can be expected.

5.2 Symmetry breaking, particle formation and the hierarchy of matter

We start with a look at the beginnings of the universe. As we write this chapter (1995), the most widely accepted scientific doctrine is that of the Big Bang hypothesis. According to this hypothesis there occurred a Big Bang, a colossal fireball, roughly 15 billion years ago, when all the matter and energy that has been observed in the universe had been concentrated in a limited region—a region smaller than a dime according to Peebles, Schramm, Turner and Kron (1994). Billions of diverse particles (the Ur particles) were present—the predecessors of the electrons and positrons, protons and antiprotons, quarks and antiquarks, neutrinos and antineutrinos, and other particles known today and still others unknown that may be unearthed by grand unification theories and experimentation. Via interaction these particles were instantaneously created and annihilated at the enormous temperatures and densities that prevailed; hence, they were indistinguishable, exchangeable one for the other—just the matter and antimatter of a soup. Concomitantly, the region of matter and energy began to expand and cool at an incredible rate. And when 'the universe had expanded an additional 1,000 times, all the matter we can measure filled a region the size of the solar system' (Peebles et al., 1994, p.53).

Also, at the time of creation, when temperatures and energies were well beyond those currently accessible in laboratories and likely always to remain so, the four basic forces that physicists have identified as of today—gravity, strong, electromagnetic and weak—and their interactions could not be distinguished; a single force prevailed. Thus, symmetry prevailed within the universe of matter and forces. As temperature dropped, density and average energy decreased and gravity separated off from the primordial force—a first break in the symmetry of the universe—and perhaps the graviton particle, which is yet to be experimentally confirmed, appeared (became built up and survived) as a distinct elementary particle. See the bottom levels of Figure 5.1. With further temperature decline and spreading of the universe, particles came to assume different identities in a kind of internal build-up and crystallization; quarks[1], leptons[2], and gluons[3] appear. The quarks and leptons were able to keep their identity under collision. Concomitant with their appearance was another phase change—the further breakdown of the symmetry of the remaining force into the strong force and electroweak force. Again, see Figure 5.1. The strong force, as carried by colored gluons, in time came to combine quarks to form hadrons such as protons[4] and neutrons.[5] Still later, with further reduction in density , temperature and average energy per particle, the electroweak force split into a weak force and an electromagnetic force (the end of electroweak symmetry).[6] Electrons[7], muons and tau-leptons, previously undifferentiated as leptons, take on specific identities, in a sense are internally crystallized from the lepton matter. Moreover, the photon, a massless field particle that travels at the speed of light and that carries the electromagnetic force, proliferated and there appeared as distinct particles the heavy W^+ (positively charged), W^- (negatively charged) and Z^o (chargeless) field particles that carry the weak force between leptons and hadrons . Once more, see Figure 5.1. Also, see Table 5.1 that records the exchange (messenger) particles and their roles, such as the role of the weak gluons in radioactive decay.[8,9]

With further cooling, nuclei and atoms could form and in time lead to the formation of stars, galaxies and planets. Later temperature drops led to the construction of both simple and complex molecules on the planet Earth—the building blocks of life. This is where, as reported in Chapter 3, the study of evolutionary biology by Dawkins and others begins.

Table 5.1 Exchange (binding) particles and their roles

Name and Symbol	Couples to	Role in Quantum Interactions
Graviton	Mass	Gravity; Cosmology; Binds Planets to the Sun, and Stars to Galaxy
Photon γ	Electric Charge	Electromagnetic Waves; Binds Electrons to Nucleus
Weak Gluons W^+, W^-, Z^0	Weak, Flavor Charges	Radioactive Decay; Breaks up Hadrons
Colored Gluons	Colored Charges	Strong Force; Binds Quarks Inside Hadrons

Adapted from Pagels (1982), p.232.

While Figure 5.1 presents developments at specific points of cosmological time, the points indicated being at best gross approximations, it is important for the analysis in the following chapters to have in mind the emergence of the hierarchical structure of matter that has been realized. We attempt to do so in an extremely crude manner in Figure 5.1. Following the discussion in Adair (1989, pp. 357-59), and assuming conservation of matter-energy generated at any limited region that experienced the Big Bang, we roughly sketch the build-up of matter, with reference to key constituents and without mention of antimatter. At the bottom are the billion units of particles in the Ur-matter. From these particles emerged the fermions, the gravitons (assuming they will be experimentally verified) and other particles. In time, as already remarked, the fermions became identifiable as quarks and leptons; and later quarks combined to form protons, neutrons and other hadrons. Still later, protons and neutrons came together to form nuclei; and further along in time, nuclei and electrons combined to form atoms, and then atoms came to interact to yield chemical compounds.

In effect a hierarchical structure came about where the billions of unidentifiable particles turned, in an irregular manner, into fewer units of

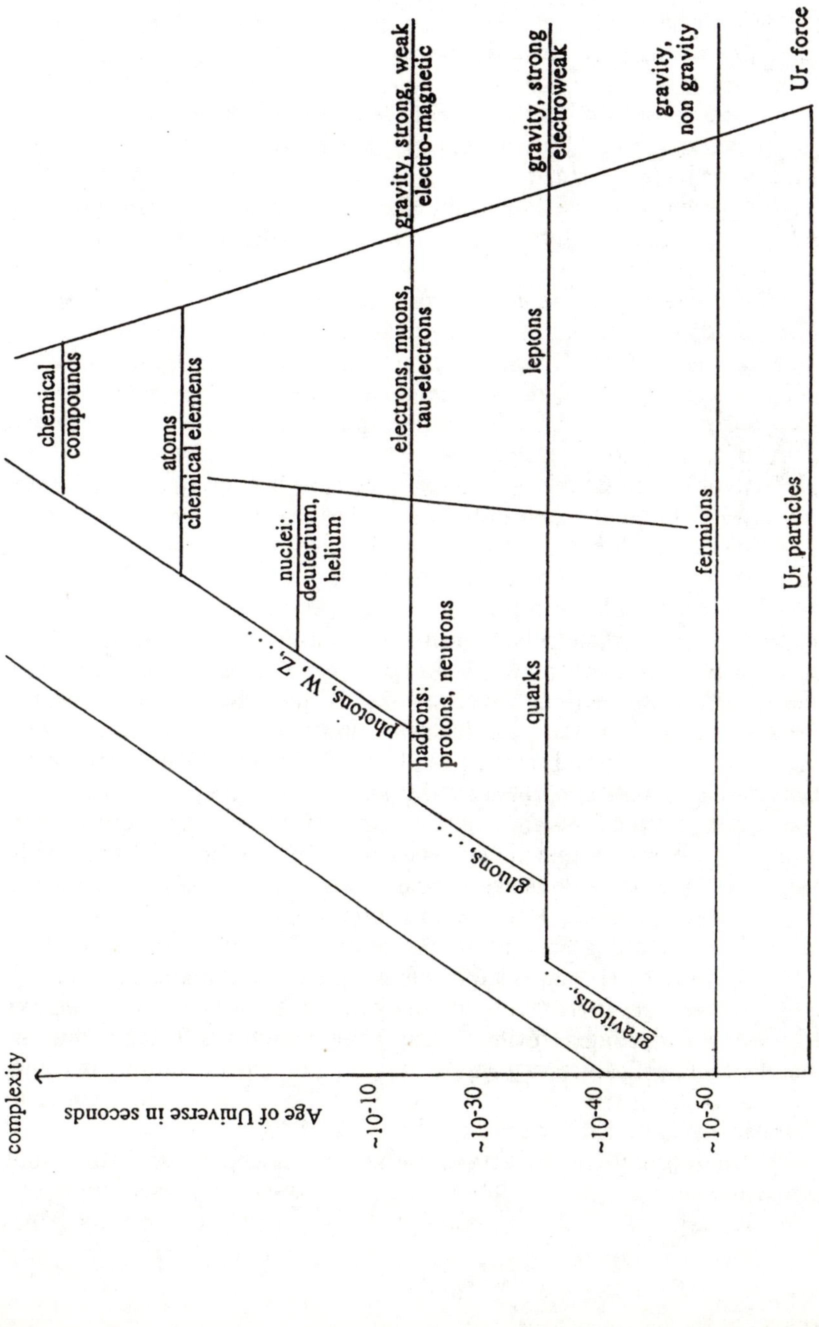

Figure 5.1 The early time path of the universe and hierarchical development

identifiable and more complex elements—the fewer as more and more complex matter evolved. Along the way, some particles that once existed were completely annihilated or disappeared during the formation of the more complex matter. Figure 5.1 of course reflects the state of knowledge in 1994, a time when particle physics was in rapid flux. (Note that the length of the several horizontal stretches in the figure should not be taken to reflect the number of units of the different matter that existed—but rather only to show the dimunition in number of units, more aggregate in nature, as time elapsed and as the process of build-up transpired.)

A few questions and additional remarks are in order. First, exactly how does the spontaneous symmetry breaking occur? To effect the spontaneous breaking of the electroweak force into electromagnetic and weak forces, a Higgs field of spinless (electrically neutral) particles is hypothesized to exist and provide mass for the resulting W^+, W^- and Z^0 particles. To effect the symmetry breaking of the unified force into the electroweak and strong forces, a field of still heavier Higgs particles is posited.

As Pagels (1982, p.233) has remarked 'the universe from its very beginning to the present may be viewed as a hierarchy of successively broken symmetries—a transition from a simple perfect symmetry at the beginning of time to the complex patterns of broken symmetries we see today.' Analytically speaking, and going backward in time (from the top to the bottom in Figure 5.1), one proceeds along the hierarchy of symmetry breaking theories to theories of greater and greater unification until the time immediately after the Big Bang when complete unification is hypothesized to prevail. To reiterate, at that time only one type of force and a single unified symmetrical field (interaction) is posited to exist.

Then there is a relatively new view of reality. The older view and one more easily corroborated by observation, is that the forces between discrete source particles (such as electrons, protons or quarks) result from the exchange of field particles (such as photons, gluons, and the postulated gravitons) emitted and absorbed by the source particles. An alternative view, and one adopted by the more abstract (intellectually grounded) quantum field theory, is that there exists a set of interacting fields with particles as quantum condensates of the fields. Force fields of photons, gluons and other particles interact with source fields of electrons, protons and quarks. As Weinberg has put it, '*the essential reality is a set of fields* subject to the rules of special relativity and quantum mechanics; all else is derived as a consequence of the quantum dynamics of those fields' (Pagels, 1968, p.239). Or as Pagels (1968, p.241) put it, the central dogmas of relativistic quantum field theory are:

1. The essential material reality is a set of fields.
2. The fields obey the principles of special relativity and quantum theory.
3. The intensity of a field at a point gives the probability for finding its

associated quanta—the fundamental particles that are observed by experimentalists.
4. The fields interact and imply interactions of their associated quanta. These interactions are mediated by quanta themselves.
5. There isn't anything else.

While to a modern theoretical physicist the essential reality may be a set of fields, to the reflective behaving unit the essential reality is what he perceives and it is not a set of fields. To him, the world is still one of objects (particle aggregates) which may be interconnected in one way or another. And from the standpoint of understanding human behavior and the society of humans, this viewpoint must prevail. Yet it is extremely valuable to ponder upon humankind as a set of fields in a quantum dynamics manner. For example, in regional science it has often been said that in appropriately defined situations the location of enterprises and the transport (movement) of commodities among them are as the two sides of the same coin. From a knowledge of transport patterns and system of transport facilities—observations on the transport field—one can deduce the enterprise location pattern (and vice-versa). Similarly, in appropriately defined situations, from the knowledge of population and idea movements—observations on movement fields—much can be learned about population distributions and the geographic pattern of idea generation. Given a modern physicist's view of reality as a set of fields whose features can be derived by the study of the quantum dynamics of these fields, the social scientist can learn much about location patterns and behavior of humans, groups and institutions by observing how physicists deduce information from field investigations.

We also should note here the extension by the physicist Bohm of field theory and related concepts to consciousness, a phenomenon to which much attention must be paid. Based in part on the interpretation of the intensity of a field at a point in space as the statistical probability for finding its associated quanta (the particles) and other aspects of both quantum and relativity theory, Bohm sets forth a doctrine of undivided wholeness, which involves a notion of an implicate order subject to unfoldment that yields the explicate order, namely the reality that man perceives. A presentation of his views is contained in the Appendix to this chapter.

Before closing this section, we should note the large number of unanswered questions and uncertainties in particle physics and cosmology. For example, from where do the Higgs particles come? Exactly what are they? They are yet to be established experimentally. Then there is the question of why particular sets of leptons and other particles exist, and why the particular masses, charges and spins of these particles. Such questions might possibly be answered once an overall unified theory incorporating gravity is achieved, and once the existence of gravity waves and gravitons are experimentally verified.

There is also the question of whether some (possibly all) particles now identified as elementary are in fact built up from new particles, preons, to be identified. Are quarks made up of subquarks, and subquarks of subsubquarks? Further, will the supersymmetry theory, an attempt to bring gravity into the unification fold, find experimental support and thus establish the existence of new subatomic particles that are sisters to existing ones—which then will belong to families that are different supersymmetric manifestations of a single superparticle?

Are there more dimensions than the four (time and the three space dimensions) that we humans perceive and currently use to describe nature? Are there for example extra dimensions that are rolled up into exceedingly tiny circles that we cannot perceive? Are fundamental particles (when conceived as oscillating string-like entities) multi-dimensional entities—to be described in ten space-time dimensions (nine space and one time) according to Grand Unified Superstring Theories?

Moreover, there are the questions regarding the origin of the universe, its expansion and contraction, the Big Bang hypothesis, and so forth.

To attack these many questions new theories are constantly being concocted. And, as in the past, a revolutionary one will in time come along to provide answers, but in doing so raise new questions, in all likelihood greater in number than the existing ones.

5.3 Stratified stability and chemical order

It is instructive to turn to the ideas of Bronowski (1970) and his notion of stratified stability, this notion being based upon our best current knowledge concerning the formation of the different atoms in stars.

As we remarked earlier, the first nuclei appeared when the temperature dropped sufficiently so that the particles, protons and neutrons, having become less agitated, could combine. The first nuclei were the lightest ones, those of deuterium (a heavy isotope of hydrogen) and helium. Later with still further temperature decline, the electrons were able to combine with nuclei to form atoms. Such formation was a major extension of the hierarchy of physical materials, or what some might consider the beginning of a new kind of hierarchy—the hierarchy of atomic elements. For via the process of nucleosynthesis, heavier atomic elements became cooked up from lighter elements—for example the carbon from the helium.

Bronowski was motivated to set forth the notion of stratified stability in order to explain why there can be direction in evolution—why increasing order, complexity and hierarchical structure emerge despite the entropy implications of the Second Law of Thermodynamics. He wants to explain: why evolution has not run backward, although in the case of viruses there has been some retrogression; why evolution does not proceed hither and thither over time; why for evolution the arrow of time is a barbed arrow; and specifically why with evolution '*there is an inherent potential for large scale order to act as a sieve* or selector on the individual chance events' (Bronowski, 1970, p. 29).

Five distinct principles form the basis of Bronowski's concept of evolution. They are: family descent, natural selection, Mendelian inheritance, fitness for change and stratified stability. The first three have already been discussed in Chapter 3. The fourth relates to the point that in order for a species to be able to change to fit into a tomorrow's environment characterized by different stresses, it must maintain its fitness for change today. Copy-erroring must be facilitated (enhanced) and/or dormant genes must exist. That is, natural selection must also take place for variability itself; mutator genes must exist. As Bronowski (1970,p.30) states:

> Indeed, we know now that there are single genes which function specifically to enhance variability. For example, there are single genes which increase the rate of mutation in several other genes at the same time. Their action could explain the tendency for genetic change in one part of an organism . . .to keep pace with change in other parts. A master gene of this kind, which increases mutation, is a mechanism that opens up the future, not by foreseeing it but by promoting the capacity for change.[10]

The fifth principle, stratified stability, is the one that is crucial for Bronowski's argument—a principle that Bronowski found lacking or inadequately appreciated in evolutionary doctrine. Clearly, evolution has involved the development of integrated systems, a fine coordination of functioning parts of an organism, as discussed in Chapter 3. Clearly, too, as these systems developed from an elementary state to an increasingly complex one, there must have been forces that precluded their disorganization or disassembly—at least precluded enough disorganization and disassembly so that we are able and do find the increasingly complex organisms and systems that have come into existence with the passage of time. Forces leading to stability must have been present, at least along certain lines of evolution, to more than offset those of instability. Bronowski, however, begins at a stage much earlier than the one at which organisms develop—at the stage of universe and star formation—and examines physical forces *per se* which operated without the intervention of natural selection.

Although today he could begin his argument with the formation of hadrons, nuclei, and the atom, he points to the evolution of chemical elements. We know that in nucleosynthesis helium (He) was built up in the stars from hydrogen (H), one summarization of the hydrogen burning process being

$$4H \rightarrow He + 2e^{+} + 2\upsilon + \text{energy}.$$

Here four hydrogens (each of mass one) are converted into an Helium particle (of mass 4), two positrons, two neutrinos, and energy (Tayler, 1972). In turn, conditions came to prevail whereby through helium burning, three helium atoms (each of mass 4) were converted to one carbon (of mass 12) and one photon plus energy (3He→1C + δ + energy), via the intermediate production of the unstable element beryllium (Be) from two helium atoms. At appropriate temperature and other conditions, oxygen (of mass 16) is formed via the synthesis of a carbon atom (of mass 12) and one helium (of mass 4) (C + He→ O + δ + energy).

At still higher temperatures, carbon burning and oxygen burning take place to yield the stable atoms neon (Ne), magnesium (Mg), silicon (Si) and phosphorous (P). For example, 2 oxygen atoms (each of mass 16) combine to yield one phosphorous atom (of mass 31) and a proton of mass one (2O → P + p).

At even higher temperatures, silicon burning and other reactions can occur yielding elements with mass numbers up to 56. The iron-peak elements of mass 56 (iron, nickel, cobalt) are associated with the end of the nucleosynthesis sequence by nuclear fusion. Reactions involving neutron capture and requiring rather than liberating energy yield the still heavier elements.

The formation of atoms in the above manner adds support to Bronowski's principle of stratified stability. In his words, the

> most telling example is the creation of carbon from helium. Two helium nuclei which collide do not make a stable element, and fly apart again in less than a millionth of a millionth of a second. But if in that splinter of time a third helium nucleus runs into the pair, it binds them together and makes a stable triad which is a nucleus of carbon. Every carbon atom in every organic molecule in every cell in every living creature has been formed by such a wildly improbable triple collision in a star (Bronowski, 1970, p. 31).

It is in this manner that Bronowski discusses how simple units form more complex units, which if stable serve as building blocks for still more complex units and so on—not only chemical units like iron but complex molecules such as hemoglobin which contain iron. (For simple illustrations in mathematical form of the transition process to a new phase of development in a social science context which generally speaking involve factors similar to those in the phenomena to which Bronowski refers, see Chapter 7A.)

Bronowski (1970, p.31) states the building up of stratified stability is present in living forms.

> Atoms build the four base molecules, thymine and adenine, cytosine and guanine, which are very stable configurations. The bases are built

> into the nucleic acids, which are remarkably stable in their turn. And the genes are stable structures formed from the nucleic acids, and so on to the subunits of a protein, to the proteins themselves, to the enzymes, and step by step to the complete cell. The cell is so stable as a topological structure in space and time that it can live as a self-contained unit. Still the cells in their turn build up the different organs which appear as stable structures in the higher organisms, arranged in different and more and more complex forms

Bronowski recalls that single mutations, as discussed in Chapter 3, are errors at random. They have no fixed direction in time; nor does natural selection provide a direction. However, the formation of stable configurations does provide direction to evolution. The sequence of more complex configurations that are stable, and therefore in general cannot be reversed, does give direction to evolution, just as the barb of an arrow prevents it from running backward. Each stable configuration represents not only a barb of an arrow but also a step which in the presence of mutation, represents a potential for sending evolution forward—and on different occasions and often in an unpredictable manner this potential is realized.

Bronowski takes up the question of the seeming contradiction between the non-entropic process involved in stratified stability and the Second Law of Thermodynamics. The Second Law, which states that all the parts which make up a system must fall progressively to their simplest states, does not in fact pertain to systems generally characterized by nonequilibrium states. It does not pertain to living systems that we have today, which are sustained by a constant net inflow of energy from the sun and wherein there are preferred configurations and hidden stabilities—such as exist at the most basic inanimate level of atomic structure and within living systems as already discussed. This may be so, even if these configurations (stable states) are rare. And when the hidden strata, configurations of stability, are one above another, as in our universe, the evolutionary process climbs them one by one. Bronowski (1970, p.34) concludes: 'What evolution does is to give the arrow of time a barb which stops it from running backward; and once it has this barb, the chance play of errors will take it forward of itself.'

Before leaving the discussion of the formation of chemical elements as phenomena supportive of Bronowski's stratified stability argument, we should note several aspects of order in the structure of chemical elements. There are the families of chemical elements—such as the alkali metals and the noble gases. Members of each family have similar properties—the members being listed in the same column in the standard periodic table. Chemical elements also have properties which occur periodically—which are associated with the number of electrons in their outermost shell and correspondingly with position along the row of the standard periodic table.

According to the shell model, the electrons surrounding a nucleus of an atom are located in successive spherical layers designated shells. One electron is in a first shell that surrounds the hydrogen nucleus, and two electrons in that first shell around the helium nucleus. In lithium, the next element in the periodic table, there are three electrons surrounding the nucleus, two in the first shell, and one in a second shell. In beryllium, boron and succeeding elements, additional electrons continue to be located in the second shell until that shell contains eight. Then for chemical elements containing still more electrons, as sodium which has eleven electrons, electrons are located in a third shell, for example, three for sodium. However, the third shell of an atom, such as zinc, can have as many as 18 electrons. The rule is that the maximum number of electrons per shell is $2n^2$ where n is the number of the shell when shells are numbered with integers beginning with 1. However, there is an overriding rule which states that the outside orbit can have no more than 8 electrons. The order behind this shell model is indicated in Table 5.2.

Thus, the orderly structure of chemical elements, as with their stratified stability, may provide, if only by analogy, insights for understanding living organisms whose structure, properties, behavioral propensities and institutional arrangements will be further examined in later chapters.

Table 5.2 Shell structure of atoms of the first 20 elements

Period	Atomic Number	Element	Number of Electrons in Shell			
			#1	#2	#3	#4
1	1	H	1			
	2	He	2			
2	3	Li	2	1		
	4	Be	2	2		
	5	B	2	3		
	6	C	2	4		
	7	N	2	5		
	8	O	2	6		
	9	F	2	7		
	10	Ne	2	8		
3	11	Na	2	8	1	
	12	Mg	2	8	2	
	13	Al	2	8	3	
	14	Si	2	8	4	
	15	P	2	8	5	
	16	S	2	8	6	
	17	Cl	2	8	7	
	18	Ar	2	8	8	
4	19	K	2	8	8	1
	20	Ca	2	8	8	2

5.4 Cosmic evolution and hierarchical clustering

The previous sections of this chapter have treated structure, order and hierarchy in areas where extensive scientific investigation has been able to be conducted to test hypotheses. In the study of cosmology, another area where order can be hypothesized, little *experimental* testing is possible; reliance must be had on observation. As a consequence, many unanswered questions about the cosmos exist and many divergent theories have been advanced. A number of theorists who accept the fireball hypothesis, believe that the existing clumpiness in the universe reflects the workings of many phenomena—for example, that stars came from the collapse of gas clouds, and galaxies from condensation within the expanding cosmic medium. Moreover, they speak of hierarchical structure. For example, Peebles et al. (1994) view the Milky Way as a knot of two dozen galaxies which are embedded in a complex of galaxies protruding from the so-called local supercluster.

Layzer (1990) another theoretical cosmologist, who does not accept the fireball hypothesis and in its stead sets up a Cold Universe model, also propounds the existence of hierarchical clustering—more particularly, of an hierarchy of nested self-gravitating systems. He emphasizes the fact that gravitation does not have a preferred interaction distance, as do the strong forces that hold atomic nuclei together. As a result, the size of an isolated planetary system, star cluster or galaxy can vary widely—a star cluster can be no bigger than a solar system, and a planetary system as large as a star cluster. Layzer (1990, pp.155-6) notes that the

> law of gravitation mandates no particular binding strength for a system of given mass. Yet astronomical systems of comparable mass do have comparable sizes and binding strengths. And when we compare the binding strengths of astronomical systems of different masses, a striking regularity emerges. . . .The binding strength increases with mass according to a simple rule: it is proportional to the cube root of the mass. For example, the masses of double stars, globular star clusters, and galaxies are in the ratios $1:10^6:10^{12}$, and their binding strengths are in the ratios $1:10^2:10^4$. This rule holds over a mass range of more than twelve powers of ten.

See Figure 5.2. Thus, after presenting his Cold Universe model, Layzer (1970; p.170) concludes that while the initial state of the Universe was wholly lacking in order, structural

> order began to emerge after the cosmic medium froze and shattered into planet-size fragments of solid hydrogen. The resulting cold gas of mutually gravitating 'particles' became unstable; density fluctuations grew and evolved into self-gravitating clusters. These, in turn, became

the 'particles' in a repetition of the process. In this way, a hierarchy of nested self-gravitating systems was gradually built up.

The Layzer hierarchical clustering and specific Cold Universe hypotheses are yet to find wide acceptance among astrophysicists and cosmologists. Nonetheless, they represent, at the minimum, a highly provocative set of ideas that can generate highly productive thinking and creative insights on hierarchical order in general, whether or not such is validated for the cosmos.

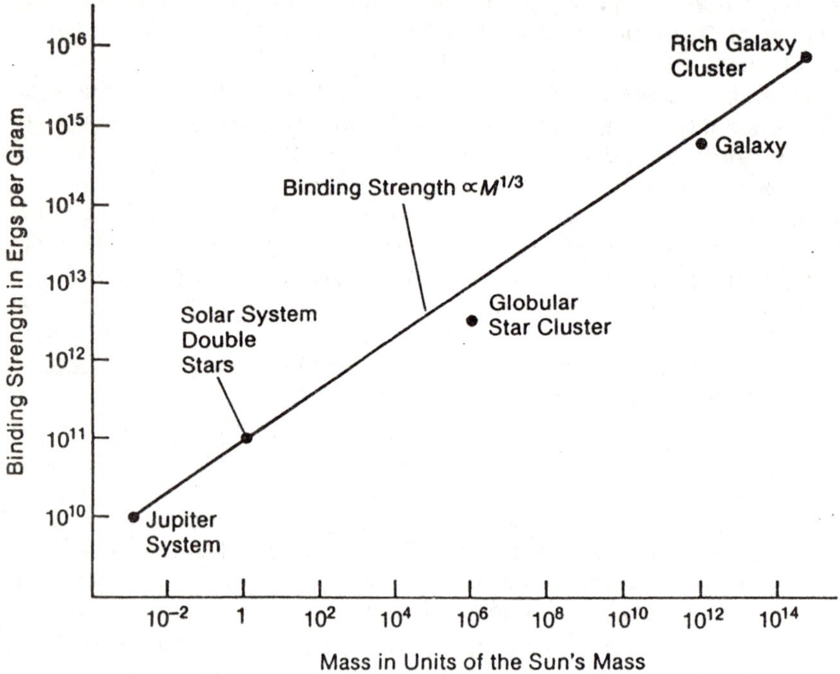

Figure 5.2 The binding strengths of self-gravitating systems
Source: *Layzer (1990, p. 156)*

5.5 Summary

In this chapter, we have examined questions of order and hierarchical structure in the physical world and continued our search for such in the biological world. We discussed hypotheses and analytical findings regarding symmetry breaking, elementary particles, the formation of nuclei and atoms—all with reference to the four fundamental forces (gravity, electromagnetic, weak and strong) and the alternative view of reality as a set of interacting fields. We found order and hierarchical structure in symmetry breaking and in the formation and construction of particles, nuclei and atoms. Extension of our discussion into the world of chemistry pointed up the presence of order in the appearance and structure of chemical elements. The existence of stratified stability in the formation of atoms of increasing mass and complexity, and the combination of fitness for change (via mutator genes) and stratified stability in the development of organisms has provided a barb for evolution, in general preventing it from running backward. Other order and hierarchical structure among chemical elements is expressed by the periodic table. There are families of chemical elements. There is order in the shell structure of electrons, and in their tier arrangement. And so forth. Finally, in the cosmos itself there is order among planets, stars, star clusters, galaxies, and galaxy clusters—all viewed by Layzer as a hierarchy of nested self-gravitating systems.

All these findings raise the question: to what extent does hierarchical structure pervade the socio-politico-economic world? Is such structure necessary for society's evolution and passage to an effective multi-cultural world government and economy? Ideally speaking, is hierarchical structure essential for a Teilhard Omega or equivalent? Moreover, as each new level is in time added to an hierarchical structure, is such necessarily associated with symmetry breaking? To these questions, we shall turn in later chapters.

The discussion of this chapter relates to order. But what about disorder? To this matter we turn in the next chapter.

Notes

1 Quarks are elementary particles. There are six kinds of quarks (designated up, down, charmed, strange, top and bottom), each having spin 1/2 and carrying one of three varieties of the color charge. The color charge is akin to electrical charge which quarks also possess.

2 Leptons are elementary particles, comprising electrons, muons, tau particles and neutrinos. They experience electromagnetic and weak forces, but not the strong quantum chromodynamic forces that quarks do.

3 Gluons are the force particles of the strong fields. Through exchange of gluons, quarks exert forces on other quarks. Also, there are forces among the gluons themselves since they carry strong interaction charges. Gluons do not carry electrical charge.

4 A proton is built from three quarks. It is a positively charged constituent of a nucleus.

5 A neutron is built from three quarks. It does not carry any electrical charge and together with a proton forms a nucleus.

6 As Weinberg (1977, p. 134) has put it:

> At temperatures below the critical temperature, the universe was as it is now: weak interactions were weak, and of short range. At temperatures above the critical temperature, the essential unity between the weak and electromagnetic interactions was manifest: the weak interactions obeyed the same sort of inverse-square law as the electromagnetic interactions, and had about the same strength.
>
> The analogy with a freezing glass of water is instructive here. Above the freezing point, liquid water exhibits a high degree of homogeneity: the probability of finding a water molecule at one point inside the glass is just the same as at any other point. However, when the water freezes this symmetry among different points in space is partly lost: the ice forms a crystal lattice, with water molecules occupying certain regularly spaced positions, and with almost zero probability of finding water molecules anywhere else. In the same way, when the universe 'froze' as the temperature fell below 3,000 million million degrees, a symmetry was lost—not its spatial homogeneity, as in our glass of ice, but the symmetry between the weak and the electromagnetic interactions.

7 An electron is the lightest elementary particle having mass. It carries a negative electrical charge.

8 The radioactive decay process can be illustrated with the Feynman diagram of Figure 5.3. There we see at the bottom a neutron composed of one *u* (up) quark with blue color and two *d* (down) quarks, one green and one red. In the beta-decay of the neutron, the weak force turns the *d* red quark into a *u* red quark and an emitted W^- weak particle. The resulting hadron (a *u* blue quark, a green *d* quark and a *u* red quark) is a proton. The W^- particle decays into an electron and an antineutrino, the latter being the antiparticle which is associated with the neutrino. A neutrino is an electrically neutral massless lepton yet to be identified experimentally.

9 Note that humans can sense the forces of gravity (in a fall from a cliff), electromagnetism (in an electric shock) and radioactivity (in the luminosity of certain dial watches). However, a human does not experience the strong force. This force takes place in the boring empty place of human sense perception that humans designate the vacuum. The vacuum is an arena of intense activity. It seethes with virtual particles—virtual electrons, antielectrons, quarks, antiquarks, and other particles which interact with each other. (A virtual particle is one with the wrong energy and momentum. It can exist for a short time, owing to the Heisenberg uncertainty principle—a principle which allows a quantum mechanical system to violate conservation of energy for a very short time dt, provided the magnitude of the product of the energy

imbalance *dE* and *dt* is less than the Planck constant.)

10 See also Layzer (1990, p.224) who distinguishes between *alpha genes* (which specify an organism's development program) and *beta genes*. The latter regulate mutation rates and rates of genetic recombination, 'suppressing variations that are likely to diminish fitness and permitting or promoting variations that are likely to increase fitness.'

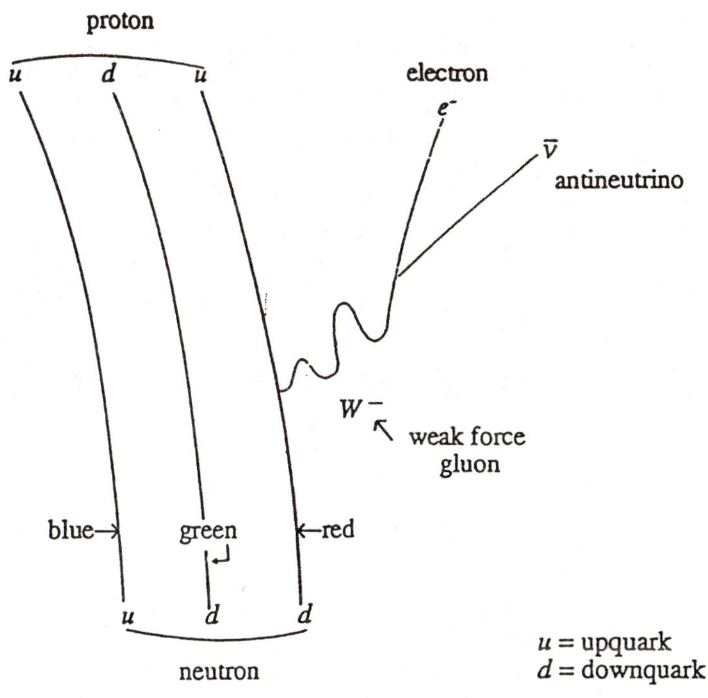

Figure 5.3 Radioactive decay: the transformation of a neutron into a proton, an electron and an antineutrino

References

Adair, R.K. (1987), *The Great Design*, New York: Oxford University Press.

Anderson, H.L. (1989), *A Physicist's Desk Reference*, New York: American Institute of Physics.

Bronowski, J. (1970), 'New Concepts in the Evolution of Complexity: Stratified Stability and Unfounded Plans,' *Zygon*, Vol.5, No.1, pp.18-35.

Cameron, A.G.W (1973), *Cosmochemistry*, Boston: D. Reidel.

Davies, P.C.W. (1989), *The New Physics*, New York: Cambridge University Press.

Gillespie, R.J., Humphreys, D.H., Baird, N.C. and Robinson, E.A. (1989), *Chemistry*, Needham Heights, MA: Allyn and Bacon.

Layzer, D. (1990), *Cosmogenesis*, New York: Oxford University Press.

Madhusree, M. (1996), 'Explaining Everything,' *Scientific American*, January, pp.88-94

Maran, S.P. (1992), *The Astronomy and Astrophysics Encyclopedia*, New York: Van Nostrand Reinhold.

Pagels, H.R. (1983), *The Cosmic Code*, New York: Bantam.

Pebbles, P.J., Schramm, D.N., Turner, E.L. and Kron. R.G. (1994), 'The Evolution of the Universe,' *Scientific American*, October, pp.53-57.

Tayler, R.J. (1972), *The Origin of the Chemical Elements*, London: Wykeham Publications.

Truman, J.W. (1973), 'Theories of Nucleosynthesis' in Cameron, A.G.W. (ed.), *Cosmochemistry*, Boston: R. Reidel.

Weinberg, S. (1977), *The First Three Minutes*, New York: Bantam.

Weinberg, S. (1994), 'Life in the Universe,' *Scientific American*, October, pp.44-9.

Weinberg, S. (1995), 'Night Thoughts of a Quantum Physicist', *Bulletin of the American Academy of Arts and Sciences*, December, pp.51-64.

Appendix 5.A: Undivided wholeness and implicate/explicate orders

5.A.1 Introduction

As noted in the preceding chapter, Pattee views the brain and sensimotor system of man as functioning coherently within the linguistic mode—only in part explainable by a complementary use of physical, computer and concurrent distributed network and related models. We have also noted Teilhard's projection of a noosystem so heavily dependent on the further development of the brain as an operational organ, *consciously* aiming at spiritual targets. Further, we have discussed the debate among evolutionary theorists as to whether the operation of the brain is by and large biologically dictated.

Other scholars with different orientations have also probed into the coherent functioning of the brain and sensimotor system. One of these is the physicist David Bohm (1981) who does so in terms of consciousness, and provides new, fruitful insights. We report on his work in the following section, paraphrasing his writings.

5.A.2 The undivided wholeness-implicate order framework

In a very broad-ranging discussion, Bohm sets forth the doctrine of undivided wholeness. To present this doctrine, begin with his views of *reality in general* and *consciousness in particular* as a dynamic (nonstatic) coherent whole unendingly undergoing movement. He maintains that often we apprehend things (objects) as static or as a series of static images; yet when we experience movement, we sense an unbroken, continuous process of flow to which we relate the series of static images formed in our mind. Actually we sense our very thinking and consciousness as a steady, uninterrupted phenomenon.

On further reflection, when we consider movement in our thinking and in an object under scrutiny, can we separate our movement in thinking from the reality of the object in movement or from reality in general? Traditionally we do so; we tend to view the brain as acting independent of reality. But Bohm would maintain that such a view cannot be maintained. Rather, the unbroken flow of reality (existence) contains both: (1) consciousness and associated reflection; and (2) the external phenomena and goings-on that we experience and perceive.

Bohm (1981, p.xv) develops his notion of *implicate or enfolded order* that is appropriate to a universe of unbroken wholeness which precludes the traditional scientific approach committed to the analysis of the world in terms of separately existent parts. In his implicate or enfolded order,

> space and time are no longer the dominant factors determining the relationships of dependence or independence of different elements. Rather, an entirely different sort of basic connection of elements is possible, from which our ordinary notions of space and time, along with those of separately existent material particles, are abstracted as forms derived from the deeper order. These ordinary notions in fact appear in what is called the *explicate or unfolded order*, which is a special and distinguished form contained within the general totality of all the implicate orders.

Being a distinguished physicist, Bohm is able to develop these notions at greatest depth when he has reference to the most advanced thinking in physics. In particular, he dissects quantum theory, currently the most widely accepted theory in physics, and Einstein's relativity theory. He finds much that is lacking in quantum theory. For example, he points up that in an experiment "the description of the experimental conditions does not drop out as a mere intermediary link of reference, but remains inseparable from the description of what is called the observed object". . . this "calls for a new kind of description that does not imply the separability of the 'observed object' and 'observing instrument'" (Bohm, 1981, p.133). Undivided wholeness of analysis is required, which he maintains is possible for quantum theory by identifying hidden variables at the sub-quantum-mechanical level and by pursuing analysis at a higher dimensional level than currently is the case.

Likewise, Bohm finds relativity theory lacking. Its dependence on the notion of a signal to correlate events that are separated in space and that are without possibility of connection through interaction is inconsistent with quantum theory. He therefore suggests that relativity theory drop its basic notion of signal and be fused with quantum theory into a new theory which retains those aspects of both quantum theory and relativity theory that are still valid. In doing so, the new theory should hold on to the nonlinear equations of relativity's field. As Bohm (1981, p.174) puts it,

> these equations could have solutions in the form of localized pulses, consisting of a region of intense field that could move through space stably as a whole, and that could thus provide a model of the 'particle'. Such pulses do not end abruptly but spread out to arbitrarily large distances with decreasing intensity. Thus the field structures associated with two pulses will merge and flow together in one unbroken whole. Moreover, when two pulses come close together, the original particle-like forms will be so radically altered that there is no longer even a resemblance to a structure consisting of two particles. So, in terms of this notion, the idea of a separately and independently existent particle is seen to be, at best, an abstraction furnishing a valid approximation only in a certain limited domain. Ultimately, the entire universe (with

> all its 'particles', including those constituting human beings, their laboratories, observing instruments, etc.) has to be understood as a single undivided whole, in which analysis into separately and independently existent parts has no fundamental status.

The resulting new comprehensive analysis, when developed, would pertain to an explicate order that emerges from the existent higher-level implicate order that remains yet to be fully revealed.

Beyond physics, his field of expertise, Bohm also finds the existence of an implicate order and undivided wholeness. Consider what happens in our consciousness as we listen to music. When a note is sounded once a composition is under way, our consciousness is aware of the sound that is produced but also contains reverberations from the sounds produced previously, and all is melded into a sensation. As a consequence over the period of time, from one moment to the next, where the moments flow into each other without finite bounds (that is are not sensed as discrete units), the simultaneous effects at each of the sequence of moments of the sound that is made by the playing of the note and contained reverberations from previous sounds generate a path of continuous sensations and an awareness of such. And this path of sensations may continue beyond the end of the composition. What is clear is that we do not retain in our mind a sequence of notes, each distinct, unique and completely independent of those that preceded it, nor do we experience a sequence of sensations, each likewise distinct and unique and completely independent of those that preceded it. Rather there exist active transformations of what went before. Moreover, the flow picked up via the ear generates a series (time-path) of emotional responses, bodily states of each (rest or unrest), muscular movements, reflective interpretations and so forth. In this way, a series of notes unfolds smoothly into several if not many levels of consciousness.

In a similar vein, Bohm considers the observation of a piece of art as an unbroken flow of consciousness at different levels, as we shall indicate when we discuss in Chapter 8 Leonardo's *Last Supper*, Kandinsky's *Small Pleasures*, and R. Isard's *Man's Extinction*.

Moreover, Bohm maintains that there were continuous flows of thought that led to Newton's discovery of the force of gravity, Einstein's recognition of the interrelation of space and time and the correspondence of energy and matter, Giotto's grasp of spatial perspective, Michelangelo's artistry in the Sistine Chapel, and Muhammad's numinous transcendent messages that came to form the Koran. All these were unfoldments of the implicate order that then came to be constituents of the explicate order.

In the above manner, and of course with much more evidence than that to which we have just referred, and with extensions to cover cosmology, linguistics and other areas, Bohm (1981, pp.212-3) sets forth his conceptual framework of undivided wholeness and implicate order of immense multidimensionality, concluding:

Our overall approach has thus brought together questions of the nature of the cosmos, of matter in general, of life, and of consciousness. All of these have been considered to be projections of a common ground. This we may call the ground of all that is, at least in so far as this may be sensed and known by us, in our present phase of unfoldment of consciousness. Although we have no detailed perception or knowledge of this ground it is still in a certain sense enfolded in our consciousness, in the ways in which we have outlined, as well as perhaps in other ways that are yet to be discovered.

5.A.3 Some evaluative and integrative remarks

Bohm's basic ideas, as discussed above, relate to sensimotor and brain activity of Pattee's linguistic mode, but also to much more. They relate to what we as humans sense, perceive, have recorded and in general know (the explicate order) and the total underlying reality (the implicate order). The classical laws of physics, the more modern quantum, catastrophe, chaos, far-from-equilibrium, punctuated equilibrium, computational and other theories, the law of supply and demand and all other economic and socio-political laws, and all data and observations associated with these theories and laws are part of the explicate order. But there remains a vast amount of reality to be unfolded, that is still enfolded. Such unfoldment will change and deepen theories in all the sciences, laws and knowledge of relationships—a view consistent with Einstein's statement in commenting upon the existing state of affairs: 'So far as the laws of mathematics refer to reality, they are not certain. And so far as they are certain, they do not refer to reality' (Einstein, 1922 in Weyl, 1949, p. 134).

Bohm (1990, p.29) insists that movement is central to reality. Humans can obtain a greater grasp of reality through acquiring greater consciousness of such movement. (Reality, however, exists independently of being known.) Further, interdependence pervades all of reality. As he puts it, 'each part of space contains waves from everywhere, which enfold the whole room, the whole universe, the whole of everything. In the implicate order everything is thus internally related to everything, everything contains everything, and only in the explicate order are things separate and relatively independent.'

References

Bohm, D. (1981),*Wholeness and the Implicate Order*, Boston: Routledge and Kegan Paul Ltd.

Bohm, D. (1990), Interview in Louwrien Wijers and Johan Pijnappel (eds.), *Art Meets Science and Spirituality*, London: Academy Group Ltd., pp.26-33.

Weyl, H. (1949), *Philosophy of Mathematics and Natural Science*, Princeton: Princeton University Press.

6 Order within seeming disorder, chaos and system transition

6.1. Introduction

In the previous chapters we have emphasized order and hierarchical structure in physical and biological systems. After the Big Bang we discussed how, according to one hypothesis, the original indistinguishable particles (Ur-particles) came to be distinguishable as elementary particles, first gravitons (still to be experimentally verified), associated with gravitation—and then later, quarks, gluons, and leptons when symmetry of the Ur force was further broken to yield the electroweak and strong forces. In time, quarks and gluons came to compose hadrons (protons and neutrons); and photons proliferated and the W^+, W^- and Z^0 particles appeared when the symmetry of the electroweak force was broken to yield the weak and electromagnetic forces. Still later, nuclei were formed and subsequently atoms, starting with the lightest and ending with the set of chemical elements manifest today. Complex molecules appeared, and then organisms and biological systems.

Concomitantly, diverse processes and radiation had led to the formation of stars, star clusters, galaxies, galaxy clusters, and the emergence of solar systems. Layzer's overall view of the total set of processes, inclusive of cultural evolution and social organization, is presented in Figure 6.1. The unknowable initial conditions and structures together with the cosmogenic processes—a situation of utter randomness—came to lead to cosmic expansion and gravitational clustering and then to chemical order and self-gravitating systems. Such order and systems led to gravitational contraction, radiation and nuclear burning which together with entropy generating processes (operating here and subsequently) formed planets, stars and galaxies. Then emerged life, leading to the production of nucleic acids and proteins. Next there appeared biological evolution generating organisms and biological populations, which later spawned cultural evolution, language, art and science with feedback effects on biological evolution.

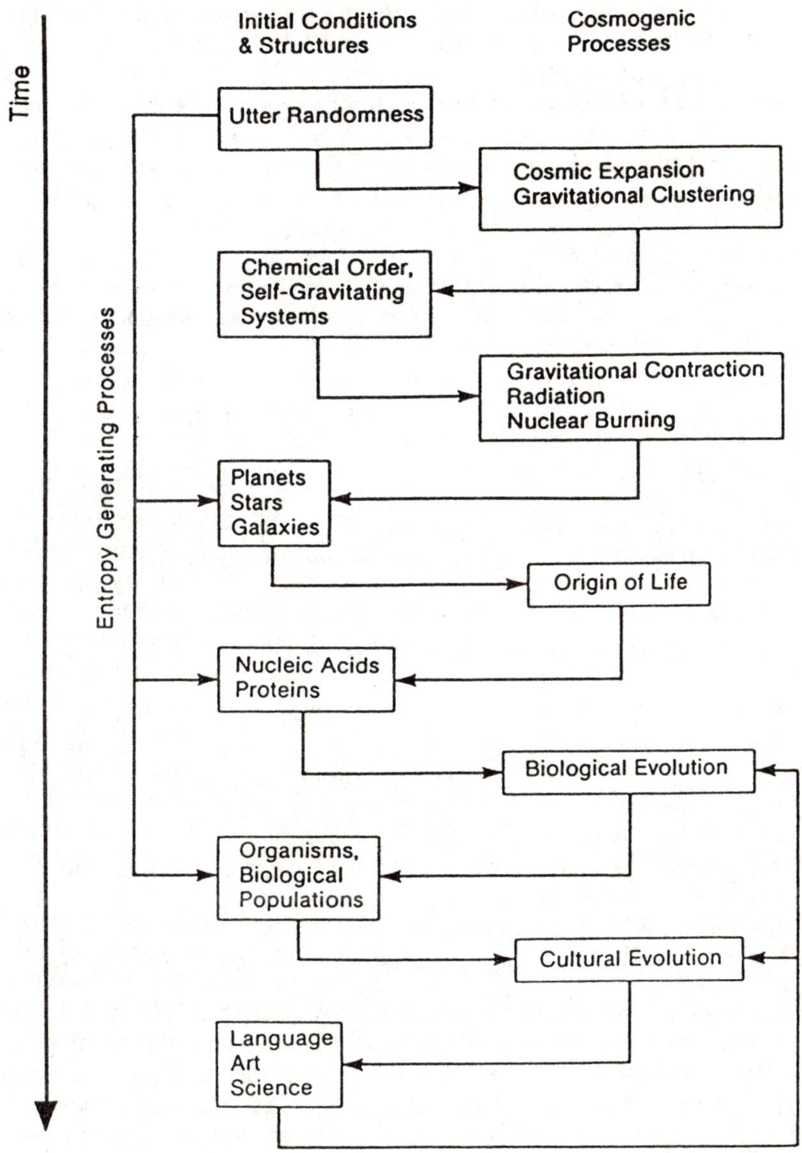

Figure 6.1 The hierarchy of time-bound order
Source: *Adapted from Layzer, 1990, p.37.*

While Layzer has been concerned with identifying order and hierarchical structure in physical and social systems, he has been exceedingly careful in pointing up the critical and inescapable effects of random elements. Other scholars in searching for order and structure in these systems have also been aware of the presence of random factors, and in recent decades much research has been conducted on these factors and associated disturbances: turbulence, stellar explosions and system breakdowns, to name a few. It is now widely recognized that understanding of order and hierarchical structure must involve, too, study of forces that generate disorder and/or constrain order as well as reestablish order—forces that can lead to the disintegration of hierarchical structure as well as to the formation of such. In this chapter we begin to examine relevant theories and empirical studies regarding these forces and associated phenomena.

The following section covers disorder, order within seeming disorder, incoherence and turbulence, periodicity and aperiodicity, deterministic chaos, strange attractors, fractal structures and related literature. Section 6.3 examines coherence, complexity, self-organization, reestablishment of order and stability, and hierarchical structure in physical and non-human biological systems stemming from the nearby- and far-from-equilibrium analysis of the Prigogine School. Section 6.4 provides a summary.

6.2 Chaos theory, fractals and strange attractors

One senses disorder. He may observe, hear, feel, smell or taste it. Yet within this seeming disorder, there still may be much if not considerable order. It is in identifying order within seeming disorder that chaos and related theory represents a major step forward in understanding reality—again as we are able to perceive and unravel reality.

To begin, consider how many biologists view the growth and decline of a population of an organism as an evolutionary phenomenon. Starting with a simple model, they might designate the population at time t to be x_t. At the next point of time, $t+1$, the population x_{t+1} would be equal to the population x_t at time t times unity plus the rate of change ρ per unit of time that is $1+\rho$ (where ρ say might be 0.1, that is ten percent). Letting $r = 1+\rho$, we would have

$$x_{t+1} = rx_t \qquad 6.1$$

But this simple statement (equation) is in general inadequate. While it might be satisfactory for a period of time when there is no pressure on resources, clearly, as discussed in Chapter 3, limited resources eventually come to restrict growth (the Malthusian effect). Hence population biologists have been motivated to introduce the effect of limited resources. One procedure sets at 1 (unity) the total population that the existing resources can support (here, unity stands for 100 percent) and subtracts from 1 the fraction of the total resources already exploited to support the existing population x_t at time t. Then by replacing r in equation 6.1 by $r(1-x_t)$, where $(1-x_t)$ represents the fraction of resources currently untapped, we have:

$$x_{t+1} = r(1-x_t)\,x_t \qquad 6.2$$

a model that brings into play the reduced amount of resources available for further growth of existing population. The nonlinear relationship 6.2 (which is a quadratic in x and recognizable as the familiar logistic equation) is plotted on Figure 6.2 where x is measured along the vertical axis and r along the horizontal. When $r \leq 1$, this means that $\rho \leq 0$ which corresponds to a non-positive growth rate, and thus there is extinction of the population when $\rho < 0$, and no change in size when $\rho = 0$.[1] See Figure 6.2. When $1 < r \leq 3$, there is steady growth of the population until its reaches after many generations a stable maximum point.[2] This resulting point is indicated for each r in this range by the corresponding value of x given in the curve designated steady state in Figure 6.2. Beyond $r = 3$, a bifurcation develops, that is, for

$$3 < r \leq 3.449490 \qquad 6.3$$

the population becomes unstable at any value of r and enters a period two cycle—that is, a period doubling. For example, for $r = 3.2$, the population will come to oscillate between 0.5130 and 0.7995. Beyond $r = 3.449490$ the population enters a period 4 cycle—another period doubling; beyond $r = 3.544090$, a period 8 cycle, still another period doubling; beyond $r = 3.564407$, a period 16 cycle, and so on. Beyond the value $r = 3.569946$, the period doubling stops and chaos takes over as indicated by the shaded region wherein population can take an infinity of values. This situation persists until the maximal value $r = 4$ is reached,[3] except that between the values of 3.569946 and 4.0 there are tiny intervals (windows), where for a given value of r, stable cycles obtain. For example, for $r = 3.83$, a three cycle period comes to exist; and for $r = 3.74$, a five cycle period. Moreover, each window when magnified reveals successive bifurcation consistent with a Feigenbaum number to be discussed below, which supports the concept of *self-similarity* at different scales, to be discussed below.

The important point to make is that there is order in the appearance of different cycles. There is a constant, 4.6692016, the Feigenbaum number, which for the critical values of r at which successive period doublings occur, defines a *scaling rule*. For example, the rule states that the *difference* between the value of r at the onset of say the 8-cycle stage and its value at the onset of the 4-cycle stage will be approximately 4.6692016 times the *difference* between the value of *r* at the 16-cycle stage and the 8-cycle stage.[4]

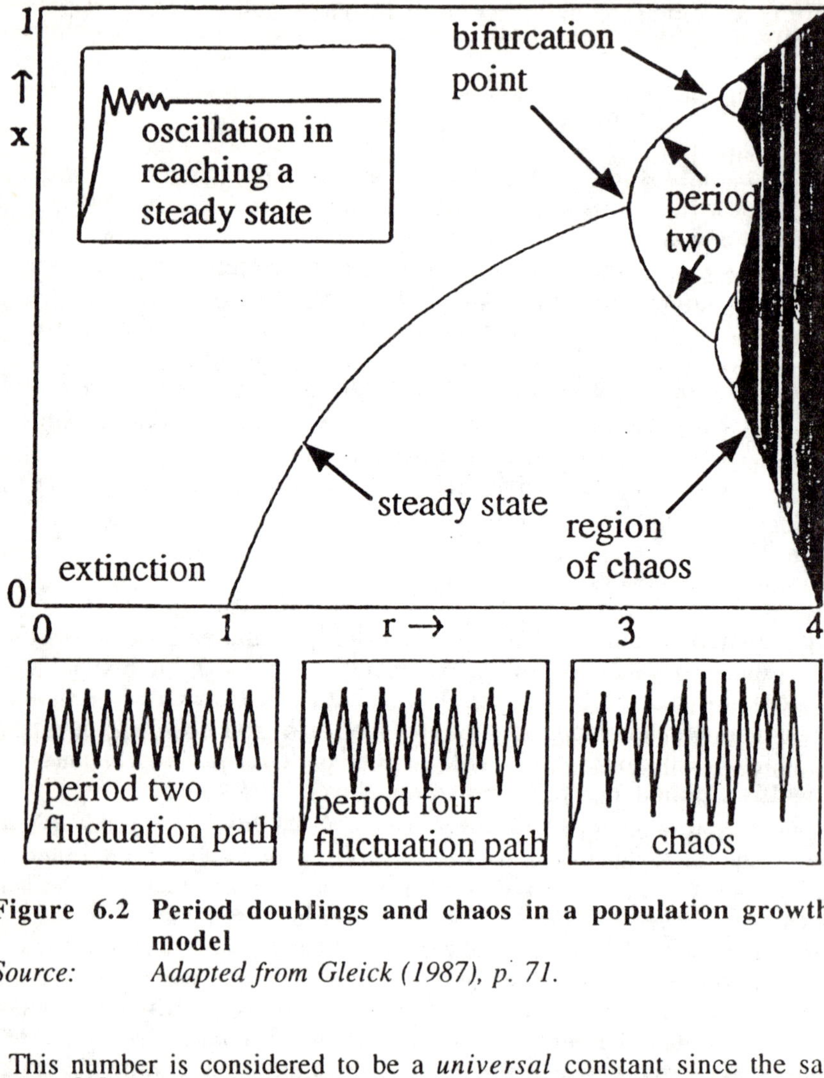

Figure 6.2 Period doublings and chaos in a population growth model

Source: *Adapted from Gleick (1987), p. 71.*

This number is considered to be a *universal* constant since the same number is found in many mathematical functions that portray phase transitions, and in various experiments in physics involving phase change.

The phenomenon of turbulent behavior in a fluid flow has been examined for many decades. One theory has considered that the time evolution of the physical parameter describing a turbulent fluid is the outcome of the joint presence of a regular force and a noisy signal that leads to a mess of disorder. Another theory, put forth by Landau (1959), has considered this parameter to be a quasiperiodic function of time involving a number of independent forces each imposing a frequency on the parameter—which suggests a number of modes (corresponding to the effective dimension of a system). Such a view, however, is inconsistent with the phenomenon of *sensitive dependence on initial conditions*, about which we shall speak later.

While the theoretical debate and experimentation regarding turbulence continues, one aspect has been clearly established experimentally, namely, period doubling (prior to the onset of chaos) á la the Feigenbaum constant with the implied frequency change. Figure 6.3, for example, shows spikes in spectra diagrams that Libchaber and his colleagues constructed to indicate the frequency of oscillations in a moving fluid. Period doubling (that is introduction of an additional oscillation (roll) in the fluid that takes place as the temperature input is sufficiently increased) is indicated by the new spikes—for example when Figure 6.3b is compared with Figure 6.3a, or when Figure 6.3c is compared with Figure 6.3b. These new spikes are predictable with the Feigenbaum number.

Moreover, the Feigenbaum number appears in other unexpected places, for example in the Mandelbrot set and the natural phenomena represented by a fractal in that set. A fractal is a geometrical figure in which an identical shape (motif) repeats itself on an ever diminishing scale—that is, where symmetry is present across scales, often designated self-similarity. For example, the logistic difference equation, as depicted in Figure 6.2, has a fractal character. A graphic picture of the Mandelbrot set which is in the complex plane is presented in Figure 6.4. We cannot go into details about this set which is a universal set of fractals that is bounded (that is, does not extend to infinity).[5] The significant point is that the intersection points of the set with the real line (the x axis) are points of tangency between circular areas whose diameters of ever-smaller size have been reduced by the Feigenbaum number.

At this point it must be noted that these and other recent discoveries of order, often unexplainable, associated with chaos and related theory do not signify that there is no disorder in real life—or any less disorder for there may not be any objective basis for measuring disorder. It does mean that there is much order within what had in many instances seemed as complete or near-complete disorder. For example, order has been found among highly irregular coastlines and geological formations (Mandelbrot, 1983), price series (Mandelbrot, 1963, 1966, 1967), and stock price indices (Peters, 1991). At the same time, fluctuations still persist once the order has been 'statistically' removed from the data—fluctuations which are still unexplainable and which therefore at our current stage of knowledge must be considered random and where possible taken into account via statistical procedure developed to treat random events and noise.

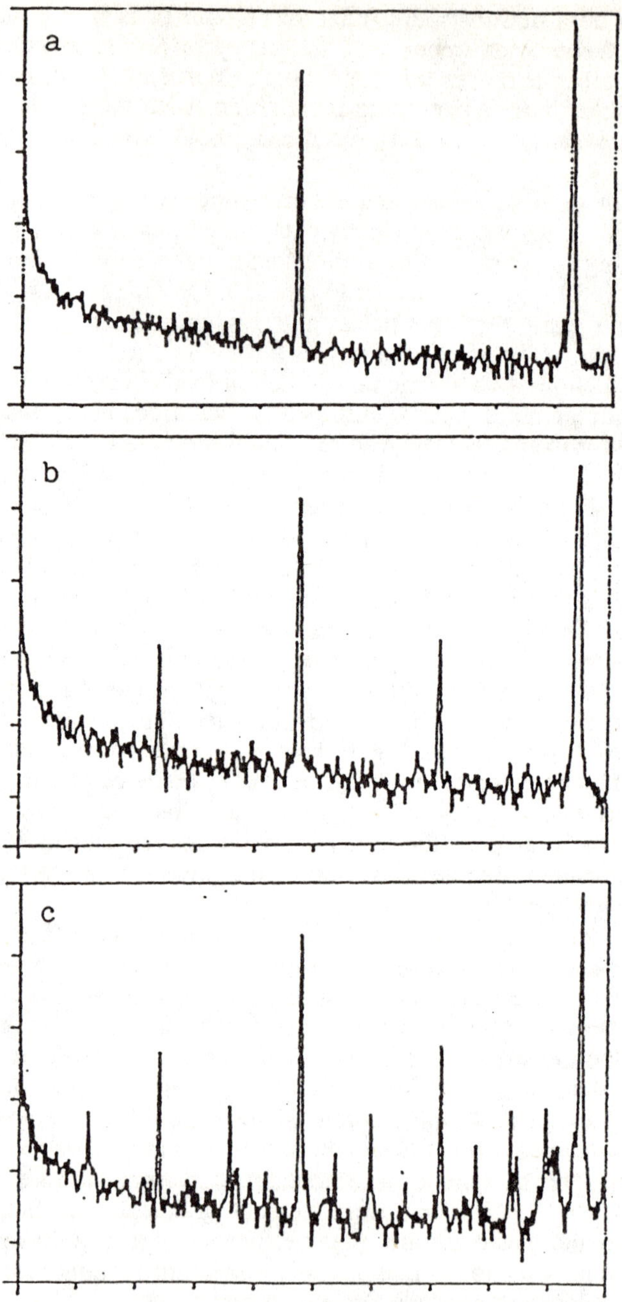

Figure 6.3 Real world data confirming theory
Source: *Adapted from Libchaber and Maurer, 1982, pp.275-78.*

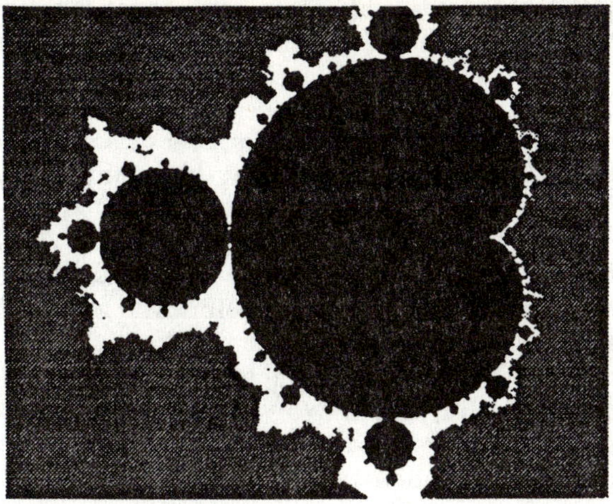

Figure 6.4 A Mandelbrot depiction *(Courtesy of John Holland)*

Another area where disorder comes to exist, but at the same time where there is order within disorder, is associated with sensitive dependence on initial conditions and the concept of strange attractor. Sensitive dependence of nonlinear dynamic systems on initial conditions was long ago recognized by Poincaré (1921). It was not fully appreciated by the scientific community until Lorenz (1963) reported outcomes of some of his simulated weather patterns—and in particular the increasing divergence over iterations when a small perturbation of an intermediate condition was introduced into a computation. See Figure 6.5 where a small change in initial conditions defining the beginning point at the extreme left of Figure 6.5—which change is unobservable—leads to patterns which at the extreme right of the figure lack any resemblance to each other. This sensitivity has now become known as the Butterfly Effect: the flutter of the wings of a butterfly in Tokyo may cause tremendous weather turbulence in New York City. Subsequently, in using a relevant set of three deterministic equations[6] to study meteorological turbulence, Lorenz (1963, 1980) found what has come to be designated the strange attractor. See Figure 6.6 for a 3-dimensional computer output of Lorenz's equations. It depicts a disorderly set of unpredictable data which come to comprise a fine structure that is bounded. The trajectory comes to form a set of loops that never intersect. No point is visited twice. The strange attractor is thus nonclassical (neither a fixed point, nor periodic, nor quasiperiodic) and has been designated aperiodic. It represents the asymptotic behavior of a chaotic system.

Because of both the disorderly array of points visited in the trajectory of the nonlinear deterministic system and sensitivity to initial conditions (remember that one can never attain 100 percent accuracy in measuring physical quantities in order to specify these conditions), one can never be able to predict the resulting time evolution—a situation of chaos. In this system there is deterministic noise, and not the stochastic noise attributed to the interplay of a large number of factors (degrees of freedom) over which one has no control in an experiment or cannot treat or account for. At best one may be able to describe mathematically the totality of all possible solutions—the bounds within which they lie and other key statistical properties. See Ruelle (1989).

As noted elsewhere (Lorenz, 1980), there is fractal structure in the Lorenz system which is continuous and three-dimensional. So also is there in another nonlinear system, the Hénon system, a simpler system (discrete and two-dimensional) and in which fractal structure is more clearly depicted. The Hénon system, which evolved from Hénon's study of the orbits of stars around their galactic center, involves two nonlinear equations.[7] When in the equations we set the two parameters *a* and *b* at 1.4 and 0.3 respectively, we obtain the set of maps in Figure 6.7. This figure when thousands or more of points are plotted depicts several successive enlargements, starting from the upper left, which tend to separate bands. Regardless of the initial conditions that are set, the bands remain the same (and again, for two different initial conditions very close to each other, the corresponding orbits diverge and become quite distinct after a relatively few iterations—sensitivity to initial conditions and unpredictability). The fractal structure—self-similarity at different scales—is clearly evident.[8]

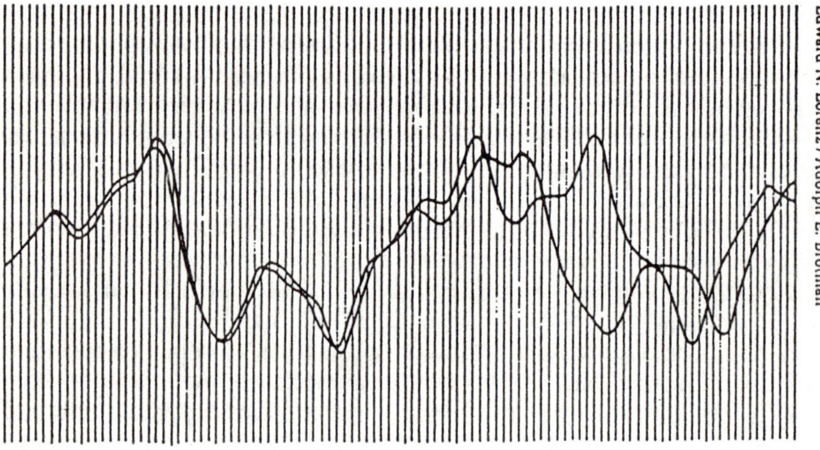

Figure 6.5 Sensitive dependence on initial conditions: weather patterns
Source: Gleick, 1987, p. 17.

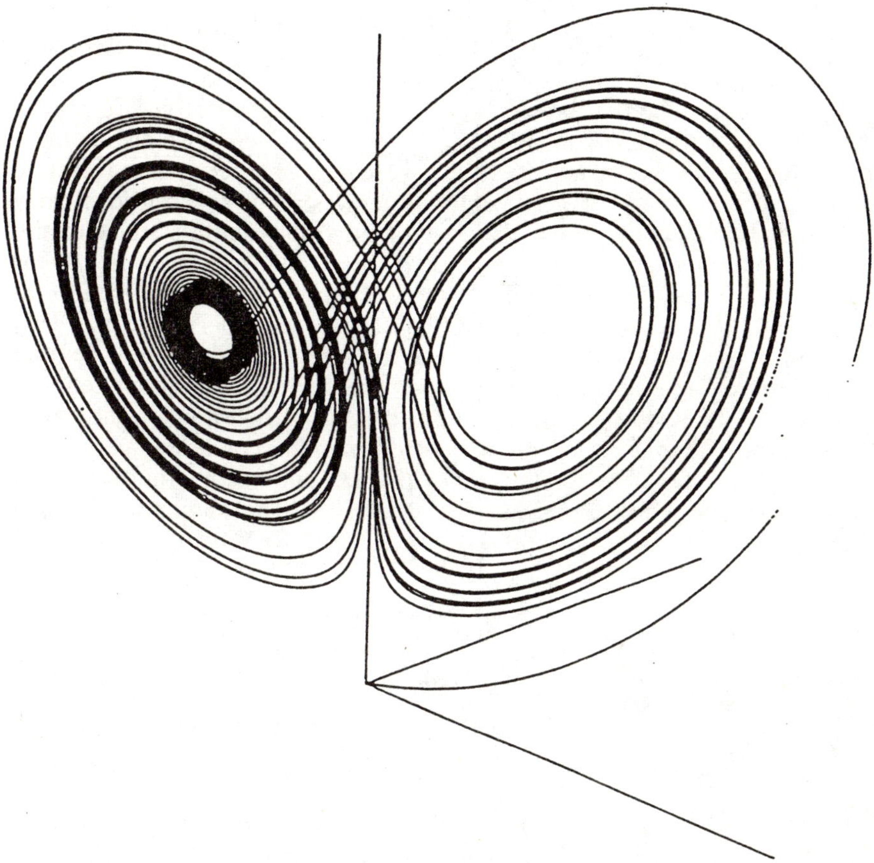

Figure 6.6 Lorenz's strange attractor: a case of deterministic chaos

Source: *Moon, F.C., Chaotic and Fractal Dynamics, Copyright 1992., p.41. Reprinted by permission of John Wiley and Sons, Inc.*

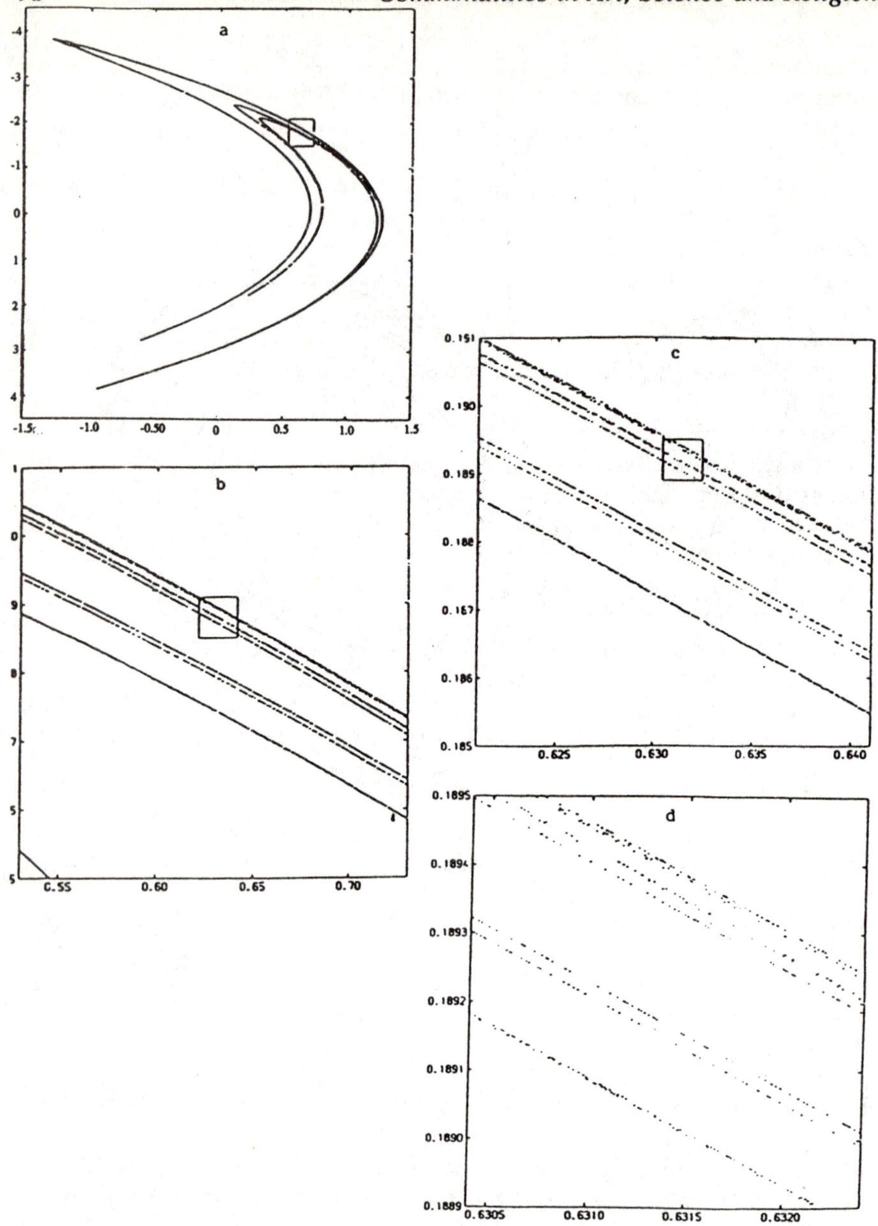

Figure 6.7 Hénon's attractor, its fractal structure and embodied self-similarity (The square block in 6.7a (upper left) is enlarged to yield 6.7b. The square block in 6.7b is enlarged to yield 6.7c. The square block in 6.7c is enlarged to yield 6.7d.)

Source: *Hénon (1976), pp.74-77*

Another useful concept emerging from chaos and related theory is fractal dimension—a concept that is of critical importance for distinguishing between trajectories reflecting the pull from the interplay of a relatively few forces that underly a strange attractor and those reflecting the pull of many more interacting forces and associated stochastic elements. This concept also promises to be of great value for distinguishing between strange attractor trajectories governed by unchanging nonlinear structures, a partial notion of dynamics, and those generated by changing nonlinear structures reflecting the full notion of dynamics.

There are several methods that have been set forth for determining the fractal dimension of a strange attractor—all currently being further developed to reduce their deficiencies. We shall therefore not discuss any of them in detail; instead we will discuss the notion and its relevance.

A straight line *per se* has one dimension even though it may lie in two dimensional Euclidean space. So also does a circle or ellipse. If the familiar Poincaré map, employed to reduce the dimensions of a time-path, were taken of the line, circle or ellipse, it would depict a point—a point typically taken to have no dimensions. Put in another way, if a periodic trajectory is enclosed within a square of two dimensions, reflecting the relation of a dependent variable on an independent variable, that trajectory will have only one dimension; and its fractal dimension is taken to be one. However, if a trajectory reflected the play of random forces, the points of the trajectory would fill the square were the trajectory observed for a long enough time. That is, the values of the previously designated dependent variable would take on all possible values permitted by the square for every possible value of the previously designated independent variable. That trajectory can then be viewed as having a fractal dimension of two. Similarly, an uneven object that needs to be described in classical 3-dimensional Euclidean space can be enclosed by a cube. The object would not fill that cube; hence it would have a fractal dimension greater than two, but less than three. Likewise would a trajectory where the dependent variable z were strongly related to two independent variables x and y. The values of z for all possible combinations of x and y would not fill the cube whose side was defined by some relevant range of these variables. However, if z were also affected by random variables, it would fill the cube the more, the greater the relative effects of the random variables. Ultimately z could fill the whole cube if z were no longer affected by x and y to any greater extent than by any other of the numerous variables at play. The trajectory would then have a fractal dimension of 3. It would constitute white noise, which like a gas, fills a volume. In a statistical sense, it would have a high degree of freedom.

In effect, the fractal dimension of an object measures the extent to which it fills its space; and of a trajectory, its clumpiness, reflecting the 'correlations' inherent in its influences. One of the characteristics of fractal objects and trajectories is that they retain their dimensionality when placed in an embedding Euclidean dimension larger than the least Euclidean dimensionality required.

Note that when the correlation measure method is used for calculating fractal dimension (where the Euclidean measure of distance between pairs of points is employed), that of the Hénon attractor is 1.21± 0.01 (for a =1.4 and b = 0.3); and of the logistic map is 0.500 ± 0.005 (for r = 3.5699456). (See Moon 1992, p.339; and Grassberger and Proccacia, 1983).

When the fractal dimension of a time-series is calculated whose structure is taken to be governed by deterministic equations, there will most frequently exist some noise (random elements) which will increase the fractal dimension dictated by the equations.

There are other relevant concepts emerging from chaos and related theory—the Lyapunov exponent (which is designed to determine whether a system is or is not sensitive to initial conditions and hence whether it is chaotic or not),[9] fractal boundary, Julia sets, and diverse invariant probability measures—which we do not have space to discuss here. Nor are we able to discuss the connections of chaos and related theory to information theory.

As already noted, chaos and related theory pertain to given nonlinear structures. As such they fail to capture the dynamics that involves structural change, linear as well as nonlinear. We now examine studies that move ahead in this direction.

6.3 Far-from-equilibrium analysis: the Prigogine approach

One approach that contributes significantly to further understanding of structural change and pure dynamics is the far-from-equilibrium analysis of Prigogine and members of his school. Asserting that equilibrium analysis pertains to closed systems only and that closed systems constitute only a small fraction of the total number of systems in the natural and social world, Prigogine focuses on nonequilibrium situations. Employing concepts from thermodynamics, he defines a closed system as one that has no exchange of energy or mass, that is has no interaction, with the outside world. Moreover, in accord with the Second Law of Thermodynamics, entropy in a closed system always increases until it reaches a maximum, resulting thereby in a homogeneous distribution of energy and matter. At that stage the closed system will be in equilibrium. When such a system is subject to a minor (weak) perturbation, often the system will return in a short time to its equilibrium state (the effects of the perturbation will gradually die out—a property of asymptotic stability).

To illustrate, we may quote from Nicolis and Prigogine (1989, p. 9):

> Imagine a layer of fluid, say water, between two horizontal parallel plates whose dimensions are much larger than the width of the layer.

Left to itself, the fluid will rapidly tend to a homogeneous state in which, statistically speaking, all its parts will be identical. A minute observer within the fluid, as in Figure (6.8), will be unable to know on the sole basis of observations of his environment whether he is within the small volume V_A rather than the small volume V_B of the fluid. All volumes that could arbitrarily be defined within the fluid will be indistinguishable; thus knowledge of the state of one of them is also knowledge of the state of any other one, or of them all. In other words, the position our tiny observer occupies makes no difference. And because all positions, all volumes, are identical, and his range of vision never extends to the fluid border, there is no intrinsic way for him to perceive the idea of space.

The homogeneity of this system extends to all its properties, and in particular to its temperature, which will be the same at all parts of the fluid and equal to the temperature of the limiting plates or, alternatively, to the temperature of the external world.

All these properties are characteristic of a system in a particular state, the state for which there is neither a bulk motion nor a temperature difference with the outside world, a state of equilibrium. (*The figure number of the above figure is the author's*).

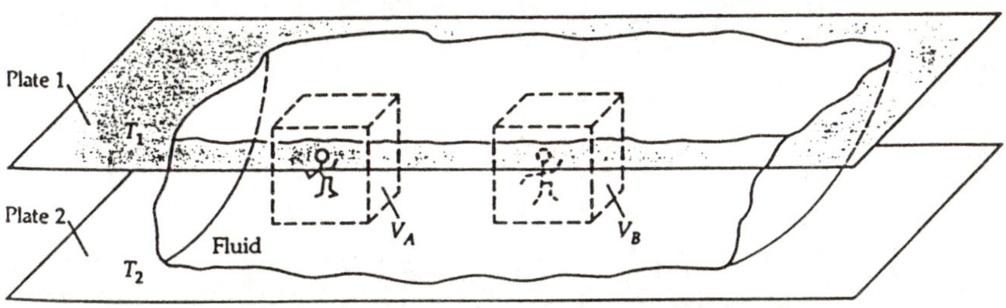

Figure 6.8 An observer contronts symmetry within a fluid (translational invariance along the horizontal direction)

Source: *Nicolis and Prigogine, 1989, p. 9*

Keep in mind that any visible element of water comprises a multitude of molecules moving in all possible directions, randomly and continuously colliding with each other—a situation that has been designated molecular chaos. Figure 6.9 is a visualization of the element when greatly magnified.

In contrast to the state of equilibrium, that of nonequilibrium, which over time can exist only in an open system, may either lose or gain energy. Loss of energy will occur when the dissipation of energy associated with entropy production exceeds the gain of energy from the external world.[10]

When energy loss prevails in a system, its order is subject to degradation, and ultimately complete disorder comes to exist, corresponding to a state of maximum entropy. However, if there is an external constraint, which can be characterized by a steady and adequate inflow of energy from the outside world, the state of maximum entropy (complete disorder) is never reached. One steady state that can be realized is a stationary state, as for instance in thermal conduction and in chemical reactions wherein the concentrations of chemicals can be held constant and thus remain time-independent. To illustrate, take a case of thermal conduction. Imagine a weak external constraint, an energy flux in the form of heat, is applied from below to the lower plate of Figure 6.8. The system is no longer allowed to remain at equilibrium. It will adopt a simple and unique state wherein heat flows from the lower to the upper plate and then passes on to the external world. Compared with the state of equilibrium, the temperature and thus the density and pressure in the fluid will no longer be uniform. However, while a state of equilibrium no longer exists, the system will reach a state of stability, namely a stationary state, where the difference between the warmer and colder regions can be practically taken to be a linear relation.[11]

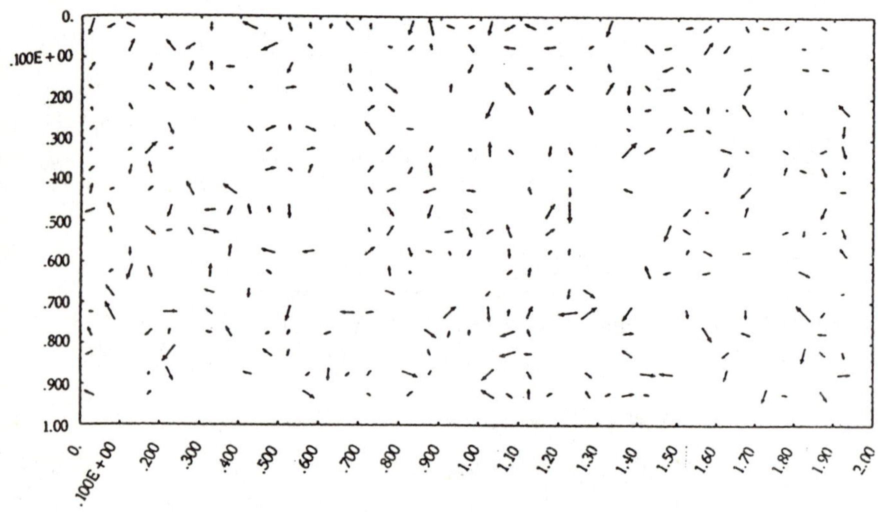

Figure 6.9 A visualization of molecular chaos
Source: *Nicolis and Prigogine, 1989, p. 7*

Stationary states like that of thermal conduction may be designated a nearby-equilibrium state. To repeat, they result from weak external constraints imposed on an equilibrium situation. Random fluctuations within them decay and disappear. These states can be said to belong to the linear domain. In contrast stand the states in the nonlinear domain. They may be reached when a marginal perturbation takes a system in an equilibrium state into a far-from-equilibrium state characterized by an entirely different internal and at times higher organization—a process which may involve a net inflow of energy from the environment. Or these states may be reached when a marginal perturbation takes a system already in a nonequilibrium state into a higher order far-from-equilibrium state. Autocatalytic or self-reproducing processes may be involved. These states may be steady and non-oscillating, or depict a regular pattern of oscillation of one or more of its elements, or be characterized by the highly irregular pattern of deterministic chaos. For example, see Figure 6.10 for a schematic representation of different oscillation patterns associated with different distances from equilibrium of the ion Br^- in the Belousov-Zhabotinski chemical reaction.

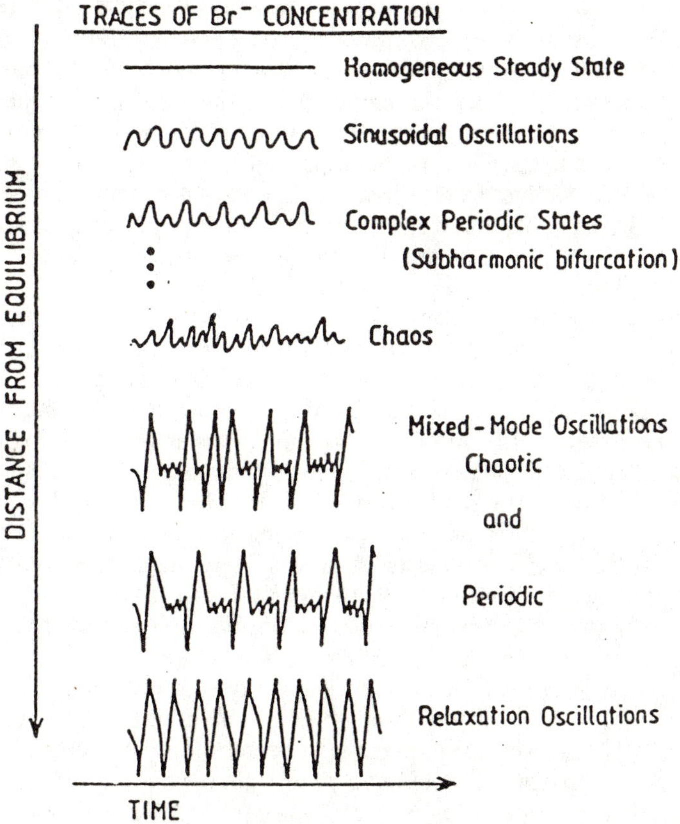

Figure 6.10 Temporal oscillations of the ion Br^- in the Belousov-Zhabotinski reaction
Source: *Prigogine and Stengers (1984), p.168*

Symmetry breaking is a phenomenon that is associated with the development of *order* in far-from-equilibrium systems. For example, if the inflow of heat applied from below to the lower plate of Figure 6.8 is increased so that the temperature difference between the lower and upper plates exceeds a critical level and results in a greater density gradient in the fluid than previously, the stationary state of thermal conduction will be replaced by a state of thermal convection. In this latter state, bulk movement takes place, structured in a series of small convection cells known as Bénard cells. See the sketch of Figure 6.11. What happens is that as a consequence of the force of gravity and the increased density gradient, perturbations upward of a volume element near the lower plate where density is lower will be sufficiently reinforced to overcome the forces opposing movement, while perturbations downward of a volume element near the upper plate where the density is higher, will be sufficiently reinforced to overcome forces opposing movement. The direction of rotation in the cells alternates, being left-handed, then right-handed, then left-handed and so on. Thus, the symmetry of the fluid in the horizontal direction has been broken. In this direction, order and coherence has emerged, which is clear when one compares Figure 6.11 with Figure 6.9 which sketches molecular chaos when the fluid is in an equilibrium state. In Figure 6.11 the tiny observer can now know where he is by noting the direction of rotation. He can obtain a concept of *distance* by sensing the change in rotation when moving along the fluid.

Many other processes go from an equilibrium state to one of far-from-equilibrium via symmetry breaking. Prigogine illustrates with the Belousov-Zhabotinski (BZ) reaction. By changing the rates at which chemicals are pumped into or out of the system so as to reduce sufficiently the residence times of the substances within the reaction vessel, one can proceed from an equilibrium state through a stationary state (the analog of thermal conduction) to a far-from-equilibrium state. In this last state a regular alternating pattern of solutions having at one time a pale yellow color and at the next time being colorless is realized. This pattern is associated with an oscillation of regular period and amplitude of excess Ce^{4+} ions and excess Ce^{3+} ions in the solution. We find that the maxima of the two different concentrations follow each other in a prescribed order. Such a process measures time, in effect constitutes a chemical clock, and permits a rigorous definition of time. Recall that in both the equilibrium and stationary states there exist constant concentrations of chemicals in the reactant vessel. Consequently, no concept of time is possible. Everything is time independent. In the new far-from-equilibrium situation, time independence no longer exists. What is found at any probe that might be instituted depends on what has been and is going on. Thus, temporal symmetry has been broken. As with thermal convection, we find once again order and coherence.

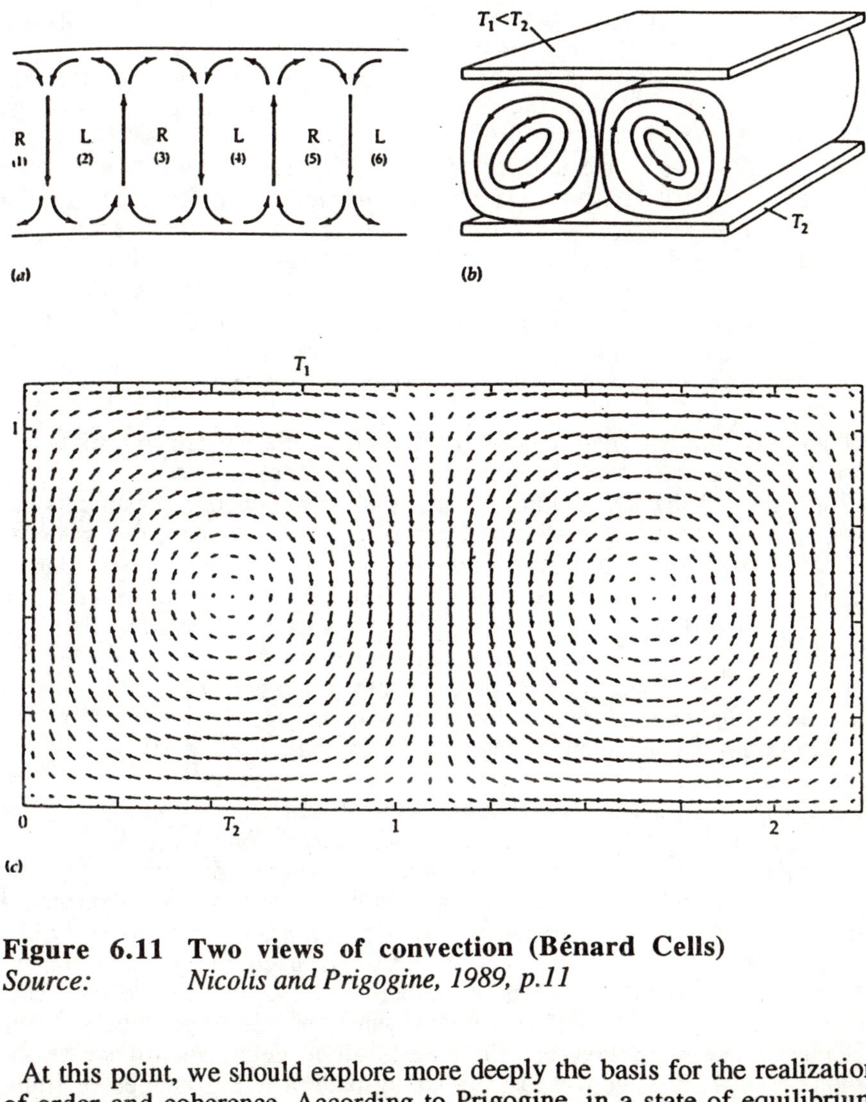

Figure 6.11 Two views of convection (Bénard Cells)
Source: Nicolis and Prigogine, 1989, p.11

At this point, we should explore more deeply the basis for the realization of order and coherence. According to Prigogine, in a state of equilibrium wherein molecular chaos exists (as in Figure 6.9), the intermolecular forces are short range. Molecules move incoherently because they cannot recognize each other over distances larger than a few angstroms, where one angstrom = 10^{-8} cm. In contrast, the molecules in the far-from-equilibrium situation of Figure 6.11 align themselves in a distinct manner. As Nicolis and Prigogine (1989, p.13) put it:

> . . . everything happens as if each volume element was watching the behavior of its neighbors and was taking it into account so as to play its own role adequately and to participate in the overall pattern. This suggests the existence of *correlations*, that is, statistically reproducible relations between distant parts of the system.
>
> . . . it is important to note the long-range character of these correlations as compared to the short range of the intermolecular forces. The characteristic space dimension of a Bénard cell in usual laboratory conditions is in the millimeter range (10^{-1} centimeter), whereas the characteristic space scale of the intermolecular forces is in the angstrom range (10^{-8} centimeter). Intermolecular forces operate up to a distance equal to about one molecule; a single Bénard cell comprises something like 10^{20} molecules. That this huge number of particles can behave in a coherent fashion, as in the case of convective flow, despite the random thermal motion of each of them is one of the principal properties characterizing the emergence of complex behavior.

Obviously, the external constraint imposed on the system modifies the natural disordered thermal motion of each individual molecule.

Similarly, in the BZ chemical reaction when the chemical elements are well stirred while residence times are sufficiently reduced, the outcome that yields regular and sustained cyclical behavior in the relative composition of Ce^{4+} and Ce^{3+} ions 'implies that its different parts act in a concerted fashion by maintaining sharp phase relationships between themselves; otherwise, destructive interference would wipe out oscillatory behavior. In other words, we expect again to observe, just as in the Bénard problem, the emergence, of *long-range correlations* induced by the nonequilibrium constraints' (Nicolis and Prigogine, 1989, p. 21).[12]

One still asks: exactly how do these external constraints operate on the natural random thermal or other behavior of molecules to produce order and coherence?—a question that still remains largely unanswered.

However, chance does play a role in determining the outcome. Prigogine views the critical point, whether a temperature difference, or a residence time, or other, as the existence of a bifurcation point—such as in Figure 6.2 where a logistic difference equation is used in studying macro population behavior. At that point, either of two paths may be followed, and chance and only chance determines which one. Correspondingly, in the Bénard problem of Figure 6.11, chance alone determines whether the rotation of the first cell in the alternating pattern is left-handed or right-handed; and in the alternating pattern of the BZ reaction, whether the first concentration is of the Ce^{4+} ion or the Ce^{3+} ion. The play of chance at this point introduces the probability factor which Prigogine later employs in providing a deeper understanding of the irreversibility and reversibility phenomena, and his conception of ever mounting evolution of order and coherence in environmental and biological organization.

As Nicolis and Prigogine (1989, p.14) put it in commenting on the onset of thermal convection as the critical temperature difference is exceeded:

> we know that the cells will appear: this phenomenon is therefore subject to a strict determinism. In contrast, the *direction* of rotation of the cells is unpredictable and uncontrollable. Only chance, in the form of the particular perturbation that may have prevailed at the moment of the experiment, will decide whether a given cell is right- or left-handed. We thus arrive at a remarkable cooperation between chance and determinism, one that is reminiscent of the duality of mutation (chance) and natural selection (determinism) familiar in biology since Darwin's era, and which was previously limited in the physical sciences to the quantum description of phenomena going on at a microscopic scale.
>
> What we have seen above is this: Far from equilibrium, that is, when a constraint is sufficiently strong, the system can adjust to its environment in several different ways. Stated more formally, *several solutions* are possible for the *same parameter values.* Chance alone will decide which of these solutions will be realized. The fact that only one among many possibilities occurred gives the system a *historical dimension,* some sort of 'memory' of a past event that took place at a critical moment and which will affect its further evolution.

The play of chance in determining which of the two or more states that will be reached when a system is undergoing transition (but not disintegrating completely) is discussed in the mathematical models of social systems in section 7A.5.

Let us once again note that other dissipative structures appear as the temperature difference between the lower and upper plate continues to increase. We may note that beyond another critical value, a new forceful manifestation of randomness appears, namely turbulence and deterministic chaos as discussed earlier. Similarly, as implicit in Figure 6.10, several types of behavior of the BZ reaction emerge under different external constraints, such as turbulence, spatial patterning, and bistability and hysteresis. Moreover, a succession of structures, beginning with a period two cycle which is then subject to a series of period doubling phenomena that ultimately lead to deterministic chaos and still later to restoration of cycle behavior, etc., as depicted in Figure 6.2, can also be encountered in a succession of far-from-equilibrium states. The critical parametric values at which the successive states appear can be, as in Figure 6.2, consistent with the universal Feigenbaum constant and self-similarity phenomena. Furthermore, far-from-equilibrium states may take the form of strange attractors whose fractal dimensions are less than the dimensions of their embedding Euclidean space.

A basic property of the processes yielding near-to-equilibrium and far-from-equilibrium states is their *irreversiblity*. As Prigogine has put it, classical physics has viewed the fundamental laws of the universe as deterministic and reversible. They are held to hold when *t* is replaced by –*t*. Of course it was observed that certain processes did not fit this scheme. Such was explained by physicists as due to their inability to fathom the complexity involved. Today, it is realized that many fundamental processes of nature are irreversible. As implied by the earlier discussion, physical systems frequently display a tendency to change spontaneously from one state to another, but not to reverse the direction of change—that is they assume states of stability of increasing order and decreasing entropy, the concomitant decrease in entropy and entropy production being exported to the environment. These states are supported by an energy or matter influx, the successive states of increasing order being provoked by permanent constraints and suitable irreversible exchanges (inflows) of progressively increasing amounts of matter and energy with the outside world. Biological organization, in a sense as *historical structures* reflecting the past joint operation of chance and determinism, clearly portrays the phenomena of steady development of far-from-equilibrium states of increasing order and coherence—states characterized by positive feedback loops and supported by increasing levels of energy and matter influx. In these organizations, the gradients of a variety of substances such as ions and metabolites provide the embodied tissues with a structure that furnishes them with information about their position with respect to their neighbors, and thus stimulates these tissues and their cells to behave in a coordinated manner. Nicolis and Prigogine (1989, p.36) conclude that 'it is therefore likely that transitions mediated by chemical substances and leading to symmetry breaking are one of the key features of life amenable to physico-chemical self-organization phenomena'.

It should be noted that the nonequilibrium analysis of Prigogine that leads to stable order and coherence within a hierachical system is not at odds with Bronowski on stratified stability. The stable systems of Bronowski may be conceived to have coherence which the normal fluctuations associated with chance have insufficient force to destroy. However, there can be fluctuations within the system that lead to increase in entropy, but only up to that maximum extent that the stability constraint of that system allows. And on rare occasions a minor fluctuation can indeed lead to a higher level of stratified stability—even a farther-from-equilibrium system with greater coherence, reflecting a further decrease in entropy.[13]

6.4 Summary

This chapter continues our investigation into order and coherence within the physical and biological world. We began with Layzer's depiction of the emergence in the world of hierarchical structure beginning with utter randomness and cosmic expansion, through the formation of stars and nucleic acids, through biological and then cultural evolution ending up with positive feedback of language, art and science upon biological, and in time, perhaps even physical evolution. But empirical evidence—namely what is perceived and interpreted by humans, whether scholars or lay persons—definitely establishes the existence of much disorder within the physical and biological world, and at the extreme, even chaos. This evidence is most clearly perceived from the examination within recent decades of nonlinear systems and their space-time paths. For example, the simple logistic model used in population studies clearly indicated, with change in the growth rate, the occurrence of a succession of bifurcations leading to period doublings and cycles and into an area of deterministic chaos (wherein population can take an infinity of values), followed by another set of cycles, and so on.

The onset of period doublings and deterministic chaos (and its exit) can be projected by the Feigenbaum number, a universal constant found in many functions that portray phase transitions (as in turbulence and star orbits). This number is also associated with appearance of self-symmetry at different scales, as is clearly revealed in the Mandelbrot set.

Order (self-organization) within disorder is also present when there is sensitive dependence on initial conditions, as with weather phenomena and the Red Spot of Jupiter. Such has led to the concept of strange attractor. Outcomes of space-time trajectories of nonlinear deterministic models used to portray weather, orbiting of stars and other physical and biological phenomena may be unpredictable, yet the totality of all possible solutions of these models may be bounded and reflect the asymptotic behavior of a chaotic system. Deterministic noise governed by a relatively few forces rather than stochastic noise (often viewed as reflecting the interplay of many factors) may be said to be present. As a consequence, one may speak of the fractal dimension of a trajectory—a dimension that is smaller than the minimal dimension of the relevant embedding Euclidean space, the fractal dimension of a trajectory being one that does not fill that Euclidean space.

Still another major advance in understanding pure dynamic phenomena through careful study of nonlinear phenomena is the nonequilibrium (nearby- and far-from-equilibrium) analysis based on notions of order, coherence and complexity of the Prigogine school. This school maintains that most equilibrium analysis (which characterizes classical physics and by analogy many other fields of study, e.g. economics, regional science and peace science) pertains to closed systems only, which account for just a small fraction of the total of existing systems. The school therefore concentrates on development of nonequilibrium analysis—analysis which comes still closer to understanding the dynamics of evolution. In doing so, it employs concepts and laws of thermodynamics and statistical mechanics, and notions of entropy production and maximization, reversibility and irreversibility, stationary state, dissipative structures, macroscopic and long-range correlations, and autocatalytic and self-reproduction processes. It extends into many other areas the path-breaking thinking on symmetry and symmetry-breaking of particle physics—illustrating spatial symmetry-breaking (as in thermal convection), temporal symmetry-breaking (as in the chemical clock emergence), and concomitant space-time symmetry-breaking in biological development. It furnishes considerable specificity to Pattee's distinction between rate-dependent and rate-independent processes and his thinking on emergence, and to Bohm's general process of unfoldment. When contrasted with the highly complex neural and connectionist models involving notions of neurons and synapses or substitute units (such as Smolensky's and Hopfield's), or with adaptive fuzzy systems (such as Kosko's), the detailed and penetrative investigations of the Prigogine school on simpler structures can provide fruitful insights into biological organization and functioning. In its deep probing into irreversible processes, and the unavoidable play of chance at points of bifurcation and elsewhere, the work of the Prigogine school adds to the relevance of Wolfram's proposed use of cellular automata models when problems are irreducible and when computational experimentation and simulation under different scenarios must be resorted to. As already remarked, the Prigogine framework, in its general employment of nonlinear systems, embraces the analyses and findings of chaos and related theory (on bifurcations, period doubling, deterministic chaos, the universal Feigenbaum constant, self-similarity, strange attractors, Mandelbrot set, fractal dimension), and it broadens Bronowski's focus on stratified stability and our understanding of how chemical order came to be. In establishing a succession of nearby- and unpredictable far-from-equilibrium states of increasing complexity, it supports the conception of hierarchical organization of Layzer and lends support to his hypothesized hierarchy of nested self-gravitation systems. More significantly, the Prigogine succession of states of increasing complexity based on the joint operation of chance and determinism, and thus the notion of biological organization as historical structures with positive feedback and self-organization, is highly supportive of: evolutionary analysis of Dawkins, the speciation thinking of Mayr, the punctuated equilibrium concept of Gould, the sociobiological frame of Wilson, and the related findings and doctrine of many others. To the scientific framework of Teilhard's vision of a religio-scientific trajectory to Omega, it adds considerable depth and breadth.

In brief, the Prigogine framework may be viewed as one that is sufficiently general to encompass many different, and in some cases, more specialized analyses of temporal, spatial, and space-time trajectories in the physical and biological world—of spatial dynamics, in effect. It provides a useful framework for a deep investigation of space-time and related trajectories in the social world, to which we turn in the next chapters of this book. However, we must not forget that in its reliance upon the joint operation of chance and determinism, there may yet be much that we now consign to the chance factor that can be unravelled as order.

Notes

1 Rigorously speaking, when $\rho = 0$, the population system is unstable and any random element (noise) having a positive effect on ρ will propel the system into the bordering stable area where $r > 1$.
2 Rigorously speaking, when $r = 3$, the population system is unstable, and any random element having a positive effect on ρ will propel the system into the bordering unstable area where $r > 3$.
3. By trying out different values of x_t in equation 6.2, one finds that x_t reaches its maximum when equal to 1/2. It then follows that $0 < r \leq 4$ if $r(1 - x_t)x_t$ in equation 6.2 is not to exceed 1, the maximum population that resources permit.
4. Formally, for the critical values of r at which successive period doublings occur,

$$\frac{\lambda_n - \lambda_{n-1}}{\lambda_{n+1} - \lambda_n} \rightarrow r = 4.6692016$$

as $n \rightarrow \infty$.

5 The Mandelbrot set is thus a subset of the Julia set which is the chaotic set of complex analytic dynamical systems.
6 The three equations are:

$$dx/dt = a(y - x) \qquad 6.3$$
$$dy/dt = -xz + bx - y \qquad 6.4$$
$$dz/dt = xy - cz \qquad 6.5$$

7 The two equations are:

$$x(t+1) = y(t) + 1 - a[x(t)]^2 \qquad 6.6$$
$$y(t+1) = bx(t) \qquad 6.7$$

Actually, the above Hénon system is a Poincaré map obtained from the successive intersections of a three-dimensional flow with a surface Σ transversal to the direction of the flow. 'A trajectory is thus replaced by a discrete set of points in Σ, and all the essential properties of the trajectory are carried into corresponding properties of this set of points in such a way that one can forget about the differential system and focus attention on the mapping T of Σ into itself' (Ruelle, 1989, p. 17). Recall that a Poincaré map is a widely used device to reduce the dimensions of a set of observations or points in order to identify nonchaotic behavior or make more explicit elements underlying its chaotic behavior. On the other hand, the Fourier transform may be useful in detecting chaotic vibrations of a system and in determining its underlying dimensions.

Some may regard Hénon system behavior as stable chaos. Another, tantalizing case of stable chaos is the Red Spot of Jupiter—a case of turbulence and coherence at the same time, a case of a self-organizing system governed by nonlinear relationships, which stays bounded and does not disintegrate. See Gleick, 1987, pp. 53-56, and the facing page of 115.

8 Note that they are Cantor-like. A Cantor set (dust) is derived by starting with a line, removing the middle third, then removing the middle third from each of the remaining segments; and continuing in like manner for many rounds until only a dust of points remains.

9. Take a one dimension map defined by an equation such as 6.2 which for a given value of *r* traces out a logistic map. Consider the initial state x_o of its trajectory and a slightly different initial state $x_o+\varepsilon$. After *n* iterations observe the divergence between the two trajectories. Do the same for many 'initial conditions' spread over the trajectory from the initial state x_o. Then the Lyapunov exponent λ may be characterized approximately as

$$\varepsilon(n) \approx \varepsilon\, e^{\lambda n}$$

where λ gives the average rate of divergence (stretching). If $\lambda < 0$, slightly separated trajectories converge (nonchaoticity). If $\lambda = 0$, the trajectory is attracted to a limit cycle. If $\lambda > 0$, nearby trajectories diverge and chaos reigns.

For a three-dimensional space, there are three relevant exponents; and if any one is positive, a volume element is expanded in some direction at an exponential rate and neighboring trajectories diverge.

10 Historically the term dissipation was associated with Lord Kelvin's observation that the simple flow of heat from a hot to a cold body, an inflow of energy, was irreversible—missed a chance to do mechanical work, and thus involved the dissipation of energy. Currently, a dissipative structure may be viewed as one which results when a system in disorder is able to transform part of the energy, mass and/or chemical potentials communicated from the environment into an ordered behavior of a new type. As will be noted below, the new regime may also be 'characterized by symmetry breaking, multiple choices, and correlations of a macroscopic range' (Nicolis and

Prigogine, 1989, p. 15).

11 Similarly, a weak external constraint, say a mass input into a mixed solution of chemicals, can yield a nonequilibrium stationary state which shares all the qualitative properties of chemical equilibrium, as in the Belousov-Zhabotinski (BZ) reaction when well stirred (Nicolis and Prigogine, 1989, p. 18).

12 Another well-known illustration of symmetry breaking—namely the breaking of *rotational symmetry*—occurs when the temperature of a piece of iron falls below 1044K. Above that temperature the iron is isotropic; below it is anisotropic and a magnetization emerges and its observable properties become characterized by a preferred direction.

13 In this statement on irreversibility, Brush (1985, p.617) interestingly notes:

> As Boltzmann himself pointed out . . . Darwin's theory of natural selection can be derived qualitatively from statistical thermodynamics. The equilibrium state of a molecular system is not simply the one of greatest disorder, but involves a balance between disorder and the forces that favor an ordered state; thus at sufficiently low temperatures water molecules will spontaneously form complex crystals. Such an ordered state, initially formed by chance, will be 'selected' to survive under certain conditions.

References

Baker, G.L. and Gollub, J.P. (1990), *Chaotic Dynamics*, New York: Cambridge University Press.

Brush, S.G. (1985), 'Irreversibility' in Besancon, R. M. (ed.), *The Encyclopedia of Physics*, New York: Van Nostrand, pp.614-17.

Cvitanovic, P. (1984), *Universality in Chaos,* Bristol: Adam Hilger Ltd.

Glansdorff, P. and Prigogine, I. (1991), 'Thermodynamics, Nonequilibrium,' in Lerner, R.G. and Trigg, G.T., *Encylopedia of Physics*, New York: VCH Publishers, pp.1256-63.

Gleick, J. (1987), *Chaos*, New York: Penguin

Goodwin, R.M. (1990), *Chaotic Economic Dynamics,* Oxford: Oxford University Press.

Hao Bai Lin (1984), *Chaos,* Singapore: World Scientific Publishing.

Hénon, M. (1976), 'A Two-Dimensional Mapping with a Strange Attractor,' *Communications in Mathematical Physics,* Vol. 50, pp.69-77.

Landau, L.D. and Lifshitz, E.M. (1959), *Fluid Mechanics*, Oxford: Pergamon.

Lauwerier, H. (1991), *Fractals,* Princeton, N.J.: Princeton University Press.

Layzer, D. (1990), *Cosmogenesis,* Boston: D. Reidel.

Lorenz, E.N. (1980), 'Noisy Periodicity and Reverse Bifurcation,' *Annals of the New York Academy of Science,* Vol. 357, pp.282-91.

Lorenz, E.N. (1963), 'Deterministic Nonperiodic Flow,' *Journal of the Atmospheric Sciences,* Vol. 20, pp.130-41.

Mandelbrot, B. (1963), 'The Variation of Certain Speculative Prices,' *Journal of Business,* Vol. 36, pp.394-419.

Mandelbrot, B. (1966), "Forecasts of Future Prices, Unbiased Markets and 'Martingale' Models," *Journal of Business,* Vol. 39, pp.242-55.

Mandelbrot, B. (1967), 'The Variation of Some Other Speculative Prices,' *Journal of Business,* Vol. 40, pp.393-413.

Mandelbrot, B. (1983), *Fractal Geometry of Nature,* New York: W.H. Freeman.

Moon, F.C. (1992), *Chaotic and Fractal Dynamics,* New York: Wiley.

Nicolis, G. and Prigogine, I. (1989), *Exploring Complexity,* New York: W.H. Freeman

Peters, E.E. (1991), *Chaos and Order in the Capital Markets,* New York: Wiley.

Prigogine, I. and Stengers, I. (1984), *Order Out of Chaos,* London: New Science Library.

Poincaré, H. (1921), *The Foundation of Science: Science and Method,* English translation, New York: The Science Press.

Ruelle, D. (1989), *Chaotic Evolution and Strange Attractors,* New York: Cambridge University Press.

Sayers, C. (1990), 'Chaos and the Business Cycle' in Saul Krasner (ed.), *The Ubiquity of Chaos,* Washington, D.C.: American Association for the Advancement of Science, pp.115-25.

7 Hierarchical social organization

7.1 Introduction

In previous chapters we have examined symmetry, symmetry-breaking and hierarchical structure in the physical and non-human biological world, at times with casual reference to the human world. Now we wish to examine such structure, regular and irregular, within this latter world. By analogy, much can be learned about the structure of the human world from what we know about physical and non-human biological systems. At the same time, our findings on the evolution (dynamics) of human organization may be able to contribute to our understanding of physical and non-human biological systems.

In the next section, we discuss types of hierarchies and interactors meaningful for our analysis. In section 7.3 we pay special attention to kinds and modes of interaction among organisms in hierarchies of both replicators and non-replicating survival vehicles with attention given to the impact on the environment, and feedback effects.

Section 7.4 notes several causal chains that have been identified and propounded as relevant by different disciplines as they examine interconnections and mutual effects of interactors at different levels of a hierarchy and among hierarchies. Here, the role of an heterarchical model is set forth as a device to encompass in a comprehensive and fairly coherent manner the diverse measurement-control networks involved and their several legitimate causal relations.

In section 7.5, we formulate some key forces and general relationships for understanding social organization. We define an operational participation potential model (related to Newtonian gravitation) to capture mutual interaction of individuals and groups, wherein aspects of measurement-control are involved to complement operational models of rate-dependent processes. At the same time, we probe the play of specialization, scale economies, joint (cooperative) actions, information flows and costs, and the concept of optimum loss of detail (information filtering) in the social system.

Lastly, section 7.6 attempts some explanation of major and abrupt transitions in social systems, setting forth a number of relevant forces at play, and seeking parallels with analytical approaches for understanding physical systems, having particularly in mind the Prigogine school's thrust. We begin to search for the why, where and when of the above transitions.

In Chapter 7A, the Appendix to this chapter, we present mathematical formulations of processes, discussed in this and previous chapters—to provide deeper probes into the play of forces that are involved.

7.2 Types of hierarchies and interactors

In each of the previous chapters we have examined hierarchies. We have done so in several different contexts, each having specific implications for analysis. We now wish to approach the concept of hierarchy in a more comprehensive manner and examine more systematically the different kinds of hierarchies that have been presented and discussed in the literature. We restrict our examination to those hierarchies and their levels meaningful for our analysis.[1] We consider hierarchies whether they are observed with humanly constructed measurement techniques (never completely independent of interconnection with what is being observed), or simply conceived for understanding the world we sense.

A first type of hierarchy is a purely descriptive, classificatory one. In biology, there has existed the study of taxonomy defined as 'the science dealing with the description, identification, naming and classification of organisms' (Random House,1947). Thus, there is the classification: kingdom (e.g. animalia); phylum (e.g. chordata); class (e.g. mammalian); order (e.g. primates); family (e.g. hominidae); genus (e.g. homo); and species (e.g. homo sapiens). In physics there is the classifications of forces and particles, as involved in the discussion of Figure 5.1; and in chemistry, of elements associated with the discussion of Figure 5.2.

In census data, cities may be classified by size (in terms of population ranges). In economics, activities may be categorized by required level of market demand for their outputs, or population by income classes. And so forth. Typically, behind each classification is at least one hypothesis, theory, or relevant concept, or in general some meaningful relationship. Explicit in the hierarchy of Figure 6.1 is the appearance in human chronological time of phenomena related to processes and structure; there is scaling involving notions of stages in evolution. Explicit in the hierarchical set of nested hexagonal market areas discussed in geography and regional science is ordering in terms of space covered, reflecting notions of scale economies. Explicit in the hierarchy of nations is scaling by levels of economic welfare (per capita income ranges) involving notions of technological development and industrialization.

For analytical purposes, it is useful to have in mind at least three basic distinctions. One is whether a hierarchy is inclusive or exclusive. An inclusive hierarchy is the biological taxonomy of the previous paragraph: all members of the homo sapiens species and all other species stemming from the genus homo belong to that genus. All members of the genus homo and any other genus stemming from the family hominidae belong to that family; and so forth. An exclusive hierarchy is one where no member of a lower level belongs to a given level. An example is that of command (pecking order), as in an army, where the levels are the general, the colonel, the captain, the lieutenant. No colonel is a general, no captain is a colonel, and so forth. However, if we examine the *chain* of command, then in a diagram all the lieutenants who come under a particular captain can be encircled to branch off of that particular captain, and all captains and their lieutenants who are under the command of a particular colonel can be encircled to fall under that colonel, and so forth.

While the classification of inclusive and exclusive is useful to facilitate theoretical analysis, in reality hierarchies are often mixtures of inclusiveness and exclusiveness. For example, in the United States, one aspect of the ranking, by level of importance, of the president, governor, mayor, and community official is in terms of the size of their constituencies (the president's constituency contains the constituency of each governor, etc.)—an aspect based on inclusiveness. However, in terms of functions those of the president are by and large distinct from those of the governor, those of the governor by and large distinct from those of the mayor, and so forth. Hence, a ranking by level of importance on this aspect must be based on the relative importance of the functions, in particular the non-overlapping ones, assigned to each official—an aspect geared to exclusiveness. Hence, the overall hierarchical ranking must take into account each of these two aspects, and perhaps others.

Another basic factor in the analysis of hierarchies revolves around the attribute of constitutiveness. A hierarchy is *constitutive* when the members of a lower level are combined to form new units that are associated with the emergence of specific functions. Cells that are combined to form tissues, tissues to form organs, and organs to form humans is one illustration of a constitutive hierarchy. Another is Simon's famous watch manufacture example. There, the 1,000 (or so) parts that are required to produce a watch are first put together to form 100 subsubassemblies of 10 parts each, then the 100 subsubassemblies are put together to form 10 subassemblies of 10 subsubassemblies each, and finally the 10 subassemblies are put together to form the final product, the watch. (See Simon, 1962; and Grene, 1988.) There are, of course, hierarchies where in one part of a hierarchy the boxes within one larger box involve a constitutive relation, and in another part, strictly aggregational. A hierarchy which does not have any constitutive property and is strictly an arrangement of convenience is simply *aggregational*—as a set of Chinese boxes, in which small boxes fit into a medium size box, and medium size boxes into a large box, with no relationship among them except embedment.

A third basic distinction among hierarchies relate to the presence or not of interaction among the units, and the degree and nature of such. The Simon case of watch manufacture is an example of a constitutive hierarchy where there is no interaction, in the sense of behavior, among the units. The constitutive case where the successive levels are macromolecules, cells, tissues, organs, humans is one where interaction among units exists.

Interaction, however, can be of different extent and nature. First, some scholars insist upon a distinction between (a) non-reflective and (b) cognitive-based, *reflective* interaction. The macro-molecules that interact to form a cellular organelle, the cellular organelles that interact to form a cell, the cells that interact to form a tissue, the tissues that interact to form an organ, etc. do not involve the *reflective* cognitive activity of a brain, although such interactions involve *control* properties of the Pattee type discussed in Chapter 4. In contrast stands the cognitive-based, *reflective* interaction among humans. In the Quaker religious order, for example, a set of interacting individuals upon reflection may form a local meeting, a geographically proximate set of local meetings upon reflection may form a quarterly meeting, and so on.

Cognitive-based reflective interaction (behavior) which is of primary concern in this chapter may be top-down, bottom-up, lateral or some combination. A pecking order, a ranking of officers in an army in terms of influence, is a top-down behavioral hierarchy. The authority structure of the Catholic church, from the Pope down through Cardinals, Archbishops, Bishops, Monsignors, Priests, Deacons, and lay persons is, by and large, another top-down hierarchy. In contrast, the Quaker religious order is an authority hierarchy which is bottom-up and lateral to a great extent. A dictatorship may be a top-down authority hierarchy; a democracy such as in the United States is much more bottom-up.

Top-down and bottom-up hierarchies may involve influence that is direct only, with no feedback. The Stalin regime may be pointed to as an example. Typically, however, indirect effects and feedback exist.

7.3 Modes of interaction and human behavior

With the above discussion in mind, which by no means covers all the possible kinds of hierarchies and their attributes,[2] we proceed to the analysis of interaction *per se*.

In Chapter 3 we discussed Dawkins' concept of extended phenotype—a phenomenon that could involve the interaction of one organism with another—for example, the reaction of a reed warbler to the red gape of a cuckoo nestling. This is still a case of interaction with an environment, although it is with an organism in the environment. It parallels the beaver's interaction that involves the construction of a dam. In Chapter 3 Dawkins' concept of meme involves more. It covers interaction among organisms (live vehicles), and from that interaction the resulting rites, practices, organization and ideas which affect the survival of the organisms (live vehicles) and thus the survival of the genes they carry. Also in Chapter 3 we began discussion of coevolution to take into account that species, as well as organisms, can be viewed as behavioral units. On all these matters more can be said.

Hull (1980, 1988), drawing upon early writings of Waddington and Mayr among others, has gone beyond the Dawkins' notions on the roles of the gene and its vehicle. First, he would discard the passive connotation of vehicle (which some scholars attribute to Dawkins despite his introduction of the meme), wherein a vehicle carries and shelters the genes, but where the genes maintain ultimate command and govern the behavior of any clever vehicle. Hull maintains that the vehicles can be and frequently are interactors, and can be aggressive interactors. Such are beavers and humans who behave to change the environment (biotic or not), thereby affecting nature's selection process and becoming a cause of differential replication. Moreover, vehicles (phenotypes) as interactors are not the only entities that behaviorally impact the environment. Hull maintains that genes, too, interact directly with their environments and do so in such a way as to bias their own replication; genes do not just ride around in vehicles. Also, chromosomes and gametes do likewise with their environments, and so do colonies, hives, and other diverse groups. Thus the fundamental concepts should be replication and interaction, replicators and interactors; and these general concepts according to Plotkin (1988) should be applicable at all levels of any hierarchical organization, biological or not. Hull (1988, p.30) considers that

> all that must actually be included in formal statements of evolutionary theory are the general characteristics of replicators and interactors and how they are interrelated. Only when this general theory is applied to particular cases is the actual informational content of the replicators and the causal situations that produced the particular adaptations relevant. Scientific theories are general. Their applications are contingent and often idiosyncratic.

Brandon (1988, p.61) also maintains that there are many interactors involved in the selection process underlying biological evolution, and he develops a hierarchy of them. He states that there are

> various levels at which selection may occur. Selection may occur among bare lengths of RNA within a 'primordial soup,' among lengths of DNA within cells, among chromosomes within cells, among organisms within (selectively homogeneous) populations, among groups of organisms within local populations, among local populations

> within communities, among species within groups of competing species (which may or may not correspond to clades), and among clades. I have argued in each case that when there is selection at a given level, the entities at that level are interactors. This should not be surprising, since my definition of levels of selection is designed to pick out levels of interaction. Thus, we have a hierarchy of interactors ranging from bare lengths of RNA through organisms to clades.

Keeping interaction and replication distinct, Brandon then proceeds to discuss hierarchies of replicators. Replicators can of course be interactors. Within the primordial soup, bare lengths of the replicator RNA must have interacted, and differences in their physical structure must have led to differences in survivorship and thus subsequent replication. Likewise, the replicators, the selfish genes (lengths of DNA) and chromosomes functioned (and still do) as interactors and affected selection. On the other hand, many interactors—for example, the beaver, homo sapiens and other sexually reproducing organisms—are not replicators although their genes are. Accordingly, Brandon conceives the hierarchy of interactors and the hierarchy of replicators to evidence different levels and composition of entities, maintaining that recognition of both hierarchies is essential for general evolutionary theory. Of course, in the functioning of an hierarchy, Brandon refers to the statistical concept of 'screening off' to help distinguish different levels. This concept is related to Simon's near-decomposability and Pattee's principle of optimum loss of detail which finds explicit recognition in several mathematical models in the Chapter 7A.

Hull's and Brandon's emphasis on interactors as aggressively active vehicles is more concretely developed by Odling-Smee (1988), who defines an interacting unit as either a phenotype or the population to which the phenotype belongs. He specifically examines any given organism, O_i $(i=1,\ldots,n)$, its relevant environment E_i, the population O_{pop} of which it is a member [which also contains other behaving organisms O_j $(j=1,\ldots,n;$ $j\neq i)]$, the relevant environment E_{pop} of that population, and the environment E in general. He investigates the changes in these variables and their interactions over time, in particular the changing mutual relationships of the behaving unit and the environment; in his mind the mutual relationships existing at any time provide the reference background (state of affairs) for the development of a general coevolutionary theory.[3] With this framework, Odling-Smee (1988, p.80) views Lewontin's concept of niche-construction as 'the modifying force that directs ecological inheritance' . . . comprising 'the outputs from active organisms to their environments.' Thus the organisms 'now become not only problem-solvers in adapting to their environment but also problem-setters who redefine their problems by modifying the environment.'[4]

An alternative framework put forth by Eldredge (1988b, p.22), which by and large recognizes both the hierarchy of replicators and that of interactors, is one that comprises (1) the genealogical hierarchy that 'is fundamentally concerned with the packaging of genetically based information in stable entities of various orders of magnitude: the various individuals that make up the hierarchy' and (2) the ecological or economic hierarchy that is fundamentally concerned with matter/energy transfer, a process of equal importance to genetic replication in all living systems. 'Put another way, it takes energy as well as structural packaging to facilitate reproduction;'. . . and, moreover, 'most of the genetic information housed in any biological system is concerned with matter/energy transfer: instructions for building the housing and running economic processes'. . . Thus, to him the processes of economics (matter/energy transfers) and genetically-based information maintenance, modification and transmission are both vital and equally important in all organismic and social behavior.

Returning to the problem-setting activity of organisms which Odling-Smee points up, we note that still another form of behavior, namely anticipatory, may be said to occur. Rosen (1985) maintains that adaptation and learning necessarily involve such behavior. Associated mechanisms, e.g. choice of an action where outcomes and environment need to be anticipated, are ubiquitous in biology and underly emergence through natural selection. Such in turn requires that the behaving unit (system) have a predictive model, however simple. Rosen thus insists that the understanding of biological behavior, which in the past has been sought by and large with the use of a pure reaction paradigm, must involve, in complementary fashion, an anticipatory paradigm. According to him, this paradigm is essential for attacking questions of social organization in the current world where social technologies involving planning and forecasting are indispensable, and where there is the omnipresence of extensive interactions among interconnected systems. In this regard, see the mathematical models in Chapter 7A which involve many interacting regions and nations. Also see Laszlo (ed.), 1991.

7.4 Causal chains, heterarchical models and transition phenomena

All the above supports the notions of Lewontin and Leven (1988) in Chapter 3 and Pattee (1991) in Chapter 4 that a system's change and the control (behavior) that governs its space-time path lie within the structure of its hierarchies, the interconnections at all levels, and between all units contained in a level, and the interactions with the environment. While a particular behaving unit or level exercises control, it can do so only at some point in the system and is dependent on what other units and levels do elsewhere. Nor by itself does the system as a whole (whenever conveniently viewed as an entity) control its development.

The modern synthetic theory of evolution is being increasingly supplanted by the above view of interaction of elements (behaving units) at different levels with relevant environments—levels within at least one hierarchical system. But there are differences among scholars over what is a relevant hierarchy, what kind of hierarchy is pertinent for analysis in a given context (situation), what levels are relevant for a given type of hierarchy, what is a relevant environment for a given level of a given hierarchy, and what types of interactions are relevant for analysis for a given level of a given hierarchy. Lying behind many of these differences are differences regarding causal chains—what are the relevant variables, independent and dependent, and the specific mathematical functions (explicit or implicit) connecting these variables.

At this point, Pattee (1991) has made a significant contribution. In line with his destinction between the complementary rate-dependent processes (as with laws of physics) and rate-independent (as in measurement and control) presented in Chapter 4, he builds upon Aristotle's distinction among causes and various differences among scholarly claims for the relevance of different causes. There is natural selection which Pattee suggests Aristotle might have characterized as *final* cause; there are genetic instructions, perhaps characterizable as an Aristotilian *formal* cause; and there is enzymatic control, perhaps denotable as an Aristotilian *efficient* cause. Still other causes have been set forth as well as various meanings of causality. To Pattee, many if not all these are relevant for a general analytical framework. He maintains that we should recognize many pertinent types of hierarchies and the causal chain models associated with them. He is then motivated to propose the concept of a *heterarchical model*—a model that can embody the simultaneous existence of a number of causal chains and can involve hierarchies that are associated with them.

As others have, Pattee (1991, p.218) observes that in most, if not all, cases a legitimate causal explanation (model) relates to a situation wherein constraints operate. For example, he notes that while genes may be said to cause protein synthesis via a chain of events, this occurs "only when imbedded in a highly constrained local cellular network of controls, and all of the molecules forming these controls are themselves causally contingent on this network. Similarly, the resulting local protein synthesized can be said to 'cause' innumerable events at all higher levels of organization, but all these events are similarly contingent on the network of controls".[5] Thus he characterizes the behavioral aspects of the biological evolution process as also involving a network of measurement-control processes—one that comprises an ensemble of behaviors (and thus constraints) and associated causal chains. This ensemble is his heterarchical model—a global model that serves to complement the specific local causal chains.[6]

A heterarchical model has certain advantages, whether or not the interconnections of the causal chains are hierarchical. As distributed causal networks, heterarchical control models do not have a defined relation to any particular local causal link. They do not need the specification of initial conditions and other detail, yet can provide statistical results. They are by nature interdisciplinary and can recognize the different approaches that each of several disciplines consider paramount. Thus, as Pattee (1991, p.220) states:

> genetics is viewed as the study of how genes control organization from protein synthesis to community behavior. Development is viewed as the study of how the individual controls genetic expression and epigenetic constraints. Evolution is viewed as the study of how environments, through selection, control genes, development of individuals, and populations. Clearly all these controls exist, but there is little evidence that they form objective levels in a causal hierarchy. Without the complementary heterarchical network view, these local causal chains will never form a coherent theory.

Moreover, heterarchical models permit a clearer picture and deeper understanding of the space-time path of social organization. With the notion of many causal chains in operation with each constrained in a consistent manner, we can conduct more insightful explanation of why society has moved in the many new directions it has taken in the past (often designated frozen accidents) and also the role of increasingly complex interactions of organisms (humans) with their relevant environments in shaping the course of evolution.

The heterarchical model conceived by Pattee in one way goes beyond the notion of a complex dynamical systems model. In the latter, the different dynamical schemes are coordinated through coupling functions where there can be both driver and follower elements. The aim is to project the behavior of the complex system where the behavior model of each component subsystem and the coupling functions are known (Abraham, 1991). Such, however, interlock only rate-dependent processes, as is evident in the models of sections 7A.3 and 7A.4 in the following chapter and in the more complex models in Isard and Liossatos (1979). They do not incorporate measurement-control processes such as in the PSIG model (Isard, 1990) sketched in Chapter 11—processes that establish autonomous control on rate-dependent physical functioning and lead to the stability and persistence of the measuring entity.[7]

As Pattee notes, heterarchical models permit a coherent theory which no local causal chain by itself generates. However, the Pattee framework needs further extension. As noted in Chapter 4, Pattee distinguishes between processes which are rate-dependent and those which are rate-independent. Laws of Newtonian physics are characterized by the former; measurement-control processes, by the latter. Attention to both types of processes are essential for coherent and adequate evolutionary analysis. Such attention is given in Pattee's concept of heterarchical models. But reality involves more than the coherent theory or framework that Pattee proposes—as is clearly evident when one attempts to recognize the abrupt bifurcations in both rate-dependent and rate-independent processes that the Prigogine school has so effectively exposed. Now that deterministic chaos—a strictly rate-dependent process—has become fully appreciated in recent years, the *dynamics* associated with it as well as that associated with (1) phenomena leading to both nearby- and far-from-equilibrium phenomena and (2) the operation of chance has yet to be embodied into Pattee's framework.

To capture such dynamics, whether induced by: (1) a snow-balling (explosive) perturbation or fluctuation in the relevant environment of a particular causal chain, ensemble of interlocked causal chains, or network, or (2) a measurement-control process at any of these places, the conceptual framework of a heterarchical model operating in a coordinative fashion needs extension. This framework has relevance for systems in stable, stationary and nearby-equilibrium situations, but not for systems in transition (via symmetry-breaking, turbulence or what-not) to another stable far-from-equilibrium state, or to decay (decomposition) or other annihilation.

Nevertheless, the Pattee framework for situations for which it is applicable coupled with Prigogine-type transitions to far-from-equilibrium situations does permit enlightening historical analysis and can generate visions of future systems (à la Laszlo, 1989, and Teilhard) or fans of possible systems (à la Isard, 1988, chap. 8). To illustrate, consider some of the historical record.

A number of scholars have interpreted past social evolution as a sequence of transitions to far-from-equilibrium states, wherein the Prigogine set of concepts—fluctuations, bifurcations and emergence of order—are directly applicable. One historian, Artigani (1981) imagines at a nearby equilibrium state the existence in the far past of many of the simplest conceivable human groups with capacity for language, technology and culture. Many arrangements of the members in these primal groups were possible, many almost equally probable so that, almost like molecules in a state of molecular chaos, knowledge about where members were at one moment of time implied very little about their next position. Being nomadic with no permanent structures, few belongings, and minimal biological bonds and social constraints, they had little sense of collective identity and little structured information in the way of tales, traditions and legends. If a group wandered into a new environment with abundant resources, its population may have likely grown, but once the environment's resources were fully exploited, further population growth within a group lacking organization would have led to at least its partial disintegration.

When a primary group began to disintegrate, or for other reasons, stragglers may have wandered into tangled, damp and malarial locales such as river valleys replete with resources for *alternative* paths of food acquisition. Many, if not most, of these stragglers may have perished, but by accident a few may have stumbled upon these new foods and paths in what initially must have been for them inhospitable environments. Flooding, disease, drought and other environmental stringencies may have compelled these individuals to adopt new forms of cooperative relationships to be added to the few already possessed. Once a successful encounter with the new harsh environment was established and an existence eked out, experience with the new way of life may have led to more fruitful exploitation of the rich resources, mounting population numbers, and requisite new organizational forms better able to cope with both predictable and unpredictable environmental hazards. An incipient civilization may have emerged—representing a successful transition to a stable, relatively far-from-equilibrium situation.

It is not necessary to go into greater detail concerning the manner in which a river valley civilization may have arisen, and areally grown over time. The rise and fall of early river valley and later civilizations are well known, although their causal analysis is still speculative and far from adequate.

In his presentation, Artigani emphasizes the role of warfare. In their efforts to survive, groups became involved in attempts to appropriate resources held by others, capture females for reproductive purposes, and so forth. These efforts led to the discovery of effective military formations—the column as an arrangement of soldiers in a file, front to back, used to move them to the battle; the line as a side-by-side front. Such arrangements were not random distributions, but only highly improbable states yet at least to a certain extent predictable—in a way analogous to the emergence of thermal convection in a Bénard cell upon increase of energy inflow. So also was the circle strategy hit upon as a defensive formation to protect a niche against attack.

Artigani goes on to develop his Prigogine-oriented interpretation of human social evolution, treating among others the organizational jumps associated with information storage and spread, ritual, myth, sex, physical and cultural structures, technological advance, consciousness and homeostasis, communication technology, language, and the balance of autonomy and control. His historical materials lend considerable support to the Prigogine doctrine.[8]

A coupling of the conceptual heterarchical model and a Prigogine-jump transition frame can also generate visions of the future. Here we may illustrate with the Grand Evolutionary Synthesis (GES) of Laszlo (1988, pp.43-6). Some of the main features of his GES are as follows:

> Autocatalytic and cross-catalytic feedback loops predominate in open dynamic systems . . . far from equilibrium in virtue of their rapid reaction rates and great stability. However, as no autopoietic reaction cycle is entirely immune to disruption, constant changes in the environment sooner or later produce conditions under which certain cycles can no longer operate. These systems encounter . . . bifurcation.

> The outcome in complex systems, as both experiment and theory demonstrate, is essentially indeterminate: it is not a function of initial conditions nor of changes in the environment. Bifurcations produce a basically random selection from among the set of alternative steady states available to the system through the amplification of some among the existing fluctuations. The new steady states assimilate the perturbations within the limits of error-tolerance of the catalytic cycles and are likely to produce increases in the complexity of system structure. There is a significant probability that the system leaps to a new plateau of nonequilibrium; that it retains a greater quantity of a more dense flux of free-energy for a longer time and thus becomes more dynamic and more autonomous in its milieu.
>
> Continuity and consistency in the processes of evolution give rise to continuity and consistency in its products. Periodic destablizations of systems far from equilibrium, coupled with the organizing effect of a constant energy flow, push bifurcating systems up the ladder of the evolutionary hierarchy, adding weaker and more flexible bonding energies to stronger and more rigid ones, complexifying system structures on each organizational level and simplifying them again on the next level, and adding highly developed control capabilities to more primitive forms of self-regulation.

As Laszlo maintains, technology has and continues to be the engine of change leading to exploitation of increasingly dense and abundant energies. In large part technology has been the driving force behind the irreversible arrow piercing through many bifurcations that involved saltatory change. Coursing along the space-time axis, one finds the following sequence of stages in the development of society: hunting-gathering→agrarian pastoral→agricultural→ preindustrial→industrial→postindustrial—this despite the colossal missteps, grossly inefficient decision making, and other potentially calamitous actions that humans and their institutions have taken.[9,10]

7.5 Key factors and general relationships in social organization

Before attempting to extend the notions of heterarchical models and the causal chains associated with interlocking hierarchies of replicators and interactors in order to incorporate Prigogine-type phenomena, let us probe at some depth into the social sciences and consider their concepts, tools, theories and models for understanding social organization. Hitherto we have examined the interaction occurring between behaving units and the environment—in a way that some would say questions the orthodox neo-Darwin synthesis. While natural selection plays a critical role, the interaction of the behaving units among themselves and with the environment, must be probed with full recognition of the feedback among the behaving units and the environment.

a. The participation potential analytic framework

Viewed from a Pattee perspective, rate-dependent and rate-independent (measurement-control) processes may occur at all points (levels) in a hierarchy, whether or not the hierarchy is regular. (We conceive an irregular hierarchy broadly to include at the extreme a one-level structure.)[11] The behavior associated with measurement and control may be that of any behaving unit in a hierarchy of replicators or interactors (as defined by Brandon (1991))—in general at any hierarchical level. It may be conscious and involve reflection on the part of the behaving unit, or it may be purely instinctive, habit-oriented or rule-determined (cybernetic). Also, it may be directly aimed at optimization (as discussed in Kauffman and Levin, 1987), or only indirectly so. In the latter case, it is associated with an organism which has survived competition, one that has not been selected against by nature.

Typically, when rate-dependent processes are discussed, it is possible to specify relevant functions, although even in the case of physical laws the empirical materials (data) employed to verify the law are based on the accepted set of concepts, theory and models that scientists inherit and the accepted manner, unavoidably subjective in part, in which measurement is conducted. For instance (as presented in mathematical form in section 7A.2 and consistent with the discussion of section 4.2), a rate-dependent process at an isolated location x at time t can be stated as:

(1) the time rate of increase of capital stock K at time t, namely dK/dt, equals the amount of goods available for investment I at t;
(2) I at t is equal to the total output of goods Y at t less consumption C at t;
(3) Y at t is taken to be equal to a constant k (a production coefficient) times the capital stock K at t, no labor or other resource being required.

However inadequate this definitional equation, it or a variant of it is widely employed as a starting point in economic and regional science analysis.

In contrast to an equation employed to model a rate-dependent process, there are few if any equations that are set forth to model a measurement-control process. To begin an attempt at such modelling, we set forth a type of relationship to capture the control (or measurement-control) process that naturally falls out of interaction phenomena. In doing so we draw upon a conceptual framework widely employed in regional science, which in turn draws heavily upon the Newtonian gravitational framework and which can be extended to incorporate the Einstein correction factors. In regional science, we speak of spatial interaction models, often designated gravity models. In its simplest form, the basic interaction I_{ij} between a behaving unit i (whether an individual, group, community, region, nation, or world region) with another unit j is defined as:

$$I_{ij} = Gm_i m_j / d_{ij} \tag{7.1}$$

where

m_i, m_j = masses at i and j, respectively, however defined
d_{ij} = distance between i and j, however defined
G = an appropriate constant

and where the exponent of the distance factor is set here at unity. In physics, when the exponent is unity, we have a measure of mutual energy; when it is 2, we have a measure of gravitational force. See Isard et al. (1970).[12]

Dividing both sides of equation 7.1 by m_i, and replacing I_{ij}/m_i by ${}_iV_j$, we obtain

$${}_iV_j = Gm_j / d_{ij} \qquad 7.2$$

Here ${}_iV_j$, being equivalent to I_{ij}/m_i, represents the interaction generated by the mass at j on each unit of mass at i—or on each individual at i if we take the above masses to represent populations. In Newtonian physics, ${}_iV_j$ is viewed as the *potential* created on a unit of mass i by the mass j at distance d_{ij}.

If we now include in the system many masses ($m_1, m_2, \ldots, m_j, \ldots, m_n$), where each mass generates interaction with or potential upon each unit of mass (individual) at i in an *additive* fashion, we can obtain the total interaction or *potential* ${}_iV$ generated by all masses on the unit of mass or individual at i by summation. Representing summation in this connection by the symbol

$$\sum_{j=1}^{n}$$

we thus have for the total potential created on the individual at i as[13]

$$_iV = G \sum_{j=1}^{n} m_j/d_{ij} \tag{7.3}$$

In a social science context, we may designate $_iV$ as *participation potential* at *i*. If we view the individual at *i* as a leader who is making a decision (or formulating a policy), then $_iV$ represents the participation potential of all masses, or groups of individuals upon this decision (or policy)—the participation potential (or influence) of any group *j* being attenuated (discounted) by its distance d_{ij} from *i*. (The notion of decision is taken to cover all thinking, reflection and other cognitive activity involved in a choice of an action including the action, 'do nothing'.) Alternatively, the individual at *i* interacts with all masses and their units and has potential influence or potentially participates in the decisions of all individuals at each *j*, the number of which we take to be m_j assuming each individual at *j* is taking a decision. (Again, this potential influence or participation is attenuated by the distance d_{ij}.) Over all decisions of all individuals in the masses $m_1, \ldots, m_j, \ldots, m_n$, the potential influence or participation potential of the individual at *i* is $_iV$ as given by equation 7.3.

In effect, the concept of participation potential for a behaving unit reflects the basic notion of mutual interaction, each decision-making unit in taking a decision being potentially influenced by all other units and in turn having potential influence on decisions being taken by all other units.

Put in another way, we have two views of accessibility: (a) accessibility of a total mass (population) to a unit decision at *i*; and (b) accessibility of an individual or behaving unit at *i* to all decisions at all *j*.

In empirical studies in the *field of regional science* mass has been measured in a number of ways. Population numbers has frequently been used as a measure of mass. But if intermetropolitan migration is to be projected (forecasted), employment or income of regions or subareas tends to be a more significant index of mass than population. Or, when the marketing problem for manufactured products is being examined, total dollar volume of retail and wholesale sales tends to be a more significant measure, etc. Clearly, the measure of mass to be employed depends on the problem to be studied, available data and related considerations. The array of possible measures of mass includes such magnitudes as total investment in facilities, commodity output, value added in manufacture, Gross Regional Product and economic opportunities. When studying *social* organizations and interaction within and among them, measures of mass such as number of families, car registrations, newspaper circulation, church attendance and school enrollment may be pertinent.

Similarly, distance has been and can be measured in a number of ways. So far, the discussion has *implied* distance *measured* in a physical manner, in particular along a straight line in terms of miles or other standard units. However, if a metropolitan traffic study is being conducted, distance in terms of travel time is at least equally important. Or, if industrial location is being analyzed, transport cost (or economic) distance is much more significant than physical. Among other possible measures of distance for use in regional science and economics are mileage along a specific transport route (waterway, highway, airline, railway, or pipeline), energy consumption in transportation, and number of intervening economic opportunities. Additionally, other general forms of distance that might be relevant are cultural, social, psychological, ideological and political; and for particular types of studies specific forms of these distances may be pertinent such as difference in the ranks of i and j in a caste system.

Note that while theoretically all individuals in a system can exert influence on a decision at i, in practice setting $d_{ij} = \infty$ can eliminate any influence from individuals at j. Thus we can replicate reality where only a limited number of individuals can influence behavior at i. Additionally, setting $d_{ij} = \infty$ can yield the reality of the individual at i having influence on a limited number of decisions at other locations j. Thus, too, this framework can handle and treat discrete sets of points, as well as a continuous set.

As already remarked, strict adherence to a Newtonian framework would require that d_{ij} be raised to a power of either 1 or 2. However, numerous empirical studies do not support such adherence. Frequently, the value for the exponent has been found to range between 1 and 2—at times falling outside these limits. See Isard et al. (1960).[14]

The above discussion points up uses of a spatial interaction, gravity-type model in regional science and economics. The widespread acceptance of the hypothesis that gravitational force was the first to break off from the primordial Ur force, its continued basic role in the physical sciences, and its recent mounting use in regional science and in economics suggest that gravity-type models can be a useful tool for the analysis of interaction and measurement-control processes in the social sciences in general and in other disciplinary areas as well. Certainly, it would seem to be relevant for analysis of hierarchies where the separation of levels may be conceived as a relevant distance. Also, bear in mind that the magnitude of the masses, distances and parameters change as interaction continues which then leads to change in interaction patterns and participation potentials; see section 7A.4 and Isard and Liossatos, 1979, chaps. 13, 14.

b. The roles of specialization, scale economies, joint actions, and information flows, costs and filtering

Examine other concepts useful for the analysis that regional scientists, economists and other social scientists conduct when they confront hierarchical structures based on interactions in the social system—assuming the existence of a stable state. To do so enriches any background framework for speculating upon the forces underlying the transition to far-from-equilibrium states or non-deterministic chaotic states and total demise.

First, there are advantages, or economies, from *specialization*. Among a set of individuals, there exist differences in attributes and capabilities at tasks. There are those that are endowed with brute strength, who can excel at moving boulders. Others are agile, and can excel at hunting and sports. Still others shine in analytical ability and excel in the organization of activities of a band of individuals or a team. Thus, sooner or later there emerges specialization in a group, whether primitive or not, that has survived, especially when owing to population growth or other factors environmental resources have become scarce and threaten to become inadequate. A two level hierarchy, at the minimum, comes to exist wherein each member of a group or team transmits information flows to the one or relatively few members who coordinate activities and where information may trickle down from the latter to the former. No formal ranking of these individuals need exist. Note that the resulting specialization that occurs becomes coupled with *economies of scale* as skill and knowledge is acquired with experience and as population increases. Each member becomes better at the task which at the start required the attribute with which he was more abundantly endowed or to which he may have been randomly assigned by a coordinator. Thereby, the gains from specialization are augmented.

As history has shown for many organisms as well as humans, occasions arise when survival of a group requires its taking a *joint action* with another group, thereby forming a *coalition*, or engaging in some other form of *cooperation*. The group may need to: defend itself against a warring tribe or nation; band together with another group to attack an enemy; or confront an environmental hazard such as the threat of a flood. Or the group may want to engage in trade. For the two or more groups involved, there emerges, almost out of necessity, another level in the hierarchy. Given that each group has a leader (or a subgroup of individuals) responsible for coordination and like activity within the group, there comes to be a need for coordination of the actions (behavior) of the leaders of the several groups. Here, in a coalition or alliance or other arrangement that has been able to survive, there typically emerges a higher-order leader who is more adequately endowed than others with ability to manage conflicts arising among the individual leaders of the several groups. He/she needs to work together with the latter in the choice of one of the options proposed (or more than one to be identified under contingency planning) or otherwise function in a similar capacity.

Carrying through this line of reasoning leads to still more levels in a hierarchy—even up to the current impending structure among nations requiring global coordination. The gains from coordination or, as some may designate them, *overview advantage*, are clearcut. But there are also costs. First, there is the cost of transmitting information up and down the hierarchy (and often laterally). Even in a small primitive group, the leader may have the same local visibility as each member; yet because of skill and knowledge gained by others from specialization he still needs to be furnished with relevant information from one or more members of the group; and there will also be costs in passing on information concerning a joint action that may be decided upon in any given situation. When a three or more level hierarchy comes to exist, typically the leader will not have visibility over all relevant localities; consequently he will need from other members situational information as well as information based on members' acquired knowledge. In general, the more levels, the greater the cost of information transmission.

Often not all desired or required information is visible. Some, if not a great deal, needs to be collected—information to be obtained by scouts or other collectors regarding the size of an enemy's attacking force, the number and quality (destructive capability) of its weapons, or the extent of rainfall in upstream areas and so forth. Thus there can be information costs not only at the bottom rung level, but also at other levels. Or a leader of a group may need to obtain reactions from members of his group to each of several joint actions being considered at the top level.

Sooner or later, given the limited capability of any mind in storing and recalling information, whether processed or not, and whether done efficiently or not, the information transmitted to the leader must be selected upon. The leader or key individual at any level finds that he cannot absorb all the information available. In general, for a hierarchy that has survived, a leader will have found that he has had to filter through the total of information available, informed others (implicitly or explicitly) about the information that he can treat or needs, or otherwise set constraints on the information that he takes into account in his coordination and other actions.

Often, information should or needs to be processed. If only a limited amount of information can be digested at the level to which an individual or group is transmitting information, it may be efficient for the recipients not only to filter out the less important information but also to process the information it has—by converting such to ratios and percentages at one extreme, to employing models at the other in order to reduce the number of variables (factors) to be reported upon. At the top level in particular, the leader will likely do simple processing, e.g. calculate rates of growth, or at times assign to his staff much more complex computations to determine the impacts of different policies on dependent variables which he considers key for society's, his group's, or his own welfare.

Costs of collecting, processing and transmitting information are factors that constrain hierarchical forms of development, but also are associated with participation potential as defined above. In general, as an individual becomes farther and farther removed in terms of relevant distance from the point at which a key decision affecting his welfare is taken—for example, as the levels of a hierarchy increase in number—his indirect participation in any decision falls; and thus his morale and consequent productivity may be negatively affected. This factor of foregone productivity operates as a cost and further limits the multi-level structure of an hierarchy. (See Isard, et al. 1970, chap.3 for further discussion.)

In addition to these factors, others have been emphasized by scholars. In his classic articles, Simon (1962, p.468) argued that hierarchical systems are 'one of the central structural schemes that the architect of complexity uses.' He contended 'that hierarchic systems will evolve far more quickly than non-hierarchic systems of comparable size.' Where time is a significant variable, as in economics when time in production may be equated with person-hours of labor, hierarchic systems are much more efficient and much less costly. Demonstrating this conclusion with his well-known watch-maker example (see pp. 109-10), he also points up the mechanism responsible for this outcome, namely the existence of stable intermediate subsystems—subassemblies or other forms—the time requirement being governed by the number and distribution of these subsystems. He then proceeds to distinguish between interactions *among* subsystems and interactions *within* subsystems. By way of analogy he points out that in physical phenomena intermolecular forces are generally weaker than internal molecular forces, and the latter weaker than nuclear forces—as we have already noted. He then proceeds to the general conclusion that interactions within a component subsystem are more intense, more numerous, and also of a shorter-run character than those among components whether or not at the same level in a hierarchy.[15] Finally, he speaks of the prominence and advantages in social systems of *near-decomposability*—loose horizontal and vertical coupling of subsystems. Each subsystem is dependent on inputs *from* other subsystems and is responsible for providing output *to* other subsystems. However, the processes any subsystem uses (its internal rate-dependent and measurement-control activities) to transform input to output can be and frequently are 'immune from scrutiny' by other subsystems. The subsystem is free to experiment, to adapt to mutations, subject only to the constraint set by the input received and output required for the system to remain integrated.

When each subsystem's internal functioning is 'sealed off' from other subsystems, near-decomposability minimizes information flows between subsystems and between levels. 'Parsimony of interactions' exists, each subsystem investing a minimum of resources in collecting and absorbing information about other subsystems. Thus, by receiving a minimum amount of information from each component of a lower-level subsystem, a given subsystem (or its leader) can treat and impose constraints,[16] and otherwise interact with a larger number of lower-level subsystems without exceeding its limited capacity to absorb and process information. For the subsystem's operation, much information about lower-level subsystems is redundant; and this is even more so when there are common elements among the inputs and outputs of these lower-level subsystems.[17]

In line with Simon's thinking, Levins (1970) speaks of the low cost and thus efficiency consequent to the low connectivity in an hierarchical system—a low connectivity that reflects an upper limit to the amount of reciprocal interactions a stable system can tolerate. As already noted, Pattee (1973) comes at near-decomposability and low connectivity from a different angle. Hierarchical controls involve a 'classification of detail'. That is, a constraint is established when the higher level classifies the detailed actions of the lower level it controls into 'significant' and 'insignificant' for its own use. It selectively neglects the irrelevant detail at the lower levels and exercises its control in response to only a very limited number of crucial pieces of information. Thus, the higher level concerns itself only with those actions and interactions on the lower level which are significant in producing an integrative function. The lower level retains its freedom in other areas. Pattee emphasizes that for the hierarchy to function effectively, the higher level must not ignore too many details or too few. It must employ a *Principle of Optimum Loss of Detail*, stating that the higher level can exercise control only by means of flexible constraints which can be neither too tight nor too loose. 'If they are too tight, we see more or less rigid bodies with no function, whereas if they are too loose, we see only boundary conditions that have no definite repeatable effect on the elements of the collection' (Pattee 1973, p 92). A balance must be maintained. The superordinate level must constrain some of the actions, or degrees of freedom of the subordinate level, but it must not leave it without freedom. It must 'lose' certain detail of the lower level actions if it is to operate efficiently. Pattee claims it is this 'simplification' of the lower level dynamics that allows a higher level function to be performed. By each level in a hierarchy being responsible for a certain degree of detail, a separability of focus is maintained, leading to an efficient specialization of function by level in a hierarchy.

The principle of optimum loss of detail as discussed in connection with vertical flows in a hierarchy can also be extended to cover lateral flows, especially when differences exist among the subsystems at any given level, often as a consequence of the different environments (physical or otherwise) that they confront. Moreover, each component may develop its own language or dialect or other variant of a language in the attempt to adjust optimally to its environment. Yet in areas in which communication is necessary for the performance of tasks, there must exist a common language and a basic set of shared-meanings. Such language and shared-meanings operate only as a set of constraints. Within these constraints each subsystem is able to develop its own separate focus (cultural, political, economic and otherwise)—i.e. to optimize in its particular environment. The outcome (commonalities and separabilities of language systems, laterally and vertically) may be said to characterize an efficient system structure (see Thayer, 1972), and one consistent with Pattee's principle of the optimum loss of detail.[18,19] Again, see sections 7A.2 and others in Chapter 7A that embody optimum loss of detail.

7.6 Dynamics forces in the transitions to far-from-equilibrium states

We now wish to attempt as best we can some explanation of the transition process in a society of human beings. In doing so, we draw heavily upon ideas in the physical sciences, and concepts, theories, and hypotheses in the social sciences, reiterating certain points covered in the previous chapters. We do so having in mind all forms of interactions among individuals and groups of all sorts including society as a whole. Thus any behaving unit (individual, group or society) may interact with any other behaving unit (individual, group, or society), the interaction of any unit with itself being excluded. We generally have in mind the concept of participation potential $_iV$, as discussed above, when we consider interaction over both time and space.

A number of interconnected factors may be said to be involved in a transition process. To unravel their interplay, we shall proceed step-by-step, starting with the consideration of one factor only and then introducing others one by one.

To begin, consider again what we have said regarding specialization. We might point to the fact that the very nature of human reproduction requires male and female to be functionally differentiated at least organically—which also leads to functional specialization re: nurturing and non-nurturing activity within the family. More significant is the specialization that evolves when groups larger than a family emerge. When individuals band together to move a boulder, kill a mammoth, defend themselves against a second band, and so forth, the need for developing a strategy, constructing a plan and performing other leadership functions arises, and such functions may be assigned to one or a small set of members of the band. Thus, at least a two-level hierarchy (formal or not) comes about. Such functional specialization mounts as experience and learning increases the capability and efficiency of each band member in the performance of his/her task, whether leadership at one extreme or pure brute force work or food gathering (preparation) at the other—one manifestation of scale economies in production, which is so prevalent and significant in today's global economy.

Consequent to successful specialization, history records major growth of population and its spatial spread. Concomitantly, markets for goods expanded permitting still greater specialization, in part based on the possibility of still greater scale economies—especially given the world's highly uneven endowment of resources. Technology, too, advanced. The wheel was discovered, as were line and column formations for military ventures, gunpowder, and more recently, electricity, nuclear fission and sophisticated communication devices. So were minerals and new energy sources. All this, and more, have led to massive concentrations of population, massive cultural developments, massive production operations and markets, massive destruction capabilities—and, generally speaking, massive interaction potentials, organized by intricate, highly interdependent hierarchical (and non-hierarchical) structures. All this, however, has been associated with an enormous stream, at times regular but mostly irregular, of perturbations, large and small—precursors of any transition, successful or not, that took place. There must have been perturbations prior to the time when several primitive bands and their leaders decided to come together and form a small village for defense purposes—thus leading to the formation of another level in the hierarchy of decision making or action taking. There must have been perturbations when villages came together to form a principality or regional grouping—and when several of the latter came together to form a nation. And certainly in the attempt to form a world government—an inevitability if the human race is to survive—the effort at establishing a League of Nations represented the amplification of a major perturbation, albeit unsuccessful.

Similarly, there were perturbations involved in the developing patterns of world commerce, the factory system, the multinational industrial giant, and the global financial community as well as in establishing and extending such world religions as Judaism, Christianity, Hinduism, and Islam.

Consider more closely the forces behind perturbations and the amplification (snowballing) of them that led to new far-from-equilibrium states that occurred in a number of the above developments. Do so against the level of specialization, technology, resource development and know-how that existed at a given point of time in a state that had a record of stability.

First, the occurrence of perturbations, large and small, is undeniable. It follows that the probability of a large perturbation (however defined) capable of being amplified (snowballed) increases with the length of time from the given point of time at which the existence of a given state is examined (assuming a Gaussian, Poisson, or other relevant distribution of perturbations for each unit of time.)

Next, if there is population growth, a natural process when resources are adequate, sooner or later the pressure upon existing resources (crowding) is likely to occur, leading to rising costs in meeting survival needs, and thus to increase in the effort at searching for ways to avoid these costs. This increase in effort enhances the probability that the amplification of any large perturbation that may occur will be able to overcome existing dampening tendencies. (The dampening tendencies can be present since an element experiencing the large perturbation and its amplification is in an environment of many other elements with which it interacts and which, via the potential ${}_iV$ that these other elements have on the given element, influences its behavior, its options, and defines its restricted action space.) At the same time, advancing technology (new input-output production functions), accumulating know-how from experience, learning and creative problem-solving, discovery of new resources, particularly new types of resources, may open up new lines of production (opportunities for gain) that will permit a successful amplification of a large perturbation. In sum, the situation may come to be ripe for change. A new behavior of an existing element subject to an amplified large perturbation may come into existence and be able to survive in an environment of many elements behaving in the old way; and the impacted element may do so profitably, with greater capability to compete and engage in cooperative relations.[20]

Our model of interdependent interacting elements involving the potential concept ${}_iV$ then suggests that the impacted element with its new behavior, having successfully combatted the dampening influences of these other elements, in turn influences their actions, especially those of neighboring elements. (Recall that the influence of a behavior at i falls off with distance and that as $d_{ij} \to \infty$, the effect approaches 0.)[21] The neighboring elements then also come to behave in the new more expedient way, tending to form a nucleation of newly behaving elements which gradually, if not rapidly, expands, and even snowballs, bringing more and more elements into its fold—until it contains perhaps the entire set of elements—all subject to the new organization that reflects the characteristics of the nucleation.[22]

Having now presented a standard social science look at transition processes and forces, let us examine some specific social forces and processes analogous with specific physical and biological forces and processes. Recall the discussion of the Bénard cell in section 6.3. There the amplification (induced by a sufficiently large temperature rise) of a large perturbation of an element can lead it to overcome the internal friction embodied in the viscosity of the fluid. This phenomena appears to occur simultaneously for all elements. But it also may take place in a succession of element changes wherein the time elapsed between the spread of the effect from one element to the next is equal to the infinitesimal time it takes an electromagnetic wave (light) to cover the intervening distance. Thus to the human eye the transition in the Bénard cell to a far-from-equilibrium state takes place instantaneously [although in the experimental studies where fluids are bounded laterally as well as horizontally and where irregularities are present in the boundaries, convection may start at subcritical levels of ΔT and spread spatially with time and temperature increase (Koschmieder, 1993)] . In contrast, today in the social world where spatial movement of objects (matter) and ideas is almost infinitely slower than that of an electromagnetic force, and is often disrupted and brought to a halt, the spatial spread (diffusion) of the effects of a major event, catastrophe, etc. via the influence associated with interaction, often taken as imitation, may be slow—where the major event may be said to correspond to a sufficiently large temperature rise.

In the Bénard cell, viscosity of the fluid imposes a frictional force on any element in its ordered movement in physical space and hence a cost in terms of an energy input to overcome that force. In society movement over space involves energy inputs or their equivalent. This is most explicit in the movement of a population unit (say a family) from one location to another. As developed elsewhere (Isard and Liossatos, 1974), this cost is equivalent to the sum of the costs over each unit of distance between any two locations A and B, and is subject to a natural transport rate analogous, for example, to energy inputs to overcome the friction in the transmission of electric power, also equivalent to transfer of mass.

Suppose in a Bénard cell experiment a small volume element of the fluid near the lower plate which is being heated is displaced upward by a perturbation. 'Being now in a colder—and hence denser—region, it will experience an upward Archimedes force that will tend to amplify the ascending movement further. If, on the other hand, a small droplet initially close to the upper plate is displaced downward, it will penetrate an environment of lower density, and the Archimedes force will tend to amplify the initial descent further' (Nicolis and Prigogine, 1989, p. 12).

In human history, the transition from a state of hunting and foraging in the Pleistocene era to one of agriculture may be said to parallel the transition from molecular chaos to the ordered, coherent pattern of molecular movement in the Bénard cell experiment. To see this, consider a hunting and foraging population in some large area such as Southern Germany where the arable land is highly productive and rather uniformly distributed. Imagine the gradual spread of agricultural know-how among this population. (The major increase in the productivity of human labor in agriculture for subsistence purposes may be said to correspond to a substantial increase in temperature.) At first we might imagine a development in this large area of an irregular pattern of population, perhaps small in number, engaged in agricultural practices. Each family would have started the cultivation of the land surrounding its domicile (cave, hut or what not), the initial distribution of families engaged in hunting having been influenced we assume by watering spots, salt licks, and other environmental features unrelated to the relative homogeneity of the fertile land.

However, as population numbers increased in this area now capable of supporting more numbers than possible via hunting and foraging activities, a point of pressure of population upon natural resources would be reached and more and more population would shift to the practice of agriculture. This in turn would lead via competition to the movement of families to a more regular spatial pattern of locations corresponding to a much more efficient distribution of farming units. (Such regular patterns have been observed by geographers, regional scientists and economists in Southern

Germany and Iowa in their pre-industrial stages of development.) The costs of movement of a family unit to a new location would correspond to the energy inputs in the Bénard cell experiment to overcome the friction associated with viscosity. The shifting around of population units in the large area would tend to increase the size of the initially small farms with a propensity to engage in a more intensive (denser) cultivation, and decrease the size of the initially large farms with a propensity to engage in a less intensive (dense) cultivation. As a result, size of farm and intensity of cultivation would tend toward an average—a consequence of competition and other forces that may be said to play the role that the Archimedes force does. (See Isard, 1956 and later location literature for detailed discussion of the competitive force.)

To proceed further with analysis by analogy, consider the tendency for amplification which takes the form of cumulation and aggregation in biological phenomena. Such occurs for example in the behavior of the amoeba, *Dictyostelium discoideum* (or slime mold), during development stages. In the unicellular stage the amoebas move in the surrounding medium feeding on such nutrients as bacteria and proliferate by cell division. As Bonner (1988, p.236) states, when discussing evolution,

> some bacteria can orient in chemical gradients, for they have receptors which react with substances that might be found in their environment: food substances such as sugars or amino acids, or noxious substances of various sorts. This suggests that receptors arose first in evolution—not in response to signals given off by other cells of their own kind, but in response to environmental cues.
>
> The next step is when cells synthesize both the receptor and the signal molecule; they become self-signaling. There is a good example of this among separate, feeding amoebae of cellular slime molds. They tend to repel one another so that they seem to be evenly spread as they feed, a phenomenon thought to provide optimal grazing when there is an even lawn of bacterial food. It can be clearly demonstrated that each cell gives off a chemical signal, and if there is a collection of cells in one spot they will slowly separate from one another by negative chemotaxis. Another good example of this is found in the development of the pigment cells in a species of salamander where these cells are evenly spaced over the body. There is excellent evidence that this spacing is also due to negative chemotaxis between the cells.

(Note that corresponding to negative chemotaxis is the force of competition at the boundary of two neighboring farmers in a spatially regular distribution of farms that repels each one from extending his cultivation into the area of his neighbor.)

Suppose now that because of an environmental change, the amoebas confront starvation conditions—a major new environmental constraint. As

Bonner (1988, pp.231,233) notes, current thinking is that after a period of time some cells begin to secrete acrasin (a chemical signal or attractant),

> sooner than the others, and those cells become the foci by attracting the cells immediately around them. The acrasin is given off by the central cells in a pulsatile fashion. As a wave of high concentration of the acrasin passes outward, it first attracts the cells that it passes over, and then induces them to secrete their own acrasin, which passes along to attract the next group of peripheral cells in the same fashion. The center of the aggregate sends off these pulses of acrasin at the highest frequency; it is the pacemaker, and all the cells become subservient to it. In the most direct sense this aggregation is an act of integration. The acrasin is a small molecule which diffuses readily, and the responding cells have specific receptors for it. In this respect it is no different from vertebrate nervous and hormonal systems.[23]

Bonner, however, notes that once a center is formed—presumably an appropriate aggregate—it produces an inhibitor (a repellent gas) which prevents other centers from forming in its vicinity.

In parallel fashion, one can analyze the tendencies for the formation of production centers serving more than one family unit. Within the above large area or region of farms that has emerged, assume that all family units have the same tastes and know-how, and in all other respects are alike, and self-sufficient, producing from the land all that they consume. Take as an initial state of this farming region a spread of farms (family units) that approximates a uniform distribution such as in Figure 7.1—a reasonable oversimplification for analytic purposes and one that is often implicitly assumed by economists when discussing stages of economic development.[24] This society, like any other system, is subject to perturbations. One subsistence farmer may have played around with a new way of producing grain, but it flopped. Another may have explored a new way of making cheese, but it too turned out to be disadvantageous. And so forth, but at some point, a farmer may have stumbled upon a new way of producing an alcoholic beverage, say beer. And it turned out to be profitable—a major perturbation. He may have found that to produce the beer that two other farmers consumed in addition to his own allowed him to receive, in bartering with these two other farmers, more grain than what he would have produced with the same amount of labor that takes him to produce beer for three households. He is better off. So may be each of the two farmers whom he would have had to persuade to barter grain for beer, for the saving in labor that obtains by not having to produce beer (after being reduced by the time spent at transporting the bartered grain to the first farmer and the acquired beer back to his farmstead) allows each to produce more grain than the amount exchanged for the beer . All three are better off. It is not necessary to go into additional detail on how other farmers learn of this new method of beer production and become beer producers; how turbulent competition involving unpredictable, irregular and inefficient entrance and exit (elimination) of farmers, and how the shifting of farmers as consumers from one producer to another may take place in an ensuing incoherent transition. The learning and change in production pattern that do result may be viewed as stemming from interactions of these farmers with each other (elements *i*, *j*, *k* . . .), and in particular as associated with changing gradients on the potential profitability of locations for production and in the potential for effective bargaining—and in general in the interaction potential at all points over the field associated with the set of elements from changes in relevant masses (production volumes) at the sites of *i*, *j*, *k* . . . These gradient changes parallel the density gradient changes in the Bénard cell, the BZ reaction, the morphogenesis of the amoeba *Dictyostelium discoideum* and other natural and biological processes referred to by Prigogine that lead to self-organization and his long-range statistical correlations, and wherein the behavior of the source element (e.g. cell, farmer) at a location is sensitive to (governed by) the gradient at its location. A regular pattern of beer producing centers results, each realizing the same profits. Interestingly enough market areas served by these centers (see Figure 7.2) come to be regular hexagons of equal size—which is also the pattern of Bénard cells that invariably results when the critical temperature required for thermal convection is passed (Koschmieder, 1995, p. 62).[25]

It is not necessary to present at length the detailed analysis that regional scientists and economists have pursued, and the additional abstractions they have required to derive more rigorously the Löschian net of hexagons in Figure 7.2 and patterns of production centers of different size depicted in Figure 7.3. The necessary abstractions parallel those in the Boussinesq approximations in the theoretical work on Bénard and Rayleigh-Bénard convections—namely, 'that the variation with temperature of the material properties of the fluid such as the kinematic viscosity, the thermal diffusivity, and the volume expansion coefficient, are negligible provided that the temperature variation is small' (Koschmieder, 1993, p.14). The Lösch hexagon derivation in Figure 7.2 results from the simple play upon the initial population distribution of two forces—transport cost as a frictional force and economies of scale from producing for seven consumers (corresponding to savings on energy input) that are sufficiently large to overcome the required transport cost.[26] Note, however, that the analysis is essentially static; it provides only an explanation of what may exist, subject to irregularities, at a point of time. It lacks any explanation of the specifics of the transition—of the space-time path of competitive (and strategic) actions that may have been taken by the given set of farmers, or other paths that could have evolved, or paths that may have in reality led to disastrous results.

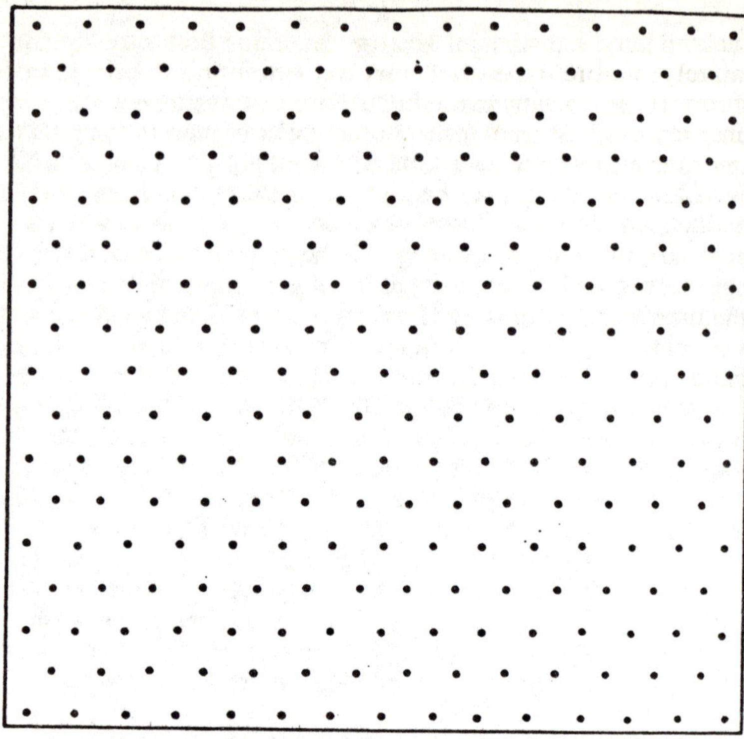

Figure 7.1 A regular distribution of settlement units (farmers)

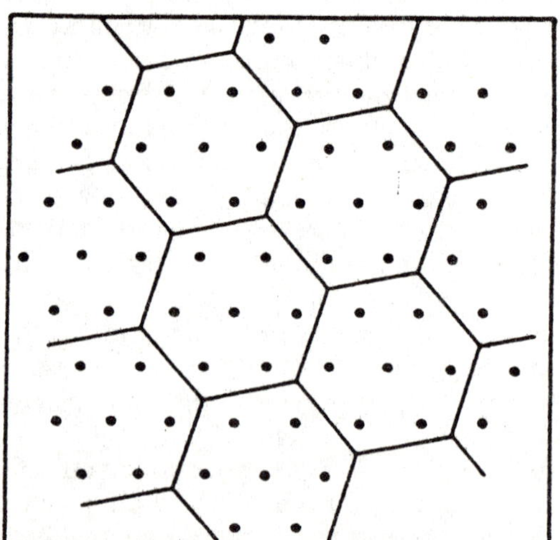

Figure 7.2 A pattern of regular hexagonal market areas

Lacking any knowledge of how the transition takes place from a relatively simple state to a far-from-equilibrium one involving the commercial production and barter exchange of just one good, we certainly cannot say much of anything regarding the transition to the production and barter exchange of a second good, of a third good, . . . , of an nth good that now characterizes the rural-urban megalopolis character of the world economy, which clearly constitutes an irregular hierarchical structure. Figure 7.3 presents a speculative depiction of structure for a set of hexagonal networks involving four classes of goods and four different size population concentrations—a primary city (at the center), metropoles, towns and villages, for a given symmetrical distribution of individuals (farming units). See Lösch (1954) and Christaller (1966) and later writing for extensive discussion on the development of such superimposed networks of regular hexagonal market areas.

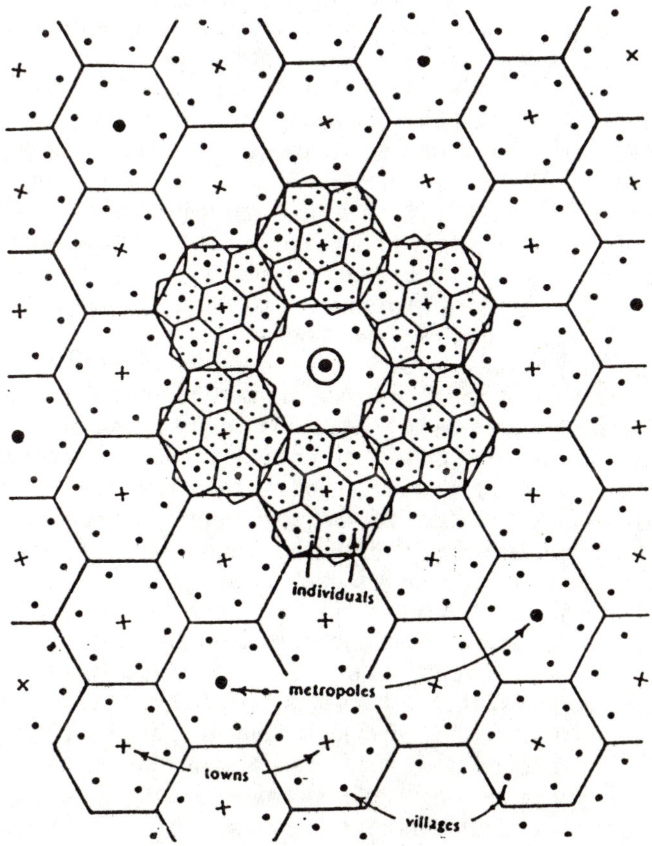

Figure 7.3 A spatial system involving a primary center and four other orders of node

Thus, while this section probes into possible relationships and forces underlying major and abrupt transitions in social systems and suggests parallels in the physical sciences, we are able to provide only a few insights regarding the why, where and when of these transitions, just as is the case for physical scientists when they confront similar transitions in physical systems. In section 7A.5 we present mathematical models involving perturbations to sharpen what insights we have.

7.7 Summary

We now wish to bring together the materials reported upon in this chapter and relate them to those of previous chapters.

We began with an examination of different kinds of hierarchies and soon focussed upon behavior and interaction within any given level and between levels of an hierarchy. There are hierarchies of replicators—genes, chromosomes, cells. These replicators interact with their environments. Likewise do the vehicles of replicators (a beaver or human being) which are not replicators. Vehicles are not passive entities; they impact the environment, too. Hence we must think of hierarchies of both replicators and interactors. If there is selection at a given level, it must be upon interactors at that level, and since both replicators and vehicles can be interactors, selection acts on both replicators and their vehicles. Thus interaction with the environment and selection occurs at the lowest level of an interactor (the gene)—where micro analysis must be pursued—and at the highest level (the species, the ecological system, the society)—where at least macroanalysis must be pursued—and everywhere in-between. Thus a species, or system involves interconnections at all levels—interconnections among the units within each level, among units at different levels, with relevant environments of units at all levels involved, and with all kinds of feedbacks. As a consequence, organisms become not only problem-solvers in adapting to their environments, but also problem-setters who define and redefine their problems by modifying their environments. And in all this, anticipatory behavior becomes essential for survival.

Given interaction at all levels, a number of possible causal chains may be proposed—by the geneticist at one extreme to the system analyst at the other. To recognize the relevance of each chain and also the constraints (network of controls) imposed on it, a heterarchical framework may be conceived—as an ensemble of models. Such a framework makes explicit the coordinative conditions that constrain the possible behavior and mutations within an existing stable system, since only those variations in a system that meet these conditions can survive. While a heterarchical framework embraces the phenomena of measurement-control processes and thus goes beyond the notion of a complex dynamical systems model with coupling functions which interlock only rate-dependent processes, it has its limitations. It does not indicate how measurement-control processes are to be explicitly incorporated, even into simple models. Nor does it shed light on Prigogine jumps—transitions to far-from-equilibrium states. The historical record all too clearly reveals such jumps—which often lead to the complexification of hierarchical structures with increasing number of levels.

In an attempt to capture some understanding of the behavior underlying measurement-control processes in a system containing many interacting units in physical space, or how these processes relate to the highly ordered connectivity properties of well-scrambled networks (as conceived by Kauffman and Levin, 1987), a participation potential concept is proposed. Two aspects of this concept were identified: the effect of the masses of interactors at all locations on decisions at a location *i*, and the participation of each behaving unit in the mass at *i*, and the mass itself, on all decisions at all locations—masses and distances being the key factors. Although this concept in both its aspects can be operationalized once definitions of masses and distance are adopted, the manner in which these aspects of interaction (which are rate-independent and associated with measurement-control processes) can be operationally fused with dynamical systems (rate-dependent) models is yet to be developed.

In addition to interaction involving measurement-control processes, Prigogine-type jumps are associated on a more concrete level with changing: (a) specialization, (b) scale, localization, urbanization and other agglomeration and spatial juxtaposition economies and diseconomies, (c) information flows and costs, (d) technology, and (e) environmental conditions. Analogies linking social system features and development—for example, its ideal hexagonal networks—to types of phenomena and forces at play in the Bénard cell experiments and the morphogenesis of the amoeba in interacting with the environment, are at best only somewhat useful in helping to explain and project the why, where, and when of social system jumps. (Even in the physical and biological sciences, Prigogine-type jumps can only in part be effectively modeled.) In Chapter 7A, we gain some deeper insights into this problem through use of mathematical models and statements—but do not resolve it.

At this point, we may examine how the materials of this chapter interconnect with those in the previous chapters. In one sense, all the materials presented in this chapter represent more detailed analysis that supports much of Teilhard's thinking. This is so whether we refer to his discussion of: (a) the genesis of increasingly elaborate organization (largely, if not entirely hierarchical); (b) the set of major increases in interiorization since the time when the first replicator came on the scene (a set of Prigogine-type jumps); (c) the superimposition in the far future of the state of noogenesis upon that of biogenesis (the emergence of another hierarchical structure); and (d) his contention that the way to Omega (or we may add, any other heavenly state) is along a deducible direction, but does not have a *deterministic* space-time path.

The materials of this chapter can easily be viewed as further elaboration of those in Chapter 3 on evolution theory, but also as further evaluation of some emerging thought not yet having widespread recognition and acceptance. Certainly Dawkins' views on the selfish gene, the extended phenotype and the meme are embraced (in the discussion of hierarchies of both replicators and interactors) and further developed. So also are: (a) Layzer's discussion of the development of complex hierarchical structures; (b) Wilson's sociobiological emphasis on how diverse species (in this chapter, we point to amoebas) adapt to the special opportunities and exigencies in their environment which enable us to gain deeper insight into principles that human societies can and have employed; (c) Lewontin's insistence that human evolution represents an integrated development of relationships between the biological and the social sets of forces; (d) Gould's and Elbridge's notion of punctuated equilibrium (a manifestation in the large of a Prigogine-type leap), and their view that the specie as well as the individual can be the unit of selection; and (f) Durham's coevolutionary theory wherein selection operates at the combined cultural/ ecological level.

The first part of Chapter 4 developed the notion of the two complementary modes of Pattee: (a) the *dynamic* mode, characterized by *rate-dependent* processes, often reflecting inexorable and universal laws, which refer to continuous and reversible time; and (b) the *linguistic* mode (also designated a cognitive-linguistic-spiritual mode) which may be equated to measurement-control activity (or to include or be included in such activity), characterized by rate-independent processes. These two modes, according to Pattee, are not reducible to each other nor integrable into a single unitary system, yet relate to a unitary *living* system—which our analysis in this chapter bears out to be the case for the social organizations of today's world and its parts.

The second part of Chapter 4 focussed on nonobservables and the emergence of novel and creative structures and behavior. To reiterate, the laws of physics are completely independent of personal experiences, which experiences may enter into measurement and much hierarchical social interaction that reveal the required initial conditions and constraints—the pre-established observables of the underlying theories. Computational models, too, have little, if anything, to offer for discovering and identifying nonobservables and describing creative thought—although they can be invaluable for studying and replicating interactions where inputs, outputs and relationships can be represented by well-defined symbols and treated as simple observables subject to rule-based operations. Yet, as discussed in Appendix A of Chapter 4 computational models, in particular, cellular automata, can serve as explicit models for a wide variety of physical and social processes—converting inputs into outputs to determine how certain social organizations and other systems behave *under specified laws*. They thus can relate to a world that any interactor or an analyst considers as a best approximation to reality as well as to hypothetical worlds whose development based upon laws and prescribed relations an analytic interactor wishes to explore. Moreover they can handle linear and nonlinear relations and move into areas too complicated for the use of traditional mathematical methods and well beyond the computational and creative thinking of the brain—as well illustrated in recent developments associated with chaos theory or in examining the implications of a number of models such as discussed in the Chapter 7A. This is particularly so when an excessive amount of computation is required. In this regard, cellular automata and other computational models permit *an analytic interactor directly to interact with them.* They allow him/her to explore implications of a particular policy option or sets of interdependent policy options, for each of a number of situations. In these ways, cellular automata and other computational models can be invaluable as tools to be exploited in an effort at understanding creative thought—certainly in sharpening up the question of the where, when, and why of Prigogine-type jumps.

Another computer-based development and type of experimentation is the adaptive fuzzy system discussed in Appendix C of Chapter 4—a system that involves the combined employment of an artificial neural network model and a fuzzy systems analysis. Such a system can provide additional insights on how the human mind confronts uncertainty, imprecision and noise in its perceptual activity, concept formation and common sense reasoning. It can help illuminate how an interactor learns from experience and how he/she adaptively develops estimates for use in the selection of a response to a problem stimulus without having to specify in precise mathematical form how outputs depend on inputs. Still such a system is abstract—still far from replicating the biological operation of the brain, its classification and learning activities, its complex pattern recognition, its coherent dynamics and generation of creative thought and decision making in the roles of an interactor within his/her relevant hierarchies.

In terms of subject matter, Chapter 5 makes an abrupt shift. However, it does cover another set of essential background material for understanding social organization and development. It does so by viewing the space-time path of the physical system, from the Big Bang on, as:

(1) a series of symmetry breakings that has led to the current set of forces and kinds of matter;
(2) the emergence of successive hierarchical organizations of *forces* (gravity and the indistinguishable aggregate of others; later to be gravity, strong and electroweak; and still later gravity, strong, electromagnetic and weak) and *matter* (leptons, quarks and gluons; later electrons, muons, tau-leptons, protons, neutrons, etc., etc.; and currently the elements of the periodic table and the kaleidoscopic variety of their compounds); and
(3) a process whereby via gravitation a sequence of clusterings of matter

> that led to the present hierarchy of nested self-gravitating systems of stars, star clusters, galaxies and galaxy clusters.

Chapter 5 provides extremely suggestive insights as to how symmetry breaking and hierarchical construction develops in human and social development. This is so whether we consider specific social hierarchies as constructed via subassembly (as Simon describes in his classic watchmaking example) or via exceedingly complex interactions among constituent hierarchical subsystems, their units and the environment, that for example characterized the Roman empire at its heights (to be discussed in Chapter 10). Moreover, the process whereby stratified stability is achieved in the formation of matter casts light on the attainment of stability and its continuance in social development.

The appendix to Chapter 5 presents thinking of the physicist Bohm, which draws heavily upon the developments in physics of field and particle theories. In attacking the question of reflection and consciousness and in setting forth insightful and provocative hypotheses and analysis regarding implicate and explicate orders, enfoldment and unfoldment, and sense of movement and continuous flow, he develops an all embracive conceptual framework. Within this framework complete interdependence exists, which is consistent with the interaction discussion and participation potential concepts presented in this chapter. Everything is related to, and contains everything, and only in the explicate order are things separate and able to be viewed as relatively independent. Whatever is judged to be valid in Bohm's statement: (1) is in keeping with the Pattee concepts of linguistic mode, measurement-control processes and the basic interaction framework in the thinking of evolution theorists covered in the previous chapters; and (2) contributes still more to a general analysis of interaction toward which the discussion in this chapter leads.

The first sections of Chapter 6 move on to recent developments in the sciences which are associated with chaos theory and the analysis of disorderly processes—processes that are concomitant and often intermixed with the orderly ones covered in this and other chapters. Layzer's hierarchy of time-bound order from the Big Bang through ecological evolution down to the human cultural systems of today where every process in the sequence of processes created structures and initial conditions that made possible new processes certainly encountered disorder along the way. Disorder must have been present in past periods of population growth when the existing conditions were of a similar nature as those that yield the areas of chaos portrayed in the May population growth model. Yet there may have been order in the succession of disorderly periods such as in the succession of areas of chaos in the May model that the universal Feigenbaum constant predicts. And there can be deterministic chaos (associated with strange attractors) and order within seemingly disorderly phenomena (such as turbulence within fluids) and structures (such as coastlines and geological formations definable as fractal structures often exhibiting self-similarity at different scales). The findings of chaos and related theory thus clearly establish that in the evolution of social structures, hierarchical or not, periods of extreme disorder can be encountered in interaction and behavioral configurations which lead to societal disintegration and collapse, as will be studied in Chapter 10.

The examination in the last section of Chapter 6 of the dynamics associated with the Prigogine school of thought—emphasis on symmetry breaking, amplification of perturbations, self-organization, jumps from one coherent stable far-from-equilibrium situation to another, and irreversibility—provides much background for the analysis of the forces for major change and sudden jumps in hierarchical structures and their interaction configurations.

Notes

1 For notes toward a formal history of the levels concept, see Sathe (1988).

2 One interesting hierarchy of relevance to research procedure *per se*, but not directly related to behavior in general, is that involving epistemological levels. See Klir (1985, 1991).

3 His framework can be considered to be a two-level space-time general equilibrium system. See Isard and Liossatos (1979, chap.5) for presentation of a somewhat parallel, but more detailed space-time conceptual framework of a social system and its interacting entities

4 See Ho (1988) for another presentation of the underlying processes nested in space and time which encompass all levels of interaction of organisms and their environments integrated in the dynamic set of experiences involved in development.

5 This statement is consistent with the earlier writing of Whyte (1965). Whyte emphasized the balance between stability and flexibility in biological systems, suggesting stability is ensured by an 'internal selection process' operating on mutations in the subsystems of an organism. Prior to the operation of external or Darwinian selection, mutations in the genotype of an organism are subjected to a test of internal coordination which determines whether they 'fit in' with the overall functional structure of the organism. Thus, 'coordinative conditions' constrain the possible mutations, since only those variations in the structure which meet the coordinative conditions—or which are physically, chemically or functionally stable—will survive the internal selection process. A successful mutation represents a change from one solution of the coordinative conditions to another which proves to be more efficient in the external competitive process.

This statement is also consistent with that of Grobstein (1973) who speaks of the process of neogenesis—the emergence of new properties or new levels of hierarchical control—illustrated by that of a linear chain of amino acids folding itself into a three-dimensional configuration to form an enzyme. Grobstein first asks what factors cause the higher level to form. He maintains that it is both the inherent properties of the individual amino acids—which lead to a certain pattern of interaction between them—as well as conditions in the surrounding environment which result in the kind of folding which will produce enzymatic activity. His next point is that the formation of the higher level results in a 'transformation' of the initial relationship between individual amino acids. Grobstein (1973, p.45) writes:

> the properties that become important at the new higher level of order do not relate equally to all of the amino acids but especially to certain of them in the new 'surface' structure of the molecule. The enzymatic property is a derivative of the amino acid sequence but it is a complex and selective derivative in which the properties of the surround play an essential role.

From a slightly different perspective on social organization, Olson (1965) has shown that a collection of rational individuals will not act in the interests of the group—even if the group interests will result in the furtherance of individual interests—unless there is some form of 'coercion' or 'separate incentive.' Again, in the language of hierarchical theorists, for an effective social entity to form, the collectivity must 'close back' on its individual members.

6 For discussion of the highly ordered connectivity properties of well-scrambled genetic regulatory networks and their extraordinarily coordinated patterns of gene expression, see Kauffman (1985). Also, see Kauffmann and Levin (1987) for a view of the functioning of these networks as involving a complex combinatorial optimization process where the conflicting constraints due to the natures of the different parts and processes or conflicting requirements due to alternative simultaneous optimization goals, limit the end result achieved.

7 À la Abraham (1991), the PSIG (The Peace Science Integrated Global) model may be viewed as a spectrum (or hierarchy) of coordinated submodels. It incorporates microscopic (e.g. industrial location), fine-grain (interregional input-output and SAM), coarse-grain (gravity), macroscopic (LINK econometric, GLOBUS and global climate) submodels and cognitive-decision (conflict management and world policy) frameworks. See Figure 11.1 and accompanying discussion on pp.366-69 below). Thus, at every point in a state space, there exists a cone of favored directions, with the cognitive-decision frameworks exercising measurement-control on whatever dynamics is embodied in the submodels. Also see Isard (1980), Isard and Anselin (1982), Isard

and Smith (1983) and Isard and Dean (1987).

8 However, some may argue that the event or disturbance that was amplified and did lead to a new far-from-equilibrium state à la Prigogine was inherent in existing social organization. For example, Dahrendorf (1959, pp.122,125) has indeed justly criticized structural-functional sociology (à la Parsons) for not recognizing adaptive self-organization. He takes issue with the view of a static social system whose parts are all 'functional' in that they contribute to its stability. He writes, 'we must find, apart from the undoubtedly important construction of functionally integrated structures, elements which are independent of these without being necessarily external to them and which determine relative stability as well as kind and degree of change of structural patterns.' Dahrendorf emphasizes the need to 'trace in the structure of society the seeds of its supersedure. The idea of a society which produces in its structure the antagonisms that lead to its modifications appears an appropriate model for the analysis of change in general.'

9 Laszlo also emphasizes autopoiesis. This concept has reference to the complex replicative processes involving renewal, repair and reproductive acts of self-creating systems where the environment, while providing energy inputs required for living, does not have any direct effect on the above acts. See Maturana and Valera (1980) and in't Veld, et al. (1991).

10 Another theory of evolution of a very general character, a dynamical replicative network theory that is applied to all levels of units ranging from the cell at one extreme to the biosphere at the other, is presented in Csányi and Kampis (1991).

11 Irregularities in hierarchical structures are implied by the spatially uneven distribution of matter and physical phenomena by type over the global space—for instance, energy resources and weather conditions.

12 A more general concept would apply exponents α and β to the masses m_i and m_j, respectively, to reflect agglomeration (deglomeration) forces present in masses—somewhat analogous to autocatalysis and the opposite. Also weights w_i and w_j should be applied. The numerator of the right hand side of equation 7.1 would then be $w_i(m_i)^\alpha\ w_j(m_j)^\beta$. If this change is made, then in the subsequent operations and equations 7.2-7.3, m_i should be replaced by $w_i(m_i)^\alpha$ and m_j by $w_j(m_j)^\beta$.

Also other types of functional forms may be considered.

13 In this formulation we must allow for the mass at i to create potential upon itself and on each unit in that mass; consequently, d_{ii} must be taken to be some positive quantity in order to avoid an indefinitely large value of ${}_iV$. In some connections, d_{ii} may be viewed as (a) the average distance of the particles of a body from the center of gravity of the body adjusted for the average distance of units in a mass from each other, or as (b) some fictitious physical distance corresponding to a

terminal charge in transportation, or a loading and unloading cost in shipping goods, or a cost of taking information out of storage, codifying it for transmission and transmitting it. For more complete discussion, see Isard (1970).

Also, the general constant G may be replaced by a set of specific constants, $G_1, G_2, \ldots, G_n$, each constant in this set being applicable to a corresponding mass.

14 Questions also arise when weights w_i are to be applied to the masses. Just as the weights of molecules of different elements are unequal, so are the weights of different kinds of interactors (people). Consider a study of interaction among individuals (masses) reflected in the volume of first-class or luxury travel. It is reasonable to expect that, *ceteris paribus*, an area with high per capita income will generate a larger volume of such travel than an area having the same population but with lower per capita income. One way to correct for such a factor is to multiply the mass, the population P_j, of each area by its average per capita income. Thus equation 7.1 becomes

$$I_{ij} = G(w_iP_i)(w_jP_j)/d_{ij} \tag{7.4}$$

and equation 7.3 becomes

$$_iV = G\Sigma(w_jP_j)/d_{ij} \tag{7.5}$$

where w_i and w_j represent per capita incomes in region i and j, respectively. But since Y_j the total income of region j equals w_jP_j, equation 7.5 becomes

$$_iV = G\Sigma\, Y_j/d_{ij} \tag{7.6}$$

which is the equation for what regional scientists designate income potential.

In a study of first-class travel it might be desirable to weight population by still another factor, say education level, or even by a third factor, say family size. In such event, w_i and w_j would represent composite weights, that is averages reflecting the relative importance of the several weighting factors.

15 For elaboration and further discussion of these points, see Simon (1973) and Augur (1989).

16 From the viewpoint of traditional political science, an optimal governmental hierarchy that addresses the problem of number and functions of different levels of government must meet the constraints of a proper *areal division of power* (Maass, 1959). For example, for the world system, this is the problem of dividing the many legislative,

administrative and adjudicative processes among the world, national, regional (state, provincial) and local levels of government. These processes clearly overlap and are not easily distinguished. Their assignment to different levels of government is exceedingly difficult if 'separation of power' and a meaningful bundle (complex) of functions for each governmental unit, and a proper set of checks and balances are to be realized.

17 Particularly interesting here are the ideas of Arbib (1972) on the way a decentralized decision-making system within an organism ensures reliability of information collected by subunits as well as the related problem of how choices are made in such a system in the face of conflicting information picked up by the localized receptors.

According to Arbib, information in a decentralized system is collected with redundancy: the system does not depend on the proper behavior of any one, or few subunits, but acts on the basis of threshold levels of information gathered by all subunits. There exists no one rigid collection-transmission pathway, but many possible pathways which must sum to a threshold level. Thus, a fundamental property of hierarchical control structures is that decisions are based on threshold effects.

Arbib points out the problem of 'resolving the potential redundancy of command' by different subunits which results when information connected by the several lower level units suggests conflicting courses of action. One model he presents depicts subsystems making tentative decisions on the basis of partial information and slowly changing them under mutual interaction until consensus is reached to commit the organism to an overall mode of action.

Arbib's statements are consistent with views earlier expressed by Weiss (1971, pp.24-25). Weiss notes that the most efficient form of behavior for a given static environment would be one in which there are 'rigidly' determined programs, 'predesigned as chains of linearly seriated steps, each step linked solely to the one immediately following —with no direct regard to the integration of the whole performance.' However, he writes that such preprogrammed behavior is not efficient in a variable environment. For adaptability, there must be freedom for components to react in a variety of ways to environmental stimuli.

18 The above discussion relates in general to a single hierarchical system. In reality, however, measurement-control and other actions taken by behaving units at one level in a given system may affect behavior of other units not only at other levels, but also in other hierarchical systems.

19 Blau (1964) in writing about social organizations, expresses similar ideas on the need for, and limit to, communication between a subsystem and the larger system. He distinguishes between 'particularistic' and 'universal' values and language. The former provide a group with a sense of identity and are 'media of solidarity' producing 'segregating boundaries between subgroups in a collectivity'; the latter are 'media of

exchange and differentiation' among subsystems. Particularistic languages ensure that subsystems remain discrete; universalistic languages ensure interaction and integration of discrete subsystems.

20 Before this point in time, any nucleation that might possibly have started to form may have not been able to grow and have disintegrated because of failure to have exerted sufficient influence on the existing non-nucleated elements.

21 It is also conceivable that simultaneous or near simultaneous perturbations in two or more neighboring elements *j, k*. . . may have impacts on *i*, the total of which may push the element *i* into a new behavior that helps provoke the transition to a nucleation.

22 À la Kauffman (1985) and Levin (1987). This nucleation process may be viewed as involving optimization by each element involved, each element having two or more options, the old option and the new one, a new one being the most efficient from the standpoint of the element. As previously noted in Chapter 2, optimization behavior is consistent with the competitive nature of humans and their curiosity—attributes that are selected for.

23 Nicolis and Prigogine (1989, pp.33,35-6) describe the process in a fuller and more technical way. They observe the following phenomena.

> First, after starvation some of the cells begin to synthesize and release in the extracellular medium signals of a chemical substance known as cyclic adenosine monophosphate (cAMP). The synthesis and emission are periodic, just as in the chemical clock of the BZ system, with a well-defined period for given experimental conditions. The cAMP emitted by the 'pioneer' cells diffuses in the extracellular medium and reaches the surface of the neighboring cells. Two types of events are then switched on. First, these cells perform an oriented movement called *chemotaxis* toward the regions of higher concentration of cAMP, that is, toward the pioneer cells. This motion gives rise to density patterns among the cells that look very much like the wave patterns in the BZ reagent . . . Second, the process of aggregation is accelerated by the ability of sensitized cells to amplify the signal and to *relay* it in the medium. This enables the organism to control a large territory and form a multicellular body comprising some 10^5 cells.
>
> Thus, the response to the starvation constraint gives rise to a new level of organization resulting from the concerted behavior of a large number of cells and enabling the organism to respond flexibly to a hostile environment. What are the mechanisms mediating this transition? Let us first observe that the process of chemotaxis leads to an amplification of the heterogeneity formed initially, when the pioneer cells begin to emit pulses of cAMP. Because it enhances the density of cells near the emission center, chemotaxis contributes, by

means of the phenomenon of relay, to an increased release of cAMP from this region, which further enhances the chemotatic movement of the other cells toward it. This constitutes what one usually calls a *feedback loop,* very similar to chemical autocatalysis or to the self-acceleration of an exothermic reaction. . .

As it turns out, a second feedback mechanism is present in *Dictyostelium discoideum* which operates at the subcellular level and is responsible for both the periodic emission of cAMP and the relay of the chemotactic signal. This mechanism is related to the synthesis of cAMP by the cell. cAMP arises from the transformation of another important cellular constituent, adenosine triphosphate, ATP, which is one of the principal carriers of energy within living cells thanks to its high-energy phosphate bond. The ATP→cAMP transformation is not spontaneous however; a catalyst is needed to accelerate it to a level compatible with vital requirements.

This catalyst is an enzyme called adenylate cyclase. It fixes itself at the interior of the cell membrane and interacts cooperatively with a receptor fixed at the exterior phase of the membrane. The cAMP that is produced diffuses in the extracellular medium through the cell membrane. It can bind to the receptor and activate it, thereby enhancing its own production. Thus, a feedback is created which can amplify signals and induce oscillatory behavior.

24 This distribution, however, is not a random distribution. (For a random distribution see Nicolis and Prigogine, 1989, p. 165.) It is viewed as having evolved from efficiency adjusting relocations of subsistence farmers.

25 In the Bénard experiment, the hexagonal pattern is in part determined by the existence of tension gradients at the upper boundary surface. When the upper boundary surface is free of tension gradients, as in Rayleigh-Bénard convection experiments, the hexagonal pattern is only one of several forms that may result. However, we view surfaces of reality to be only occasionally lacking in tension gradients and irregularities and take the situation of existing tension and irregularities as much more likely, and thus the hexagonal pattern to be by far the most prevalent in non-experimental situations.

26 In reality of course, the empirical materials reveal irregularities in the observed hexagonal nets reflecting topological variation, the existence of a finite number of transport routes (because of scale economies) and different transport modes (land, water, etc.), variation in the fertility of land as well as farming abilities and many other factors—all of which give rise to gradients comparable to the tensions and other gradients in the Bénard cell convection.

References

Abraham, R. H. (1991), 'Complex Dynamical Systems Theory: Historical Origins, Contemporary Applications' in Laszlo, E. (ed.), *The New Evolutionary Paradigm,* New York: Gordon and Breach.

Arbib, M. (1972), *The Metaphorical Brain*, New York: Wiley-Interscience.

Artigani, R. (1991), 'Social Evolution: A Nonequilibrium Systems Model' in Laszlo, E. (ed.), *The New Evolutionary Paradigm,* New York: Gordon and Breach.

Auger, P. (1989), *Dynamics and Thermodynamics in Hierarchically Organized Systems,* Oxford: Pergamon Press.

Bonner, J.T.(1988), *The Evolution of Complexity,* Princeton, N.J.: Princeton University Press.

Brandon, R.N. (1988), 'The Levels of Selection: A Hierarchy of Interactors' in Plotkin, H. C. (ed.), *The Role of Behavior in Evolution*, Cambridge, MA: MIT Press, pp. 51-73.

Bronowski, J. (1970), 'New Concepts in the Evolution of Complexity,' *Zygon,* Vol 4, pp.18-40.

Christaller, W. (1966), *Central Places in Southern Germany,* Englewood, N.J.: Prentice-Hall.

Csányi, V. and Kampis, G. (1991), 'Modelling Biological and Social Change: Dynamical Replicative Network Theory' in Laszlo, E. (ed.), *The New Evolutionary Paradigm,* New York: Gordon and Breach.

Eldridge, N. (1988a), 'Niche-Constructing Phenotypes' in Plotkin, H. C. (ed.), *The Role of Behavior in Evolution*, Cambridge, MA: MIT Press, pp. 73-133.

Eldridge, N. (1988b), 'The Evolutionary Context of Social Behavior' in Greenberg, G. and Tobach, E. (eds.), *Evolution of Social Behavior and Integrative Levels*, Hillsdale, N.J.: Lawrence Erlbaum Associates, pp. 19-30.

Greenberg, G. and Tobach, E. (eds.), (1988), *Evolution of Social Behavior and Integrative Levels*, Hillsdale, N.J.: Lawrence Erlbaum Associates.

Grene, M. (1988), 'Hierarchies and Behavior' in Greenberg, G. and Tobach, E. (eds.), *Evolution of Social Behavior and Integrative Levels,* Hillsdale, N.J.: Lawrence Erlbaum Associates, pp. 3-18.

Grobstein, C. (1973), 'Hierarchical Order and Neogenesis' in Pattee, H.H. (ed.), *Hierarchy Theory*. New York: Braziller.

Ho, Mae-Wan (1988), 'Genetic Fitness and Natural Selection: Myth or Metaphor' in Greenberg, G. and Tobach, E. (eds.), *Evolution of Social Behavior and Integrative Levels,* Hillsdale, N.J.: Lawrence Erlbaum Associates, pp.85-112.

Hull, D.L. (1980), 'Individuality and Selection,' *Annual Review of Ecology and Systematics,* Vol. 11, pp.311-32.

Hull, D. L. (1988) 'Interactors versus Vehicles' in Plotkin, H.C. (ed.), *The Role of Behavior in Evolution,* Cambridge, MA: MIT Press, pp.19-50.

in't Veld, R.J., Schaap, L., Termeer, C.E.M. and Van Twist, M.J.W. (eds.) (1991), *Autopoiesis and Configuration Theory: New Approaches to Societal Steering,* Dordrecht: Kluwer Academic.

Isard, W. (1956), *Location and Space Economy,* Cambridge, MA: MIT Press.

Isard, W. et al. (1960), *Methods of Regional Analysis: An Introduction to Regional Science,* Cambridge, MA: MIT Press.

Isard, W. et al. (1970), *General Theory: Social, Political, Economic and Regional,* Cambridge, MA: MIT Press.

Isard, W. and Liossatos, P. (1974), 'On the Natural Transport Rate and Some Parallels from Physics,' *Papers,* Regional Science Association, Vol.32, pp.95-100.

Isard, W. (1980), 'A Definition of Peace Science, the Queen of the Social Sciences, Part II,' *Journal of Peace Science,* Vol. 4, No. 2, pp.97-132.

Isard, W. and Anselin, L. (1982), 'Integration of Multiregional Models for Policy Analysis,' *Environment and Planning A,* Vol. 14, pp.359-76.

Isard, W. and Smith, C. (1983), 'Linked Integrated Multiregional Models at the International Level,' *Papers,* Regional Science Association, Vol. 51, pp. 3-19.

Isard, W. and Dean, W. (1987), 'The Projection of World (Multiregional) Trade Matrices,' *Environment and Planning A,* Vol. 19, pp.1059-66.

Isard, W. (1988), *Arms Races, Arms Control and Conflict Analysis,* New York: Cambridge University Press.

Isard, W. (1990), 'Progress in Global Modeling for World Policy on Arms Control and Environmental Management,' *Conflict Management and Peace Science,* Vol. 11, pp.57-94.

Jantsch, E. (1975), *Design for Evolution,* New York: Braziller.

Kauffmann, S. (1985), 'Self-organization, Selective Adaptation and Its Limits: A New Pattern of Inference in Evolution and Development' in Depew, D.J. and Weber, B.H. (eds.), *Evolution at a Crossroads,* Cambridge, MA: MIT Press, pp.169-207.

Kauffmann, S. and Levin, S. (1987), 'Towards a General Theory of Adaptive Walks on Rugged Landscape,' *Journal of Theoretical Biology,* Vol. 128, pp.11-45.

Klir, G.J. (1985), *Architecture of Systems Problem Solving,* New York: Plenum Press.

Klir, G.J. (1991), *Facets of Systems Science,* New York: Plenum Press.

Koschmeider, E.L. (1993), *Bénard Cells and Taylor Vortices,* New York: Cambridge University Press.

Laszlo, E. (1988), *Evolution: The Grand Synthesis,* New York: Simon and Schuster.

Laszlo, E. (1991), *The New Evolutionary Paradigm,* New York: Gordon and Breach.

Layzer, D. (1990), *Cosmogenesis*, New York: Oxford University Press.

Levins, R. (1970), 'Complex Systems' in Waddington, C. (ed.), *Towards a Theoretical Biology*, Vol. III, Chicago: Aldine.

Lewontin, R.C. and Levins, R. (1988), 'Aspects of Wholes and Parts in Population Biology', in Greenberg, G. and Tobach, E. (eds), *Evolution of Social Behavior and Integrative Levels*, Hillsdale, N.J.: Lawrence Erlbaum Associates, pp.31-52.

Lösch, A. (1954), *Die räumliche Ordnung der Wirtschaft*, Jena: Fischer, 1940. Translation by Woglom, W.H. and Stolper, G. *The Economics of Location*, New Haven: Yale University Press.

Maass, A.A. (1959), *Area and Power*, Glencoe, Ill: Free Press.

Mayr, E. (1982), *The Growth of Biological Thought*, Cambridge, MA: Harvard University Press.

Nicolis, G. and Prigogine, I. (1989), *Exploring Complexity*, New York: W.H. Freeman.

Olson, M. (1965), *The Logic of Collective Action*, Cambridge, MA: Harvard University Press.

Odling-Smee, F.J. (1988), 'Niche-Constructing Phenotypes,' in Plotkin, H.C. (ed.), *The Role of Behavior in Evolution*, Cambridg, MA.: MIT Press, pp.73-133.

Pagels, H.R. (1983), *The Cosmic Code*, New York: Bantam Books.

Pattee, H.H. (1973), *Hierarchy Theory*, New York: Braziller.

Pattee, H.H. (1991), 'Measurement-Control Heterarchical Networks in Living Systems,' *International Journal of General Systems*, Vol. 18, pp.213-21.

Plotkin, H.C. (1988), *The Role of Behavior in Evolution*, Cambridge, MA: MIT Press.

Prigogine, I. and Stengers, I. (1984), *Order Out of Chaos*, London: New Science Library.

Random House (1987), *The Random House Dictionary*, New York: Random House.

Rosen, R. (1985), *Anticipatory Systems*, New York: Pergamon Press.

Salthe, S.N. (1988), 'Notes Toward a Formal History of the Levels Concept' in Greenberg, G. and Tobach, E. (eds.), *Evolution of Social Behavior and Integrative Levels*, Hillsdale, N.J.: Lawrence Erlbaum Associates, pp.53-64.

Simon, H. (1962), 'The Architecture of Complexity,' *Proceedings of the American Philosophical Society*, Vol. 106, pp.462-82.

Simon, H. (1973), 'The Organization of Complex Systems' in Pattee, H.H. (ed.), *Hierarchy Theory: The Challenge of Complex Systems*, New York: George Braziller.

Thayer, L. (1972) 'Communication Systems' in Laszlo, E. (ed.), *The Relevance of General Systems Theory*, New York: Braziller.

Whyte, L.L. (1965) *Internal Factors in Evolution*, New York: Braziller.

7A. Some relevant mathematical space-time development models and examination of the jump (transition) problem

7A.1 Introduction

In this chapter, which is presented as an Appendix to chapter 7, we attempt to state more precisely some of the processes discussed in previous chapters, and in particular chapter 7. We do so by using mathematical models from economics and regional science which are based upon work in the physical sciences. This approach may be bypassed by the nonmathematical reader; no basic idea will be missed.

In section 7A.2 we treat dynamical system models (involving rate-determined processes) which by and large are non-spatial. We start with an extremely simple one. We then introduce in turn: (1) an element of control via optimization; and (2) the more explicit measurement-control process in a multilevel hierarchical system after first examining this system in a non-dynamical manner. If economic activity were to occur at all locations with resources everywhere the same, spatial symmetry would be preserved—no location would be identifiable. Temporal symmetry, however, would be broken by the rate-determined processes in these models.

In section 7A.3 we move to discrete spatial models. We introduce simple spatial interactions via commodity shipments among many one-point regions. This leads to the continuous spatial models of section 7A.4. There we treat non-field flow analysis, wherein flow through any location at a given point of time is analyzed as an aggregate variable. This then enables us to proceed to develop a field description of spatial interaction. Into this description we can introduce for consideration: (1) behavioral interaction in space-time via a multi-ombudsperson framework; and (2) the concept of participation potential. However, the problem of developing a unified model composed of both rate-dependent and rate-independent processes remains unresolved. When we introduce into any continuous model a commodity requiring a resource which is unevenly distributed, the spatial symmetry of production is broken.

In section 7A.5, we examine models incorporating perturbations (noise). With the use of accepted urban economics and regional science theory, we develop a simple dynamical system, Thom-type model of urban growth. To it we add noise and attempt to construct rationales for the why, when, and where of abrupt, Prigogine-type jumps to new far-from-equilibrium regimes. We ask what a stochastic, Master Equation approach and then recent developments in noise theory can add to the speculative probability-type rationales we set forth. Finally we examine the introduction of a new variable as a perturbation of a multination situation, the basic why, when and where remaining unresolved.

Section 7A.6 provides a summary.

7A.2. Non-spatial models

a. Single level (non-hierarchical)

A first dynamical systems model (which illustrates Pattee's rate-determined process) might be an elementary macro model of economics for an isolated region or nation. Let

$K(t)$ = capital stock (the state of the system) at time t
$I(t)$ = investment (the input) per unit of time at time t
$C(t)$ = consumption per unit of time at time t
$kK(t)$ = production (output) per unit of time at time t which is taken to be a linear function of capital (no labor or other resource being required), k being a relevant constant production coefficient.

Taking all investment to represent the time rate of increase $\dot{K}(t)$ in capital stock, and positing no depreciation of capital and all output in our point economy to be either consumed or invested, we have

$$\dot{K}(t) = I(t) \tag{7A.1}$$

$$C(t) = kK(t) - I(t) \tag{7A.2}$$

Given an initial stock of capital K_0 at the start of a time period t_0, and any feasible time path of investment, $t \rightarrow I(t)$, we can determine the time path $t \rightarrow C(t)$ of consumption over that time period.

This model has been extended to cover (in vector form) many state, input, output, and production variables (as in Leontief's dynamic input-output framework). For complete discussion of the assumptions and other aspects, see Isard and Liossatos (1979), chap. 3.

Now introduce into the model a first control element. Assume a performance function $W(C)$, which, as in economics and regional science, can be designated a utility or a welfare function. Let $w(C(t))$ = welfare per unit of time from $C(t)$, a current welfare; and consider a simple undiscounted summation

$$W(C) = \int_{t_0}^{t_1} w(C(t))dt \tag{7A.3}$$

The optimal control problem is to identify a consumption plan $t \rightarrow C(t)$, as the input to the problem (replacing the previous input $t \rightarrow I(t)$), that maximizes the overall utility or welfare, namely $W(C)$. In doing so we may combine equations 7A.1 and 7A.2 to form

$$\dot{K}(t) = kK(t) - C(t) \tag{7A.4}$$

Assuming $w'(C) > 0$, namely positive marginal utility (welfare) from a unit of consumption, and $w''(C) < 0$, namely diminishing marginal utility, we derive, using the Hamiltonian approach, the following conditions for utility (welfare) maximization (the time argument being omitted in the following equations):

$$\frac{dw(C)}{dC} = p \tag{7A.5}$$

$$\dot{p} = -kp \tag{7A.6}$$

Here p is a costate variable. At the particular point of time t being considered, p represents the price of a unit of output which can either be consumed or added to the stock of capital and thus also represents the price of capital. Equation 7A.5 states that this price must equal marginal utility, the value of the last unit of consumption. In Equation 7A.6, $\dot{p}$ is the decrease in the price of a unit of capital (at the same point of time) from the fact that the total potential value of a unit of capital diminishes because of its use at t. Equation 7A.6 states that this decrease equals the amount of output this unit produces at time t (namely k) times the value of that last unit of output.

A third (a terminal) condition is that the price of capital at t_1, the end of the time-period, is equal to whatever scrap or other value a unit of capital will have. For this simple model we have taken this value to be zero.

The above model is easily extended to cover depreciation of capital, labor as an input in a nonlinear production function, and limited government production and distribution. For example, when depreciation of capital and labor (L) are included, the welfare and production functions must be rewritten, respectively, as $f(C,L)$ to cover the disutility of labor, and $F(K,L)$ to cover the productivity of labor. The problem then is to identify time paths of both consumption and labor to

$$\max_{(C,L)} \int_{t_0}^{t_1} f(C(t), L(t))dt + \Phi(K(t_1)) \qquad (7A.7)$$

(where $\Phi(K(t_1))$ represents the value of the remaining units of capital at t_1, the end of the time period), subject to

$$\dot{K} = F(K,L) - C - \mu K \qquad (7A.8)$$

where μ is the rate (positive) of depreciation of a unit of capital. At year t, equation 7A.8, a local income identity, reads: the net increase in capital stock equals output less consumption less depreciation (the amount of capital used up in production). The conditions for maximization are

$$f_C(C,L) = p \qquad (7A.9)$$

$$pF_L(K,L) = -f_L(C,L) \qquad (7A.10)$$

$$\dot{p} = \mu p - pF_K(K,L) \qquad (7A.11)$$

In equation 7A.9, $f_C(C,L)$ represents the marginal utility of consumption which, as in equation 7A.5, equals the costate variable p, the price of a unit of output. In equation 7A.10 $F_L(K,L)$ represents the marginal productivity of labor, and $f_L(C,L)$ the marginal disutility of labor which is a negative quantity. This equation states that the value of the marginal productivity of labor (that is, the value of the output realized by adding a unit of labor) is equal to its disutility. In equation 7A.11, $F_K(K,L)$ represents the marginal productivity of capital. Equation 7A.11 states that the price of capital declines at time t because, once the last (marginal) unit of capital is used at time t, its potential for producing value from output at subsequent times for the rest of the time period has fallen by the value of what it produced at time t—all this adjusted for physical deterioration (depreciation).

A much more rigorous and full treatment of this model is presented in Isard and Liossatos (1979), chaps. 4 and 5. See these chapters also for the inclusion of government production and distribution, and explicit treatment of investment.

b. Multilevel (hierarchical) with measurement-control: non-dynamical

Thus far, each model has involved implicit controls. A statement at the start specifies the initial capital stock K_0, the fixed labor supply L, the value of the capital stock at t_1 and the objective to be achieved, to which the rate-dependent growth process must comply. If a price mechanism associated with market control is embodied, a rule must be prespecified, namely that price rises (falls) when demand exceeds (is less than) supply.

We now develop a model of a very simple hierarchical system, one however with a more explicit control process—a model that is more in keeping with Pattee's rate-independent, measurement-control activity. We develop it in full so that the more complex control models that follow can be more easily grasped. We draw heavily upon the work of Mesarovic, Macko and Takahara (1970).

Imagine a 'two-region' system S transforming inputs E into outputs designated $y \equiv (y_1, y_2)$. These outputs result from the behavior and interaction of the following subsystems: a decision-making unit D_1 for region 1, a decision-making unit D_2 for region 2, a coordinator, and an overall process P summarizing production activities (see Figure 7A.1). The overall process P is an input-output relation; the inputs are u_1 and u_2 (capital, labor, energy or some combination of these or other resources) and the outputs are y_1 and y_2. The overall process is the resultant of the two subprocesses (#1 and #2) and their interactions (couplings) represented by the box designated 'Field.' Subprocesses #1 and #2 represent production activities in regions 1 and 2, respectively, and are controlled by the respective decision-making units, who select u_1 and u_2. In effect, the decision-making unit D_1 together with subprocess #1 represents region 1, and the decision-making unit D_2 together with subprocess #2 represents region 2.

Concretely, let us analyze the system S, where we take E fixed (and hence to be ignored in the discussion that follows) and where the subprocesses 1 and 2 are given, respectively, by

$$y_1 = u_1^{3/2} - v_1$$

$$y_2 = u_2^{3/2} - v_2 \qquad (7A.12)$$

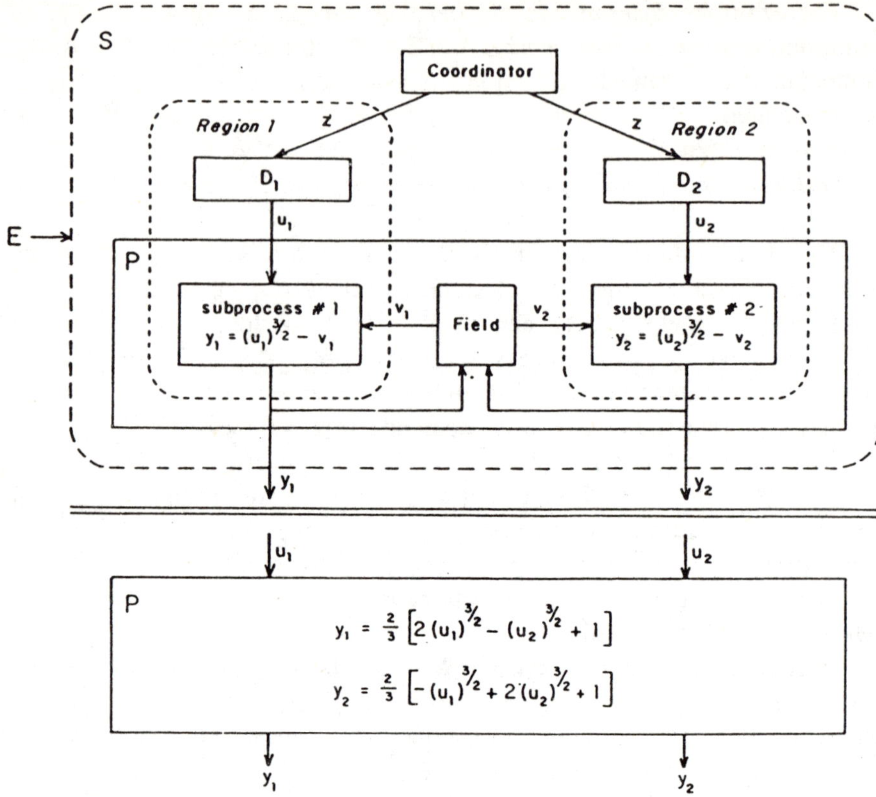

Figure 7A.1 Commodity flows and production in a system

where y_1, y_2 are outputs of two different products, u_1 and u_2 are inputs of labor (energy) and v_1 and v_2 are interaction terms (interface inputs). The latter couple the two subprocesses. (The variables are multiplied by appropriate constants to give consistent units.) More specifically, take (for illustration purposes only)

$$v_1 = \frac{1}{2}\, y_2 - 1$$

$$v_2 = \frac{1}{2}\, y_1 - 1 \qquad (7A.13)$$

represented by the box 'Field' in Figure 7A.1. Thus where $y_2, y_1 < 2$, v_1 and v_2 are negative, so that the interaction of the two sub-processes involves net positive externalities.

For example, the positive effects that may result from the flow of information about production experience may dominate the negative effects from pollution diffusion. But when $y_2, y_1 > 2$, v_1 and v_2 are positive, so that the interaction involves net negative externalities (pollution effects dominate). Using (7A.13) to eliminate v_1 and v_2 in (7A.12), we obtain the *overall process*

$$y_1 = u_1^{3/2} - \frac{1}{2} y_2 + 1$$

$$y_2 = u_2^{3/2} - \frac{1}{2} y_1 + 1 \tag{7A.14}$$

from which we may derive via simple algebraic operations

$$y_1 = \frac{2}{3} (2u_1^{3/2} - u_2^{3/2} + 1)$$

$$y_2 = \frac{2}{3} (-u_1^{3/2} + 2u_2^{3/2} + 1) \tag{7A.15}$$

as is indicated for the system in the lower part of Figure 7A.1.

Assume that the decision-making unit of each region i, $i = 1,2$, has a payoff (objective) function G_i dependent on the region's input and output, the interface input to it from interaction, and variables $z = (z_1,z_2)$ set by the coordinator. More specifically, take[1]

$$G_1(u_1,y_1,v_1,z) = 2y_1(v_1 + 1) - u_1^3 + 2z_1(v_1 + 1) - z_2y_1$$

$$G_2(u_2,y_2,v_2,z) = 2y_2(v_2 + 1) - u_2^3 + 2z_2(v_2 + 1) - z_1y_2 \tag{7A.16}$$

In order to see the meaning of these equations we can rewrite them, taking into account (7A.13) as

$$G_1(u_1,y_1,v_1,z) = y_1y_2 - u_1^3 + z_1y_2 - z_2y_1$$

$$G_2(u_2,y_2,v_2,z) = y_2y_1 - u_2^3 + z_2y_1 - z_1y_2 \tag{7A.17}$$

We may view the first term on the RHS of each equation as a 'utility' from consumption of goods, the second term as disutility from labor, and the third and fourth terms as penalties and subsidies imposed by the coordinator.

Posit that the overall two-region system functions like an organized whole—the optimal performance of the whole is of importance and concern to all its parts. Accordingly, let there be an overall performance function G depending in general on inputs and outputs. For example, consider

$$G(u,y) = 2y_1y_2 - u_1^3 - u_2^3 \tag{7A.18}$$

which we can interpret as the system welfare, taken here to be the sum of the welfares of the two regions [as defined by (7A.16)], provided v_1 and v_2 are given by (7A.13).

Accordingly, the following decision problems exist. The overall decision problem is to max $G(u,y)$ subject to (7A.15). The decision problem of region i is to max G_i, $i = 1,2$ subject to its subprocess constraint in (7A.12). As already indicated, the coordinator's decision problem is to select z so that when all regions respond by choosing their optimal decisions, system optimization automatically results.

With respect to region 1, setting $\partial G_1/\partial u_i = 0$ yields the first-order condition for optimization

$$3(v_1 + 1 + \frac{1}{2} z_2)\, u_1^{1/2} = 3u_1^2 \tag{7A.19}$$

This states that the labor input should be increased until the marginal utility stemming from the production of an additional unit of labor, which is given on the LHS, comes to equal the disutility of that labor, given on the RHS. Setting $\partial G_1/\partial v_1 = 0$ yields another first-order condition,

$$2y_1 - 2(v_1 + 1) = -2z_1 - z_2. \tag{7A.20}$$

This states that the desired level of interaction should increase until the net utility via increase in production from increase in this level, namely $2y_1 - 2(v_1 + 1)$, comes to equal the net utility lost because of penalities (in terms of goods taken away), namely $-2z_1 - z_2$, imposed by the coordinator. Simple algebraic operations on (7A.19) and (7A.20) yield as the solution of the optimization problem of region 1

$$v_1 = z_1 \tag{7A.21}$$

and

$$u_2^{3/2} = \frac{2z_1 - z_2 + 2}{2}. \tag{7A.22}$$

By symmetry, the solution of the problem for region 2 is obtained by interchanging the indices. We thus derive

$$v_2 = z_2 \tag{7A.23}$$

and

$$u_2^{3/2} = \frac{2z_2 - z_1 + 2}{2}. \tag{7A.24}$$

Note that we allow each region to view v_i, $i = 1,2$, as a variable, so that v_1 and v_2 in (7A.21) and (7A.23) are *desired* interface inputs; to indicate this in what follows, we write them as v_1^d and v_2^d, respectively. However, the regions can apply only u_1 and u_2 to the production process. They merely report the v_1^d and v_2^d, —the actual v_1 and v_2 hereafter designated v_1^a and v_2^a, being determined by (7A.13).

Next consider the coordinator's job. It is to select z so that the differences $/v_i^d - v_i^a/ \quad i = 1,2$ disappear, that is that

$$v_1^a = v_1^d, \quad v_2^a = v_2^d. \tag{7A.25}$$

For this purpose we can imagine that the coordinator has a payoff function

$$\pi = z_1 /v_1^d - v_1^a/ + z_2 /v_2^d - v_2^a/ \tag{7A.26}$$

Restricting z_1 and z_2 to negative values, it follows that the coordinator will select $z = (z_1,z_2)$ so that the conditions (7A.25) are realized. Specifically, the coordinator may announce z. The regions respond by selecting u_i and v_i^d. They report the v_i^d to the coordinator and apply the u_i to the production process. The coordinator then observes directly the actual interactions v_1^a resulting from the production carried on by the regions, and if the conditions (7A.25) are not met, he selects another z. This process is continued until the actual interactions become equal to the desired ones.

Now examine whether in fact system optimization is automatically achieved when the regions and the coordinator all maximize. Given z, the regions apply u_1 and u_2, as defined by (7A.22) and (7A.24). This yields outputs given by (7A.15). These outputs give actual interactions by (7A.13). So we obtain

$$v_1^a = \frac{1}{3}(-u_1^{3/2} + 2u_2^{3/2} + 1) - 1$$

$$v_2^a = \frac{1}{3}(2u_1^{3/2} - u_2^{3/2} + 1) - 1. \tag{7A.27}$$

Using (7A.22) and (7A.24) we then obtain

$$v_1^a = \frac{1}{6}(-4z_1 + 5z_2 + 4) - 1$$

$$v_2^a = \frac{1}{6}(5z_1 - 4z_2 + 4) - 1. \tag{7A.28}$$

Applying the condition (7A.25), where v_1^d and v_2^d are given by (7A.21) and (7A.23), respectively, we obtain

$$10z_1 - 5z_2 = -2$$

$$-5z_1 + 10z_2 = -2, \tag{7A.29}$$

which yield the unique solution

$$z_1 = z_2 = -\frac{2}{5} \tag{7A.30}$$

from which we then obtain

$$v_1 = v_2 = -\frac{2}{5} \tag{7A.31}$$

$$u_1^{3/2} = u_2^{3/2} = \frac{4}{5} \tag{7A.32}$$

and

$$y_1 = y_2 = \frac{6}{5} \tag{7A.33}$$

The selection of these $\boldsymbol{z}$, $\boldsymbol{u}$, and $\boldsymbol{v}^d$ corresponds to maximal behavior, since the associated Hessian matrix can easily be shown to be negative definite. These results also optimize system welfare, because when we select u_1 and u_2 to *max* $G(\boldsymbol{u},\boldsymbol{y})$ subject to (7A.15), we obtain the first-order conditions

$$-25u_1^{3/2} + 20u_2^{3/2} = -4 \tag{7A.34}$$

$$20u_1^{3/2} - 25u_2^{3/2} = -4, \tag{7A.35}$$

which are satisfied by (7A.32), and the associated Hessian matrix is negative definite.

Note the similarity and difference between the above two-region system problem and the typical two-region game-type conflict situation. [See, for example, Isard et al. (1969, pp. 278–89).] In the latter, the problem is to suggest to the two regions that they adopt a guiding principle to follow in their behavior to lead them from an inefficient equilibrium point to a Pareto-optimal point on the efficiency frontier. The role of coordination that such a guiding principle might provide is limited in the sense that the welfare of the system comprising the two regions is ignored. In the former problem, the role of coordination is a full role in the sense that the system is explicitly recognized and the objective of maximizing its welfare is introduced.

Generalizing the above illustration (and therefore encompassing dynamics—to be treated later—as well as statics), we can consider n subprocesses described by

$$y_i = P_i(u_i, v_i), \qquad i = 1, \ldots, n. \tag{7A.36}$$

which parallel (7A.12). Each subprocess is directly controlled by a corresponding region, each region having a payoff function $G_i(u_i, y_i, v_i, z)$, which its decision-making unit maximizes given z. Further, the n subprocesses interact in many different ways, and the effect upon i of all these interactions can be summarized by expressing v_i as a function h_i of all inputs and outputs:

$$v_i = h_i(u,y), \qquad i = 1, \dots ,n, \tag{7A.37}$$

where $u = (u_1, \dots ,u_n)$ and $y = (y_1, \dots ,y_n)$. These equations generalize (7A.13). The v_i can thus be viewed as field variables, because they embody the net effect of the actions of the entire system on i. Still more, we posit that the collection of subprocesses (7A.36) together with their interactions (7A.37) determine an over-all process

$$y = P(u), \tag{7A.38}$$

which generalizes (7A.36). This signifies that if we insert v_i from (7A.37) into (7A.36), we obtain (7A.38). Finally, we require that there be a system welfare (performance) function $G(u,y)$. Then, the overall optimization problem is to maximize $G(u,y)$ subject to (7A.38).

In oversimplified terms, the coordination procedure described is applicable provided the overall welfare (performance) function $G(u,y)$ can be expressed as an increasing function Ψ of the welfare (performance) functions of the regions, that is,

$$G(u,y).= \Psi(G_1(u_1,y_1,v_1,z), \dots ,G_n(u_n,y_n,v_n,z)), \tag{7A.39}$$

where the y_i's are computed from (7A.36) and the v_i's from (7A.37), where (7A.38) holds, and where Ψ is an increasing function of its arguments. For a more precise statement, see Isard and Liossatos (1979), chap. 12.

Note further the generality of the above model. The coordinator receives from each region i information on its desired interface input $\overset{d}{v_i}$ and observes (receives information on) its output y_i. Given the latter, he then can calculate $\overset{a}{v_i}$. In all this he need not know u_i which may be a single input, or in the more general problem a vector of inputs. The overall process then involves loss of detail.

Simon's watchmaker example discussed on pp. 109–10 illustrates a hierarchical structure in which the attainment of efficiency is based upon the loss of detail, although it involves a hierarchy of more than two levels. Any one subprocess at a given level is concerned with only a small number of inputs and constraints, and passes along to the next subprocess information about its output only, the information about its inputs and constraints being lost to that next subprocess. See Isard and Liossatos (1979), p. 276 for a relevant figure.

The above simplifed model, involving both rate-dependent processes and measurement-control in an instantaneous sense of a static model, is easily extended to incorporate Pattee's principle of optimum loss of detail. One way is to introduce a filtering operation involving the use of resources as inputs into each region's production operation. At each region, the extent of filtering is established when the marginal gain in efficiency in decision making at the next level equals the marginal cost of resource inputs—an operation consistent with Levins' low cost via low connectivity condition for an efficient complex system. A second way is to recognize explicitly that the setting of constraints and the (instantaneous) up and down transmission of information utilizes resource inputs. Hence, the question arises: how much setting of constraints and transmission should take place? Recognizing the cost of resource inputs, the optimal welfare of a society is not what would be achievable were this cost zero, but a smaller amount of welfare defined by the condition that its marginal increase of welfare just equals the marginal resource cost of achieving it.

The above two ways are presented in rigorous mathematical form in Isard and Liossatos (1979), pp. 277–81. In other kinds of situations there are still other ways as well as combinations of ways in which the principle of optimum loss of detail can be attained.

c. Multilevel with measurement-control in a dynamical system framework

Thus far we have not treated a dynamical system framework with explicit measurement-control. In a conceptual manner, the extension to cover such a framework, which approaches more closely a unitary system as viewed by Pattee, is easily achieved. Again consider the case of two regions (1 and 2) and a coordinator (see Figure (7A.1). They are concerned with optimization over a period of time $[t_0, t_1]$. Each region specializes in the production of a single product, that is, a distinct subprocess (production) is associated with each region. Instead of Equation 7A.12 we now have a differential equation for each subprocess, which for simplicity we take to be linear:

$$\dot{y}_i(t) = u_i(t) + v_i(t) - a_{ii}y_i(t) \qquad y_i(t_0) = y_{i0} \qquad i = 1,2 \tag{7A.40}$$

for $t_0 \leq t \leq t_1$. In order to make the following statement for a two-region system more compact, from here on we let $i = 1,2;\ j = 1,2;\ j \neq i$. Hence whenever $i = 1, j = 2$ and when $i = 2, j = 1$—or when $j = 1, i = 2$ and when $j = 2, i = 1$.

In (7A.40), $y_i(t)$, $i=1,2$, may be viewed as some stock of social wealth (goods, health, and other tangibles and intangibles), let us say *social stock*, at time t. $\dot{y}(t)$ is its time rate of change. $u_i(t)$ is a control variable of the ith decision-making unit, summarizing its decision regarding the use of labor (or any other type of resource) per unit time. When multiplied by appropriate dimensional constants, $u_i(t)$ also can be expressed in units of social stock and thus can measure the amount of social stock produced from labor per unit time. $v_i(t)$ is the net effect per unit time of interaction upon production, and may be either positive (say from flow of information, or a sense of rivalry) or negative (say from pollution diffusion). Finally, the social stock depreciates over time, the constant a_{ii} measuring its rate of depreciation, the amount of depreciation per unit time being $a_{ii}y_i(t)$. y_{i0} is the initial value of the stock at t_0, which is taken as prespecified. Thus (7A.40) states for each time t that the rate of change in social stock of region i is equal to the sum of the output from the application of labor (energy) per unit time, *plus* the output (either positive or negative) from the presence of interaction (summarized by the interface input), *less* the amount of depreciation (a loss) per unit time. Observe that for each selection of control input $t \rightarrow u_i(t)$ and an interface input $t \rightarrow v_i(t)$, each equation uniquely determines a time pattern of social stock. Symbolically, (7A.40) is equivalent to

$$y_i = P_i(u_i, v_i) \qquad i = 1,2; \quad j = 1,2; \quad j \neq i \tag{7A.41}$$

where the symbols u_i, v_i, and y_i represent time patterns. So (7A.41) has the same form as (7A.12) or (7A.36).

The couplings between the two subprocesses are

$$v_i = a_{ij}y_j; \equiv h_i(u,y) \qquad i = 1,2 \tag{7A.42}$$

paralleling (7A.13) or (7A.37). The coefficient a_{ij} is the coupling coefficient, which is taken to be constant, indicating the effect of a unit of social stock of region j on the rate of output of social stock of region i. Using (7A.42) to eliminate v_i, we see that the *overall process* is

$$\dot{y}_i = u_i + a_{ij}y_j - a_{ii}y_i \quad y_i(t_0) = y_{i0} \qquad i = 1,2; j = 1,2; j \neq i \tag{7A.43}$$

For any given control input, namely $t \rightarrow \mathbf{u}(t) \equiv (u_1(t), u_2(t))$, (7A.43) uniquely determines a time pattern of social stocks, $t \rightarrow \mathbf{y}(t) \equiv (y_1(t), y_2(t))$. That is, (7A.43) determines $\mathbf{y}$ as a function $\mathbf{P}$ of $\mathbf{u}$:

$$\mathbf{y} = \mathbf{P}(\mathbf{u}), \tag{7A.44}$$

where $\mathbf{u}$ and $\mathbf{y}$ represent time patterns, which function is of the form (7A.15) or (7A.38).

In addition to a production process, there is a decision-making unit within each region i. This unit desires to optimize a payoff (objective) function G_i by manipulating u_i subject to the constraint of its production process. However, we assume that each decision-making unit operates as a part of an organization in which there exists a superior coordinating-type decision-making unit. Thus the payoff function of the ith region depends not only on the variables u_i, v_i, and y_i pertaining to its subprocess, but also on the parameters imposed by the coordinator, $t \rightarrow z(t) \equiv (z_1(t), z_2(t))$—which we now designate as the *coordination input*. For example, let the payoff function, or more generally the performance function, be

$$G_i(u_i, v_i, y_i, z) = \int_{t_0}^{t_1} [u_i^2 + (a_{ij}r_j - v_i)^2 + \bar{q}_i(r_i - y_i)^2 + z_i v_i - z_j a_{ji} y_i] dt. \qquad (7A.45)$$

(In what follows, we no longer specify $i = 1,2;\ j = 1,2;\ j \neq i$ when the argument implies such.) In (7A.45) the time arguments are omitted from the integrands; r_i and r_j are large positive constants playing the role of targets (desired levels of stocks); and the G_i are essentially cost functionals. In this equation the integrand is the total cost per unit time. The third term, $\bar{q}_i[r_i - y_i(t)]^2$, is a cost associated with not reaching the target r_i, where $|r_i - y_i(t)|$ is the deviation from that target and $\bar{q}_i$ is a cost coefficient. When $y_i(t) = r_i$, this cost is zero, but when the target is set high enough, it can be achieved only with the application of control inputs (labor). Such inputs, however, are costly; their cost is taken to be $[u_i(t)]^2$, the first term in the integrand; its coefficients is set at unity, so that $\bar{q}_i$ indicates the relative importance of these two costs. The second term in the integrand is a cost that the decision maker of region i associates with each level of v_i. In this particular situation we assume that he knows $a_{ij}r_j$ and knows that there is a cost, $a_{ij}r_j^2$, imposed upon region i, which is reducible by the extent of interaction v_i; so the decision maker of region i desires that $v_i = a_{ij}r_j$. Observe that if the interaction is the correct one, namely $v_i = a_{ij}y_j$, then this term becomes a cost imposed on region i because region j does not reach its target. The fourth and fifth terms are subsidies and penalties imposed by the coordinator, which, as discussed before, are regulations of the system. We construct the payoff definition of region j in a similar fashion.

We posit that the overall two-region system functions as an organized whole, the optimal performance of the whole being of importance and concern to all its parts. To convey this idea we posit an overall performance function G depending in general on the time patterns of control inputs and social stocks. For example,

$$G(u,y) = \int_{t_0}^{t_1} [u_1^2 + u_2^2 + q_1(r_1 - y_1)^2 + q_2(r_2 - y_2)^2]dt. \tag{7A.46}$$

where time arguments are omitted from the integrand. Here q_1 and q_2 are positive constants, indicating the relative importance of the cost terms. The first two are, as before, costs from control inputs (labor); the last two terms, costs from failing to reach targets. If we take

$$\bar{q}_i \equiv q_i - a_{ji}^2 > 0 \tag{7A.47}$$

then we can show that if the interactions are the correct ones, as given by (7A.42), the overall system cost functional is the sum of the cost functionals of the two units:

$$G(u,y) = G_1(u_1,a_{12},y_2,y_1,z) + G_2(u_2,a_{21},y_1,y_2,z). \tag{7A.48}$$

As before, there are four decision problems. The overall decision problem is to minimize $G(u,y)$ subject to (7A.43) or (7A.44). The decision problem of region i is to minimize $G_i(u_i,v_i,y_i,z)$ subject to (7A.40) or (7A.41). Thus, the coordinator's decision problem is to select z so that when all regions respond by choosing their optimal decisions, system optimization automatically results.

Consider the optimal-control problem of region 1. We construct the current Hamiltonian

$$H_1 = u_1^2 + (q_{12},r_2 - v_1)^2 + \bar{q}_i\,(r_1 - y_1)^2 + z_1v_1 - z_2a_{21}y_1 \\ - p_1(u_1 + v_1 - a_{11}y_1). \tag{7A.49}$$

The minimum principle with reference to u_1 (the controllable input) and v_1 (as if it were a controllable input)—that is, $\partial H_1/\partial u_1 = 0$ and $\partial H_1/\partial v_1 = 0$—yields

$$p_1 = 2u_1 \tag{7A.50}$$

$$p_1 = -2(a_{12}r_2 - v_1) + z_1 \tag{7A.51}$$

assuming interior optima. By the nature of the costate variable in the Hamiltonian, $p_1(t)$ is the imputed price of an additional unit of $y_1(t)$ (capital stock at time t). As discussed elsewhere, $p_1(t)$ is a value that region 1 assigns to a unit of additional output (capital stock) at time t because of anticipated cost savings over the time period $[t,t_1]$. This being the case, (7A.50) states that at each t the control (labor) input should be increased up to the point where its marginal cost (disutility), $2u_1(t)$, equals the imputed price $p_1(t)$ of $y_1(t)$. In a similar manner, Equation (7A.51) states that at each time t the desired level of interaction (interface input), namely $v_1(t)$, should be increased up to the point where its marginal net cost, $-2[a_{12}r_2 - v_1(t)] + z_1(t)$, falls to the level of the imputed price $p_1(t)$. Note that by elimination of p_1, (7A.50) and (7A.51) yield

$$2u_1 = -2(a_{12}r_2 - v_1) + z_1, \tag{7A.52}$$

which in a sense represents a substitution principle between two inputs, namely, that their marginal costs must be the same, a basic allocation principle of optimization.

In addition to (7A.50) and (7A.51), minimization of the cost functional of region 1 requires that Hamilton's equation be satisfied, namely

$$\dot{p}_1 = \frac{\partial H_1}{\partial y_1}, \tag{7A.53}$$

which yields

$$\dot{p}_1 = -2\,\bar{q}_1(r_1 - y_1) - z_2a_{12} + a_{11}p_1, \quad p_1(t_1) = 0. \tag{7A.54}$$

The meaning of this equation is best seen if we multiply both sides by dt and integrate to obtain

$$p_1(t) = \int_t^{t_1} \{2\,\bar{q}_1[r_1 - y_1(t')] + z_2(t')a_{21} - a_{11}p_1(t')\}dt' \tag{7A.55}$$

In the integrand, the first term is the reduction in cost due to being one unit closer to the target r_1, a marginal cost saving. The second term, $z_2(t')a_{21}$, is the subsidy or penalty applied by the coordinator to a unit of $y_1(t')$. The third term, $-a_{11}p_1(t')$, represents a depreciation cost. Together the three terms in the integrand yield current net returns, or net cost savings, or net gains per unit of capital stock at t'. Hence (7A.55) states that the level $y_1(t)$ of capital at t should be increased up to the point where the cumulative sum of its current net returns over the time period $[t,t_1]$ comes to equal its imputed price at time t—an investment-type principle.

For region 2, we obtain the parallel optimality conditions.

By solving (7A.40), (7A.50), (7A.51) and (7A.54), we obtain for region 1 the optimal inputs $\{u_1,v_1\}$ as functions of the coordination input z. To show this dependence we write these inputs as $\overset{z}{u}_1,\overset{z}{v}_1$, which collectively are the optimal decisions of region 1 for given z. Similarly region 2 obtains $\overset{z}{u}_2,\overset{z}{v}_2$ for given z.

Next consider the coordinator's problem. His strategy is based on the fact that each subunit can select $\overset{z}{u}_i$ and $\overset{z}{v}_i$ but can only apply $\overset{z}{u}_i$. The interactions that result from the application of $\overset{z}{u}_1$ and $\overset{z}{u}_2$ are in general different from those desired. Using (7A.42) and (7A.44), the actual interactions (interface inputs) are

$$\bar{v}_i^z = a_{ij} y_j^z = a_{ij} P_j(u_i^z, u_j^z) \tag{7A.56}$$

Hence the coordinator's job is to choose z so that

$$v_i^z = \bar{v}_i^z, \quad i = 1,2 \tag{7A.57}$$

As shown in Mesarovic et al. (1970, pp. 178–81), a solution $\hat{z}$ to (7A.57) exists, and in addition

$$\hat{u}_i = \hat{u}_i^z, \tag{7A.58}$$

where $\hat{u} = (\hat{u}_1, \hat{u}_2)$ is the optimal control for the overall problem. We let $\hat{y} = (\hat{y}_1, \hat{y}_2)$ be the resulting capital stocks obtainable from (7A.44). The relation (7A.58) states that the optimal controls selected by the units are the optimal controls for the system. It necessarily follows from (7A.52) that, for example,

$$\hat{y} = 2\,\hat{u}_1 - 2a_{12}(\hat{y}_2 - r_2). \tag{7A.59}$$

Hence when the overall process as well as the subprocesses are described as dynamical systems involving time patterns of inputs and outputs, the formalism for a two-level, hierarchical control system, developed in the discussion of the example covered by equations (7A.12–7A.35), applies directly to the system just discussed. Only the techniques of computation change. In the former case we have techniques of static optimization, while in the latter we apply optimal-control theory. Accordingly, the general formulation of equations (7A.36–7A.39) gives the essence of the method, which applies to all problems regardless of the specific representation of processes (algebraic equations, differential equations, difference equations, integral equations, etc.).

Needless to say, the two region (subsystem) model above that involves linear differential equations as laws governing the processes, has obvious limitations. Few situations of social life are describable by linear dynamical systems and have solutions that can be written explicitly in analytical form. Moreover, stochastic elements are always present; so decision problems become much more complex than mere optimization problems because they must take account of uncertainties. Further, we have assumed that the subunits (regions) transmit 'correct' information regarding their short-run optimization; however, in practice regions may not do so. They may 'bluff,' purposely mislead, and be concerned with optimization over the middle run and/or long run. Hence the coordinator's problem must be restated, especially with regard to the uncertainty concerning the truth of the information received. But to restate the model in terms of stochastic and incomplete information makes the identification of a solution much more difficult. Nonetheless, posing the problem as we have done in a simple linear, deterministic, nongame fashion provides valuable basic concepts and insights for analysis of the nonlinear, multilevel control problem.

For example, the central planning models discussed in Isard and Liossatos (1979), chaps. 5 and 6 can in principle be restated to recognize explicitly a multilevel framework. At the highest (supremal) level is the ombudsman, who plays the role of the coordinator. At the lowest (infimal) level are the many consumers, producers, and traders in each of the Z regions, operating on respective 'local' processes to maximize their utilities, profits, or payoffs. Also, at the lowest level (*from a control standpoint*) are regional and local governments and the central planner, maximizing their respective functions. However, it should be borne in mind that in reality the hierarchy is more complicated and frequently comprises more than two levels. In centrally planned economies, such as in socialist countries, there is at the top the political leader together with the central planning staff, who may make a few critical decisions, as on prices of basic commodities or the allocation of capital and other scarce resources. These decisions are handed down to, among others, regional governments and regional planners, who may be able to react by allocating capital to producers, consumers, and others in their region. Also, they may be able to modify the central planning staff's prices by adding subsidies or penalties. Then there may be local governments and local planners who, upon receiving information on decisions of regional governments and regional planners, can react to these at the local level. At the lowest level are numerous individual consumers, producers, traders, and others, who operate to maximize their utilities, profits, or payoffs.

Also note that a typical general equilibrium model (see Isard et al., 1969, chap. 11) can be viewed in a rather anemic way as a two-level control system. The supremal unit (coordinator) is the fictitious market participant. Infimal units are firms, consumers, traders, etc. By the coordination activity of the market, social welfare (the overall performance function) is maximized (see Mesarovic et al., 1970).

As already noted, the analysis of a two-region system can be generalized to a system of *n* regions (subsystems). The restriction to linear differential equations may be dropped and the interpretation will thus pertain to general nonlinear laws of change and nonlinear performance functions. See Isard and Liossatos (1979), pp. 295–98 and 362–71 for mathematical statements which then model rate-dependent processes, elements of measurement control and loss of detail in still more involved interacting hierarchical subsystems.

7A.3 Discrete spatial models

Consider a system of many regions, $J,K,L = A, \ldots ,Z$, each being a single-point economy. There exists a single (composite) good which may be shipped between any pair of regions. Our initial problem is to maximize system welfare for a multiregion system that has no regional objectives. We thus have as the current welfare function

$$C \rightarrow f(C), \tag{7A.60}$$

where

$$C = (C^A, C^B, \ldots ,C^J, \ldots ,C^Z) \tag{7A.61}$$

and $C^J(t)$, $J = A, \ldots ,Z$ is the consumption per unit time in region J at time t. The production function for each region, where labor is fixed, is

$$Y^J = F^J(K^J), \quad J = A, \ldots ,Z, \tag{7A.62}$$

where $Y^J(t)$ and $K^J(t)$ are, respectively, the production per unit time at time t and the capital stock in region J at time t. Taking into account imports, exports, and transport cost, the local income identity for region J is[2]

$$\dot{K}^J = Y^J - C^J + \sum_{K \neq J} s^{K \rightarrow J} - \sum_{L \neq J} s^{J \rightarrow L} - \sum_{L \neq J} \sigma^{J \rightarrow L} s^{J \rightarrow L}, \tag{7A.63}$$

where

- $s^{J \rightarrow L}(t)$ is the amount of the good shipped from region J to region L per unit time (exports) at time t,
- $s^{K \rightarrow J}(t)$ is the amount of the good shipped from region K to region J per unit time (imports) at time t, and
- $\sigma^{J \rightarrow L}$ is the fraction of a unit of export used up in effecting the shipment from J to L. We take $0 < \sigma^{J \rightarrow L} < 1$ for all $J \rightarrow L$. In effect, we adopt the convention that the region of export is responsible for transportation.

The problem is to

$$\max_{\{C, s^{J \rightarrow L}\}} \int_{t_0}^{t_1} f(C(t))dt + \Phi(K(t_1)), \tag{7A.64}$$

where

$$K = (K^A, K^B, \ldots, K^Z) \tag{7A.65}$$

subject to (7A.63) and initial conditions

$$K^J(t_0) = K_0^J, \quad J = A, \ldots, Z, \tag{7A.66}$$

and nonnegative shipments,

$$s^{K \to L} \geq 0, \quad K,L = A, \ldots, Z, \quad K \neq L, \tag{7A.67}$$

and where the consumption vector and shipment matrix play the role of control variables.[3] Forming the Hamiltonian

$$H = f(C) + \sum_J p^J [F^J(K^J) - C^J - \sum_{K \neq J} s^{K \to J} - \sum_{L \neq J} (1 + \sigma^{J \to L}) s^{J \to L}] \tag{7A.68}$$

we apply the maximum principle. If we take C and $\{s^{J \to L}\}$ as control variables, we obtain the usual *consumption principle*

$$\frac{\partial H}{\partial C^J} = 0 \Rightarrow \frac{\partial f}{\partial C^J}(C) = p^J, \quad J = A, \ldots, Z, \tag{7A.69}$$

assuming an interior solution.[4] We also obtain the Kuhn-Tucker conditions

$$\frac{\partial H}{\partial s^{J \to L}} \leq 0, \quad s^{J \to L} \frac{\partial H}{\partial s^{J \to L}} = 0. \tag{7A.70}$$

From (7A.68) we obtain

$$\frac{\partial H}{\partial s^{J \to L}} = p^L - p^J - \sigma^{J \to L} p^J \equiv \tau^{J \to L}, \tag{7A.71}$$

where, in accord with standard terminology in the regional science literature, $\tau^{J \to L}$ represents net gains from trade per unit good shipped from J to L. Thus the maximum principle, with shipments as control variables, yields, in view of (7A.63)

$$\tau^{J \to L} \leq 0, \quad s^{J \to L} \tau^{J \to L} = 0. \tag{7A.72}$$

Observe that (7A.72) corresponds to the normal *transportation principle* and yields the traditional equilibrium conditions of interregional trade under pure competition when p^L and p^J are interpreted as market prices. The first half states that net gains from trade in region J (price in import region L less price in export region J less transport cost) cannot be positive; the second half states that when shipments are positive, gains from trade are zero.

The Hamilton equations $\dot{p}^J = -\partial H/\partial K^J$, $J = A,B, \ldots, Z$, imply

$$\dot{p}^J = -p^J \frac{dF^J}{dK^J}(K^J), \quad p^J(t_1) = \frac{\partial \Phi}{\partial K^J}(K(t_1)). \tag{7A.73}$$

The first part of (7A.73) is the normal investment principle for each region J; as before, $p^J(t)$ is the imputed price of capital in region J at time t. The second part states that at t_1, the imputed price of capital must equal the welfare value attached to a unit of capital stock in place at t_1.

The above model corresponds to a central-planning model for a system of Z regions wherein the central planner determines consumption and import-export plans so as to maximize the system welfare for the planning period. Behavior is fully governed by the central planner's objective. In effect, this model pertains to a system consisting of a single decision-making unit with a single goal. The behavior of the many micro units in the system is not considered, since the model treats aggregates only.

This simple spatial interaction model involving commodity shipments and only implicit control (with the initial capital stock, the terminal value of capital stock and the objective given) can be significantly extended. It can be modified to encompass: (a) investment and a general equilibrium framework embodying many commodities, many consumers, many producers, and many traders in each of many regions; (b) limited government production and distribution; and (c) central planning motivated to maximize the social welfare over all individuals in the entire system of many regions, each region a point in space. This extension, however, requires explicit measurement-control processes of the type discussed in section 7A.2b where a coordinator imposes subsidies and penalties or other inputs (as variables) upon the regions (subsystems)—and thus a still more complex hierarchical structure and set of interactions.

In this extended central planning model, these coordination inputs are provided by an ombudsperson who controls a basic set of variables that enter the central planner's social welfare function and that insure that the distribution of goods to each consumer in each region is in accord with the bundle of goods the consumer demands to maximize his/her own utility. The model, as developed in Isard and Liossatos (1979), also has: each producer maximizing his/her profits; each shipper, his/her gains from trade; and each government, the welfare of its constitutents via taxation and distribution of government produced goods. The formal derivation of necessary first-order conditions involving the use of a Lagrangian for each behaving unit is presented therein.

7A.4 Continuous space models

a. Non-field flow analysis

Thus far we have treated discrete space—a set of points in space, each a locality or region (subsystem) at which is concentrated all production, consumption, investment, export-import activity and population of a locality or region. We now wish to consider continuous space in order to employ more powerful tools of analysis, especially those developed in physics. Analysis in continuous space does not preclude the treatment of discontinuities since δ-functions, for example, can describe point concentrations as densities over continuous space.

As in Isard and Liossatos (1979), we find it convenient to designate the regions $A,B, \ldots ,J,K, \ldots ,Z$ respectively as $1,2, \ldots ,J, J+1, \ldots ,N$, where N is some positive integer. Assume these regions are ordered along a straight line (the x-axis).[5] This set occupies the interval $[0,B]$ $(B > 0)$. Consider a partition of this interval, $\{x^1,x^2, \ldots ,x^{N+1}\}$, where $x^1 = 0 < x^2 < ,\ldots, < x^{N+1} = B$. Then we take $[x^1,x^2]$ to be the first region (region A), $[x^2,x^3]$ to be the second region (region B), ... $,[x^J,x^{J+1}]$ to be the Jth region (region J), ... $,[x^N,x^{N+1}]$ to be the Nth region (region Z). We let $\Delta x^J \equiv x^{J+1} - x^J$, $J = 1,2, \ldots ,N$. If $max_J\, \Delta x^J << B$, then, approximately, we can look upon the discrete set of regions as a continuum of regions occupying the interval $[0,B]$, each region becoming a point in $[0,B]$. It is this continuum that we have in mind when we allow $\Delta x^J \rightarrow 0$ for all J.

Consider any magnitude $Q^J(t)$, where J refers to the region $[x^J,x^{J+1}]$. We posit that $Q^J(t) \rightarrow 0$ as $\Delta x^J \rightarrow 0$ in such a way that $Q^J(t)/\Delta x^J$ approaches a definite limit, which we denote by $Q(x,t)$, where x replaces x^J. *Then* $Q(x,t)$ is the local density of Q. With these concepts, $K^J(t)$, $C^J(t)$, and $Y^J(t)$ become $K(x,t)$, $C(x,t)$, and $Y(x,t)$, respectively, that is, densities of capital, consumption, and output at (x,t).[6] Further, by the convention already established it is seen that the magnitude $s^{J\rightarrow L}$ corresponds to shipment from $[x^J,x^{J+1}]$ to $[x^L,x^{L+1}]$. Thus we posit that as Δx^J, $\Delta x^L \rightarrow 0$, $\Delta s^{J\rightarrow L} \rightarrow 0$ in such a way that $s^{J\rightarrow L}/\Delta x^J \Delta x^L$ approaches a definite limit, which we call $s(x,x',t)$. Here $x' = x^L$ and $x = x^J$. So, dividing both sides of (7A.63) by x^J,[7] letting $\Delta x^1, \ldots ,\Delta x^J, \ldots ,\Delta x^N$ each $\rightarrow 0$, and appropriately transforming summations to integrals we obtain

$$\dot{K}(x,t) = Y(x,t) - C(x,t) + \int_0^B s(x',x,t)dx' - \int_0^B [1 + \sigma(x,x')]s(x,x',t)dx', \qquad (7A.74)$$

where $\sigma(x,x')$ takes the place of $\sigma^{J \to L}$, and where the dot $\cdot$ indicates the partial derivative with respect to time, i.e.,

$$\dot{K}(x,t) = \frac{\partial K}{\partial t}(x,t) \tag{7A.75}$$

In parallel with (7A.62) we assume

$$Y(x,t) = F(K(x,t),x), \tag{7A.76}$$

where the explicit dependence of F on x indicates the possibility of spatial variation in the production function. From (7A.60) we observe that current system welfare $W(t)$ is a function of $C^A(t), C^B(t), \ldots, C^Z(t)$. Accordingly, in our new context, $W(t)$ must be a function of the map $x \to C(x,t)$, $0 \leq x \leq B$. That is, $W(t)$ is a functional defined on the spatial distribution of consumption. Here we posit that this functional is an integral, and we write

$$W(t) = \int_0^B f(C(x,t)dx). \tag{7A.77}$$

We interpret $f(C(x,t))$ as the current welfare density at point x. Further, we posit that current welfare density at x has the standard properties of utility (positive and diminishing marginal welfare with increase of C).

Additionally, we must keep in mind that at time t_1, the end of the planning period, there may exist at any location x a stock of capital $K(x,t_1)$ and that the distribution of capital over all locations may have a value which is a functional of this distribution. For simplicity, we take this functional to be expressed as an integral, the value of the capital stock at location x being $\Phi(K(x,t_1))$.

We can now formulate the optimization problem as

$$\max_{\{C,s\}} \left[\int_{t_0}^{t_1} \int_0^B f(C(x,t))dxdt + \int_0^B \Phi(K(x,t_1))dx \right] \tag{7A.78}$$

subject to (7A.74), (7A.76),

$$s(x,x',t) \geq 0, \qquad x,x' \in [0,B], \tag{7A.79}$$

and prespecified initial conditions

$$K(x,t_0), \qquad x \in [0,B]. \tag{7A.80}$$

We can now proceed to set up the Lagrangian and derive the first-order conditions. See Isard and Liossatos (1979), pp. 109–10 for their presentation.

At this point we wish to proceed from a model that makes explicit each of the diverse flows between any single location and all others to a model that consolidates all these flows into a single local variable. Put otherwise, we want to replace the interactions of location x with every other location x' as shown in Equation (7A.63) with a single interaction of x with its neighborhood $x + \Delta x$ (for $\Delta x \approx 0$). Such an approach to spatial interactions will be called a local or field-theoretic description.

To do so, we first assume that there is one-directional trade, namely, trade in the positive x-direction (this assumption will be dropped later). This means that $s(x,x',t) = 0$ for any $x' \leq x$. We now introduce another concept, namely, the flow $U(x,t)$ of goods through location x at time t per unit of time. This flow may be viewed as consisting of all exports from all originating locations up to and including x to all terminating locations extending from x up to and including B. That is,

$$U(x,t) = \int_{x'=0}^{x} \int_{x''=x}^{B} s(x',x'',t)dx'dx'' \tag{7A.81}$$

Note that by (7A.81), we have $U(0,t) = 0$, $U(B,t) = 0$, and $U(x,t) \geq 0$ for $0 \leq x \leq B$ and $t_0 \leq t \leq t_1$. Also note that if we take the partial derivative of (7A.81) with respect to x, we obtain[8]

$$\overset{x}{U}(x,t) = \int_{x}^{B} s(x,x'',t)dx'' - \int_{0}^{x} s(x',x,t)dx', \tag{7A.82}$$

where the first term on the RHS represents exports from x to all other locations x'' (which by definition must be to the right of x if exports are to be positive), and the second term represents imports into x from all locations x' (which by definition must be to the left of x if imports are to be positive).

With this new concept it becomes convenient to restate our notion of transport cost. Instead of assuming that the transportation required by a shipment is obtained at the market of the region of export (as we do in the model in the previous paragraphs), or at the market of the region of import (as is sometimes done in other models), we adopt an equally good assumption for a continuous-space model. We assume that in making a shipment from x to x' over the distance $x' - x$, society supplies the transport inputs at x, x', and each intervening location, and in particular supplies at each of these locations just the amount of transport inputs used up to effect the flow of goods through that location.

Accordingly, if $U(x,t)$ is the flow of goods through x at t, we posit that $\sigma U(x,t)$ is the quantity of goods that society takes out of the output $Y(x,t)$, for purposes of transport, that is to be used up in effecting the flow $U(x,t)$. So $\sigma U(x,t)$ is the transport inputs (in terms of goods) per unit length and time, where σ is a positive constant having dimensions of 1/length. Assume that our single homogeneous good is measured in tons. Then σ is dimensionally weight/(length × weight), or tons used per unit of length per ton passing through a location. Further, take the unit of length to be a mile. Then σ has units of tons per ton-mile. Thus σ is a *'physical transport rate'* in terms of weight of goods used up per ton-mile of transport, which concept parallels a 'money transport rate' in terms of dollars per ton-mile. Whenever we refer to σ as a transport rate, we shall have in mind the definition of σ as a physical transport rate.

If this new convention had been used in the model in the previous section, we should have had precisely the same first-order conditions with τ redefined as

$$\tau(x,x',t) = p(x',t) - p(x,t) - \int_{x}^{x'} \sigma p(\bar{x},t)d\bar{x}, \qquad (7A.83)$$

where the integrand $\sigma p(\bar{x},t)d\bar{x}$ is the transportation cost over the distance $d\bar{x}$ incurred by (assigned to) the location $\bar{x}$ when society ships one unit of good from x to x'.[9] Obviously the integral in (7A.83) represents the total transport cost in shipping one unit from x to x'.

With this notion of local responsibility for transportation of all flows through a location and with the relations (7A.76), (7A.81) and (7A.82) in mind, we can replace (7A.74) with

$$\dot{K}(x,t) = F(K(x,t)x) - C(x,t) - \overset{x}{U}(x,t) - \sigma U(x,t). \qquad (7A.84)$$

With the new equation of motion it becomes necessary to restate the problem of (7A.78). See Isard and Liossatos (1979), p. 112. Also, we can recognize flows in both directions from any x, where U^+ is in the positive direction and U^- in the negative. Such type of flows may also be introduced into a macro central planning model encompassing micro behavior with regard to many commodities on the part of many households, many producers, many shippers, and limited government activity. Once again, an ombudsperson is required to perform measurement-control functions in order to have a macro model responsive to maximizing behavior of individual households. In effect, the ombudsperson alters a basic variable to insure that the macro variable $C_h(x,t)$ pertaining to the total consumption magnitude of every good h at location x and time t comes to be equal to the total $\Sigma b_{hi}(x,t)$ of demands for that good h by all households i at location x at time t. See Isard and Liossatos (1979), pp. 134–40 for a full statement of the model.

b. Field description of spatial interaction

The macro model in the previous section may be claimed to capture only partially the dynamics of growth (evolution, or development) in space-time. It focuses on balance at the local level obtained from the equality of supply (local production plus imports) and demand (consumption plus investment plus exports), an equality resembling a local income identity. But specifically what are the forces that determine imports and exports and their changes in a system that exists over space—where the system can be envisaged as a continuum of local economies (subsystems), one at each location.

Unlike the forces at work internally (as in production and as reflected in an equation like 7A.74), these latter forces must depend on conditions in other local economies—economies that provide the imports and economies which absorb the exports. Hence, these latter forces relate to the interactions among local economies.

At time t take two neighboring local economies x and x' with capital stocks $K(x,t)$ and $K(x',t)$. If these two stocks are different, say larger at x than at x', then there is a force that tends to equalize them. Letting the distance separating them $\Delta x \rightarrow 0$, the difference may be stated as $dK(x,t)/dx \equiv \overset{x}{K}(x,t)$. The resulting force will be proportional to $\overset{x}{K}(x,t)$. It may reflect the higher rate of marginal productivity at x' than at x when the use of capital at both locations has run into diminishing returns, which in turn reflects what some may designate the greater development potential at x'. It therefore will tend to increase at time t the time rate of flow $(dU(x,t)/dt) \equiv \dot{U}(x,t)$. Hence, recognizing (1) l to be a factor of inertia (a delay in response), (2) $\dot{U}(x,t)$ to be a development force at play in the interaction of the local economies at x and x', and (3) ξ to be a factor which may be viewed as an elasticity coefficient reflecting the change of the development gradient in response to a unit of capital gradient, we have the relation

$$l\dot{U}(x,t) = -\,\xi\,\overset{x}{K}(x,t) \tag{7A.85}$$

If we now add a frictional type force $vU(x,t)$ (where v is a frictional-type coefficient) to equation 7A.85, we obtain a 'two equations of motion model' (where x,t is suppressed)

$$\dot{K} = F(K) - C - \overset{x}{\dot{U}} - \sigma U \tag{7A.86}$$

$$l\dot{U} = -\xi \overset{x}{\dot{K}} - \nu U \tag{7A.87}$$

This *two equations of motion* model is developed in Isard and Liossatos (1979), chap. 6.[10] There also is presented its embodiment in a central planning macro model with many micro behaving units (producers, shippers, and households) and an ombudsperson providing coordinating inputs. Because of externalities (positive and negative) resulting from the existence of a development potential gradient, the central planner must also be involved in measurement-control processes in order to impose upon the individual producers and shippers penalties and subsidies to take into account the positive and negative externalities.

c. Behavioral interaction: A multi-ombudsperson framework in space-time

As emphasized in the main text, behavioral interaction is a basic factor in evolution and hierarchical development. Thus far, our models have considered: (1) a time development path for a point region with a coordinator involved in measurement-control processes and imposing inputs at certain places in the system to achieve system goals; (2) a space-time development path for a multiregion system, each region being a point but trading goods with other regions, also with a coordinator concerned with a system's goals; and (3) a space-time development path for a multiregion system in continuous space. In the last model, which contains a central planner, an ombudsperson and many microbehaving units in each region, we employed the notion of a continuous flow of goods, the $U(x,t)$ variable—a variable subject to change in both time and space. However, while there was an ombudsperson in each region concerned with measurement and control, there was no direct behavioral interaction among the ombudspersons. Their behavior was connected only indirectly through the $U(x,t)$ variable.

We now wish to consider *direct* behavioral interaction as emphasized in section 7.3 which specified both a *hierarchy of interactors* as well as a *hierarchy of replicators*. Such suggests that the concept of participation potential ${}_iV$, as developed in section 7.4, be employed. Thus, for a continuous space-time model, we can consider: (1) the participation potential of all ombudspersons in the system in determining the coordination inputs imposed at (x,t) on behaving units at (x,t); and (2) participation potential of the ombudsperson at (x,t) in determining all coordination inputs imposed at all locations x' at time t. However, to apply this potential we would need to associate a mass with the ombudsperson at each (x,t) and discount it by distance to each and every x'. One of course could consider population at (x,t) as a measure of mass, an approximation of which could be: (1) the labor force $L(x,t)$ at (x,t), a magnitude which would be explicit in the production function say $F(K(x,t), L(x,t))$ of any non-simplistic model; (2) some weighted combination of $K(x,t)$ and $L(x,t)$; or (3) simply the production potential at (x,t).

If one were to take labor force, as the measure of mass, then participation potential could be given by

$$ {}_iV(x,t) = k \int_0^B \frac{L(x',t)}{|x - x'|}\, dx', \qquad i = 1,2 \tag{7A.88} $$

k being a positive constant.

Alternatively, we may define ${}_iV(x,t)$ so that all behaving units in the system participate in decisions (actions) at (x,t), and where all the behaving units at (x,t) participate in all decisions in the system. But while it is tempting to proceed to define and develop the notions of ${}_i\overset{x}{V}(x,t)$ and ${}_i\dot{V}(x,t)$, it should be kept in mind that coordination inputs represent rate-independent processes and therefore ${}_iV(x,t)$, ${}_i\overset{x}{V}(x,t)$ and ${}_i\dot{V}(x,t)$ cannot be introduced into equations of motion, such as (7A.86) and (7A.87). Yet some insight might be gained from such an inconsistent introduction.

7A.5 Models with perturbations and noise

a. Preliminary transition (major jump) analysis

We now introduce perturbations and noise into the analysis. We need to do so, whether they occur in connection with rate-dependent processes or rate-independent (measurement-control) processes. As already discussed, a perturbation when amplified can lead to a jump to a new far-from-equilibrium state. This is so whether the transition is associated with symmetry breaking among forces and particles, the formation of a new chemical element à la Bronowski's stratified stability, or a new level in an hierarchical structure.

In discussing isolation mechanisms, Bonner notes, as does Prigogine (see pp. 131–3), that when the amoeba (Dictyostelium discoideum) cells spaced evenly over a lawn of food come to confront a condition of starvation, because, say, the environment has suddenly become a less abundant source of food, they converge to form an aggregate—a center constituting a new multi-cellular organism. In effect, there occurs a major transition from one stable state to another via a perturbation whose amplification is fostered by decrease in the fertility of the environment—and we shall concoct conceptually a parallel type of transition in a social human world.

However, it is interesting to note that Bonner observes that once a sufficient aggregation of cells takes place to form a center, that center produces an inhibitor which prevents other centers from forming in its vicinity. Given that any lawn of bacterial food is not unlimited, the repulsion of cells by centers already formed must force some cells to aggregate at or close by the lawn's boundary. The centers that materialize in the less fertile and/or less extensive area may be less stable, and subsequent perturbations within it may be more likely to be subject to amplification; and thus attainment of still another-far-from equilibrium state may be realized—a possible result consistent with saltation doctrine.

When appropriate conditions and constraints obtain, the transition to a new far-from-equilibrium state is well illustrated in an example developed by Isard and Liossatos (1979, chap. 11) in connection with discussion of Thom's catastrophe theory. Imagine a 19th century city on a fertile plain where initially resources are plentiful. Its population numbers, which we take to be our state variable x, increases through normal reproduction and in- and outmigration. Permit there to be three types of forces at play that affect the city's welfare denoted by W. There are agglomeration forces—somewhat similar to autocatalysis and crosscatalysis in biology—that reflect scale economies and advantages from diversity and geographic specialization that come from population growth. Let this force be $\alpha x^2/2$.

There are deglomeration forces, or diseconomies, associated with size due to increase in pollution, crime, traffic congestion, impersonality, etc. Let this force be $-\gamma x^4/4$. When α is significantly positive relative to γ, which henceforth we will set at unity, deglomeration forces do not come to dominate until x reaches a large enough magnitude. Let there be a third force, which is strictly proportional to x—associated simply with population numbers—namely βx. Thus we have the function:[11]

$$W(x,\alpha,\beta) = -\frac{x^4}{4} + \frac{\alpha x^2}{2} + \beta x \qquad (7A.89)$$

The graph of each of the three terms involving x, which correspond to the three types of forces, is given in Figure (7A.2) for a particular structural framework where $\alpha = 3$ and $\beta = 5$. For W of (7A.89) to be a maximum, we have as a first-order condition

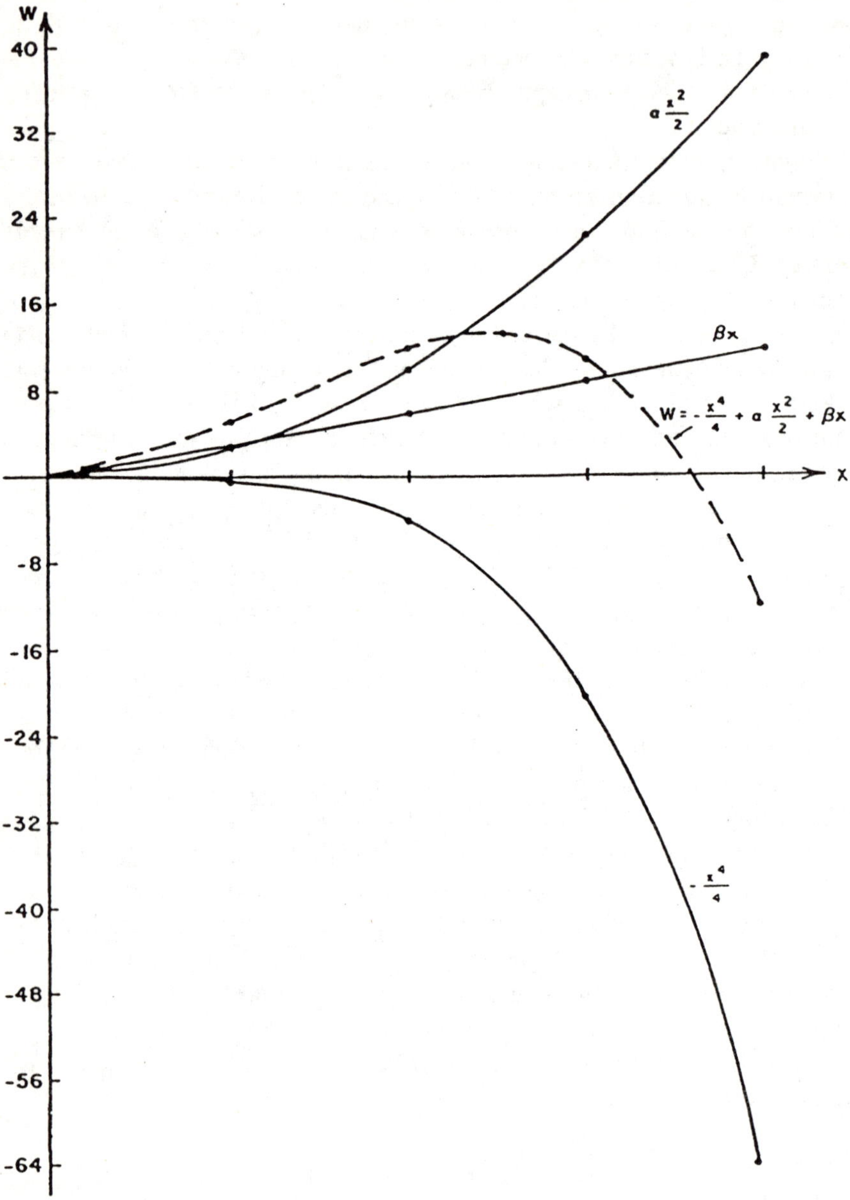

Figure 7A.2 The terms in a welfare function

$$\frac{\partial W}{\partial x} = -x^3 + \alpha x + \beta = 0 \qquad (7A.90)$$

which implies that the solutions of (7A.90) are the equilibrium states of

$$\dot{x} = -x^3 + \alpha x + \beta \qquad (7A.91)$$

To study solutions to (7A.90), it is instructive to plot β against x for different values of α. We do so in Figure (7A.3). We see for $\alpha = 0$ that there is for each a unique x, given by $x = \sqrt[3]{\beta}$. Also, for $\alpha < 0$ there is a unique x for each β. However, for $\alpha > 0$, for each β within the interval $[-2(\alpha/3)^{3/2}, 2(\alpha/3)^{3/2}]$ there are three distinct solutions for x, although for values of β outside that interval there is only one.[12]

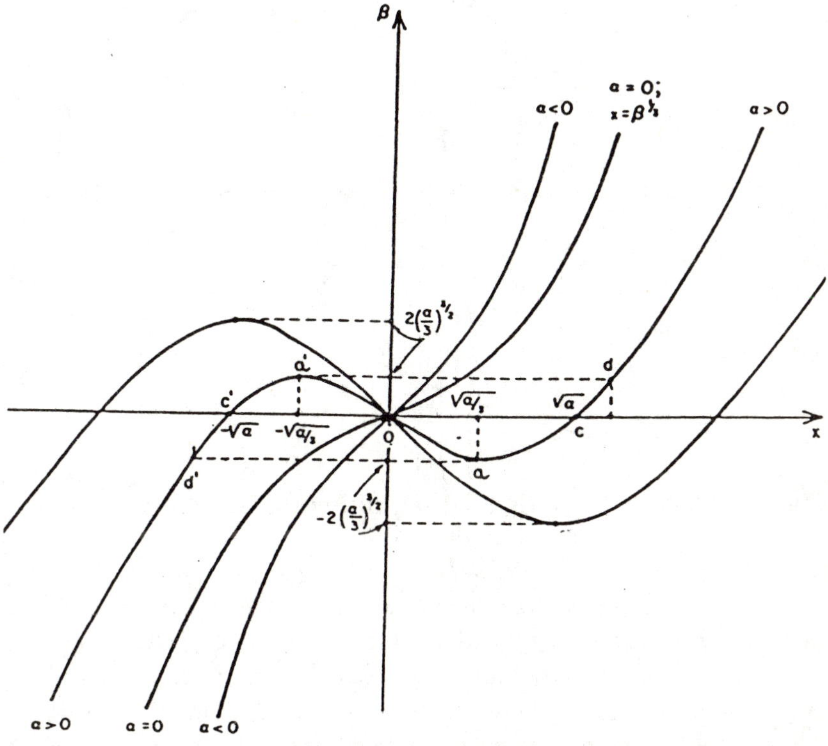

Figure 7A.3 Variation of equilibrium states x with β for selected values of α

Of critical interest is the stability of any equilibrium point, inferred from the sign of the second derivative. From (7A.90), we have

$$\frac{d^2W}{dx^2} = \alpha - 3x^2. \qquad (7A.92)$$

Hence asymptotic stability characterizes all states for which $/x/ > \sqrt{\alpha/3}$, since for these values $d^2W/dx^2 < 0$; and instability characterizes all states for which $/x/ \leq \sqrt{\alpha/3}$, since for these values $d^2W/dx^2 \geq 0$.[13]

The implications of these conditions for stability can be clearly seen by examining one of the solution curves of (7A.90) in Figure (7A.3) indicating multiple equilibrium states for some range of β. For example, we reproduce in Figure (7A.4) that solution curve indicated by $\alpha > 0$. And

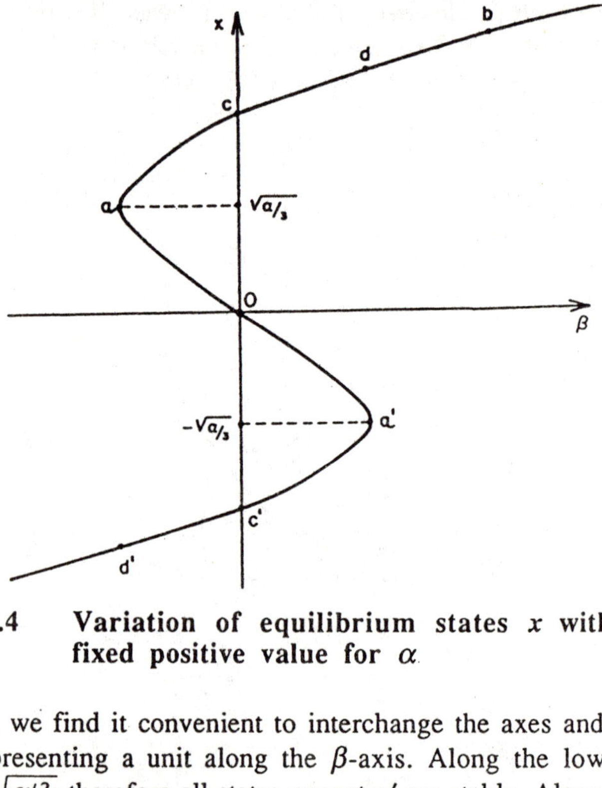

Figure 7A.4 Variation of equilibrium states x with β for a fixed positive value for α.

in doing so, we find it convenient to interchange the axes and triple the distance representing a unit along the β-axis. Along the lower branch $b'c'a'$, $/x/ > \sqrt{\alpha/3}$; therefore all states except a' are stable. Along the upper branch acb, $/x/ > \sqrt{\alpha/3}$; therefore all states except a are also stable. Along the intermediate branch $a'0a$, $/x/ \leq \sqrt{\alpha/3}$; therefore all states are unstable. Hence, on the basis of this analysis alone, the system cannot stay, even under the smallest perturbations, at any equilibrium state on the intermediate branch. It then follows that for the range of β for which multiple solutions exist, any shift away from the lower branch that does not tend to return (regress) to the lower branch must end up on the upper branch, and vice versa. However, *for all small perturbations*, the property of asymptotic stability for all equilibrium points on the lower and upper branches (except for a and a') ensures that no jump from one branch to the other will occur. At a and a', small perturbations cause jumps to the other branch.

We obtain further analytical insights, when our welfare function is of the specific form (7A.89) if we plot the values of W for the equilibrium states [solutions to the equations (7A.90) of Figure (7A.4)]. We obtain Figure (7A.5). There, only points along the two curves $a'b'$ and ab correspond to welfare maxima. The points along the curve aa' correspond to welfare minima, the points a and a' being inflection points and corresponding to the values $\beta \pm 2(\alpha/3)^{3/2}$. Further, we see that the welfare maximum associated with any equilibrium solution along the stretch from d' to c' is greater than the welfare maximum associated with the corresponding equilibrium solution along the stretch from a to c. Also, the

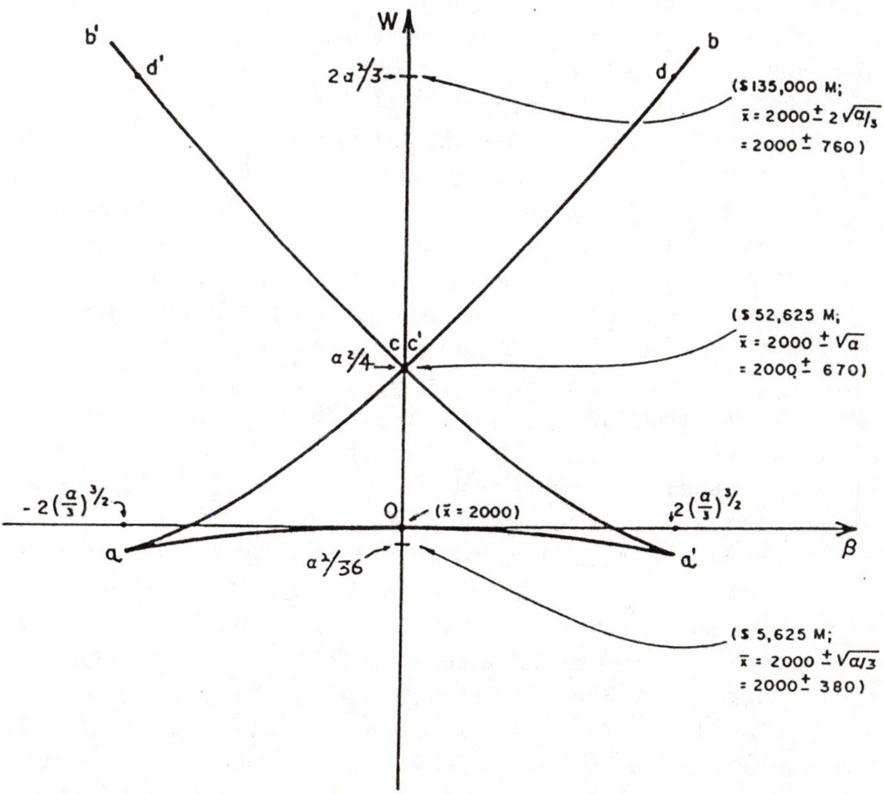

Figure 7A.5 **Variation of W with β for equilibrium states**

welfare maximum associated with any equilibrium solution along the stretch beyond c and up to d is greater than the welfare maximum associated with the corresponding equilibrium along the stretch beyond c' up to a'. At c and c' the welfare maxima are the same, being $\alpha^2/4$. To illustrate these points, we have indicated at the side in Figure 7A.5 selected dollar values for the welfare function and the corresponding state values (populations) where $\alpha = 450{,}000$ and $x = \bar{x} - 2000$, $\bar{x}$ being actual population.[14]

On Figure (7A.5), we see that an equilibrium state on the stretch from d' to c' is preferable, *from a welfare standpoint*, to its corresponding state on the stretch a to c; and an equilibrium state on the stretch beyond c to d is preferable to a state on the stretch beyond c' to a'. At c and c' the equilibrium states are equally valued. Accordingly, if the system is moving *in macro time T* along the lower branch, it is 'logical' for it to shift to the upper branch once it passes c'. But *on the basis of local stability considerations*, the system under small perturbations will continue to move along the lower branch up to point a', and only then will it jump to the higher branch. Refer also to Figure (7A.4). Correspondingly, on the basis of local stability considerations, if the system is on the upper branch and moving in macro time T in the direction from b to c to a, it will continue to move along the upper branch even after it passes c until it reaches a, at which point it will jump to the lower branch. Thus the local stability approach contradicts the logical approach based on a welfare criterion. This contradiction is due to the fact that local stability has nothing to say about global stability, that is, about what happens to an equilibrium state under large perturbations. Accordingly, on the basis of our welfare rationale, we can posit that all states between c' and a' on the lower branch, and all states between c and a, are stable under infinitesimal perturbations, but unstable under sufficiently large (noninfinitesimal) perturbations. We may designate these equilibrium states as *metastable*.

Observe that this situation parallels that of phase transitions in thermodynamics (from say liquid to vapor), where the phase transition should take place at points c and c' if the potential (which corresponds to our social welfare) is to be optimized. However, both thermodynamics and our particular approach thus far developed ignore micro states, that is, the states and behavior of individual molecules and the corresponding multitude of individual behaving units in a social system—whose behavior does not conform to the averages of the macro model. Hence we must make it explicit that equation (7A.89) constitutes a macro description of a large social system containing many small behaving units (consumers, producers, traders and others). Each of these units presumably maximizes some payoff and is a little system in itself.

Exactly how does the phase transition take place? Here we may quote Callen (1960), who refers to a thermodynamic system involving the macro variables volume, pressure, and temperature, and whose statements supplement those of Prigogine in Chapter 6. We may

> . . .inquire about the manner in which the system, slowly brought to the state *c′*, becomes aware of the existence and attractiveness of the competing state *c*. We may well visualize that the system at *c′*, faced with the unpleasant result of an increasing Gibbs function [*decreasing welfare*] if it ventures to alter its state slightly, returns from this slight exploratory excursion to the locally preferable state *c′*. How then is it to learn that far removed at *c* is an *equally attractive state*? To answer this question, we must appeal to the perspective afforded by statistical mechanics. According to statistical mechanics, the micro state of a system in equilibrium is not static, but the system perpetually undergoes rapid fluctuations. It is by means of these fluctuations that the system probes and explores to discover the state with the *maximum* value of the appropriate potential. To be specific, with reference to our particular system brought to the state *c′*, the system is not quiescent on the microscopic scale, but rather there are local fluctuations in density from point to point, the average density being that corresponding to the point *c′*. Of course, small fluctuations in density are most probable, and large fluctuations, such as that required to take a small portion of the system to the state *c*, are very rare indeed. So the system may sit for a very long time at *c′*, until eventually a large spontaneous fluctuation occurs, taking a small portion of the system to *c*. This fluctuation, unlike others, does not decay, for the portion of the system brought to *c* finds its new state equally satisfactory to its initial state. Such a stabilized fluctuated portion of the system then becomes a nucleus for the growth of the second phase. . .
>
> In most practical instances a system brought to the state *c′* does not have to wait for the eventual appearance of a spontaneous fluctuation to take it to *c*, for artificial fluctuations are often induced by outside mechanisms. Thus a mechanical shock or tap on the system sets up longitudinal waves, with regions of condensation and rarefaction. The region of condensation is a region of small molar volume; it is consequently a region in which the system may be brought into the state *c* and in which it forms a phase nucleus (Callen, 1960, pp. 162–3).[15]

In parallel fashion, we can conceive the micro state of our social system in equilibrium to be not static, but perpetually undergoing rapid fluctuations. Its numerous firms, consumers, traders, organizations, governments, and planners are not average-type behaving units. Rather there are numerous differences within the population of each of these and other classes of behaving units; they can differ greatly in their intensity and extent of interactions with other units and the environment. Thus some of any class are always exploring—almost invariably, however, in small or smallish ways. Before the system comes to c' in macro time T, we may imagine that these explorations, even the occasional large ones, all regress to the equilibrium state. They do, too, when the system is at c', except for the very large fluctuations. But if the system sits for a long time at c', eventually there will occur large spontaneous explorations by some units, especially under the stimulus of random shocks from the external environment. These take one or more behaving units to c. Since at c the situation is as good as at c', those units arriving at c remain there and form the nucleus for growth of numbers at c—and beyond c if the system continues to move in the same direction.

However, this kind of interpretation is not fully satisfactory. As Callen suggests, one needs in physics to resort to statistical mechanics in order to derive a fuller explanation of transition phenomena. But this should mean that generally speaking we ought to employ probabilistic concepts in order to describe the behavior of the system when more than one attractor is present. The deterministic equation *per se* is not sufficient.

To motivate an alternative approach, consider the system at an equilibrium state on the lower branch. With respect to macro time T, let the system just cross d' from the left. As it does so, a second attractor (equilibrium state) on the upper branch becomes relevant for the system. We may then imagine that there enters, for the system, a certain amount of indeterminacy as to the equilibrium state toward which the system will regress from any random perturbation. That is, of the two and only two possible equilibrium states toward which the system will regress, there must be a positive probability for each—even though it may be extremely small for one. This indeterminacy persists until state a' is reached, provided the system has remained on the lower branch. At a' there is no indeterminacy, as there exists only one attractor, namely d. However, it may be presumed that as the system passes from d' to c', the probability of regressing from a random perturbation to an equilibrium state on the upper branch increases. Such a pattern may still be considered to be roughly consistent with thermodynamic (Maxwell) conventions and the

Callen interpretation quoted, if the probability of occurrence of a random perturbation large enough for the system to regress to the upper branch remains very small until the system reaches the immediate neighborhood of c'. Here we may hypothesize that the probability sharply increases and almost reaches unity as the system comes close to, crosses and just passes c'. This argument, however, is only a reasonable speculation, as it is in thermodynamics. There is nothing in the equation of motion and our definitions of stability that definitely implies this interpretation or hypothesis. As in physics, in order to make more definite statements about the behavior of the system confronting two or more alternative attractors, we can no longer rely on deterministic equations of motion to define the trajectory of the system. We must make additional assumptions that explain so-called 'random' behavior in terms of well-defined probabilistic laws.

b. Noise and stochastic approaches embodying deterministic aspects

At this point, we should note the progress in formulating a problem that encompasses both deterministic and probabilistic behavior through the use of both a Master Equation and a Fokker Planck equation. First, consider the use of a Master Equation alone (see Isard and Liossatos, 1979, chap. 11). Assume a large system where a key property is dependent on the size of the system. Also, posit that the dynamic process governing the path of this property (its magnitude) is Markovian, that is where the future and past are statistically independent if the present is known, or put otherwise, where the future is dependent on only the present and not the past. Thus, where at any time t there exists a probability distribution of states sharply peaked around the state of maximum probability, which is the mean, we become interested in an equation of motion pertaining to the mean. We become so interested because (1) we can take this equation of motion as a good approximation to the equation of motion governing the state in a purely deterministic model, and (2) such an equation of motion has the advantage of being more 'realistic,' since it is embedded in a stochastic process. Put otherwise, when we represent a system by deterministic equations like (7A.91), we in essence ignore fluctuations. Hence our representation is not 100% accurate. However, such representation can be viewed as valid when it refers to the expected or mean values. If an equation like (7A.91) is to refer to the actual instantaneous values of x, a random force term must be added to its RHS. Equivalently , however, we can associate with (7A.91) a Master Equation, as developed in Isard and

Liossatos (1979), pp. 326–8. So to fully describe the process, we need to solve the equation of motion of the probability distribution, namely

$$\frac{d}{dt}\langle x\rangle_t = \alpha\langle x\rangle_t + \beta - \langle x\rangle_t^3 \tag{7A.93}$$

where $\langle x\rangle_t$ is the mean value of x at time t. In particular, from the Master Equation we can derive the equation of motion for the mean square deviation around any equilibrium state. If the equilibrium state is stable, the mean square deviation approaches a steady value. However, if the equilibrium state is unstable, then the mean square deviation grows exponentially with time, a phenomenon often associated with the period of turbulence. If and when the system reaches the basin of another attractor, the mean square deviation regresses to another steady value. In sum, the Master Equation allows us to study more closely, at least to some extent, the sequence of fluctuations not only during the normal development of a system with a stable equilibrium state, but also during the period of transition of the system from one stable state to another, or during its collapse.

Next, consider recent developments in noise theory which exploits a combined Master Equation and Fokker–Planck equation approach. One simple way to do so is to assume that the stochastic fluctuations in a system originate in the environment and directly impinge on only one element, say the α parameter of equation (7A.91), as in Isard and Lobo (1996). Accordingly at time point t,

$$\alpha_t = \langle\alpha\rangle + \sigma\xi_t \tag{7A.94}$$

where the bracketed term represents the mean value of α, and where ξ_t is Gaussian White Noise and σ is the strength, or amplitude, of the noise.

To accommodate the effect of stochastic fluctuations on the evolution of the state variable x, we replace equation (7A.91) with the following stochastic differential equation:

$$\frac{\partial x}{\partial t} = \langle\alpha\rangle x + \beta - \gamma x^3 + g(x)\sigma\xi_t. \tag{7A.95}$$

When the $g(x)$ term in the above equation is a constant, the noise affecting x is termed *additive*, as where the influence of environmental fluctuations is independent of the state of the system; when $g(x)$ is not a constant and the effect of the stochastic fluctuations does depend on the state of the system, the noise is termed *multiplicative*. The state variable x has thus become a random variable, whose time evolution—and in particular the value it takes at a specific point of time—can only be described probabilistically by means of the probability density function $p(x,t)$.

As in Doering (1991), we can study the effects of different levels of noise on the behavior of equation (7A.95) by means of simulations of time series involving extremely rapid fluctuations of an environment acting upon a steady state. To do this we discretize the continuous equation (7A.95) using the procedure outlined in Lasota and Mackey (1985). The discrete-time analogue of (7A.95) is then given by:

$$x_{t+1} = x_t + \Delta x_t = x_t + (\langle\alpha\rangle x_t + \beta - x_t^3)\,\Delta_t + g(x_t)\,\sigma\Delta\tilde{W}_t \tag{7A.96}$$

where $\tilde{W}_t$ represents a Wiener process.[16]

Assuming additive noise, in particular $g(x) = 1$, and setting $x_0 = 1.04668$ at $t = 0$ (the steady state for one particular situation examined), with $\langle\alpha\rangle = 1$, $b = 0.1$, $\gamma = 1$, and $\sigma^2 = 0.1, 0.15, 0.20,$ and $0.25,$ we obtain the output depicted in Figure (7A.6). As can be expected, at low levels of noise ($\sigma^2 = 0.1$) the value of x undergoes small fluctuations around its deterministic steady-state with an occasional jump to a negative value. When the noise amplitude is increased to 0.15, the time series can look significantly different; not only may the deviations from the steady state increase in magnitude, but also the system may suddenly jump into the range of negative x values centering around a second steady-state value. In effect, the system jumps between the wells of the potential function corresponding to the welfare equation (7A.89). Such jumps are more prominent in Figure (7A.6c) ($\sigma^2 = 0.2$). When the noise amplitude is increased even further, the system jumps still more widely and can become explosive.

For multiplicative noise, where in our model $g(x) = x$, we obtain a set of graphs which, for a given steady state equilibrium point, show fluctuations of increasing magnitude with increase in the magnitude of noise and which, too, may become explosive.

The effects of varying noise levels on the behavior of equation (7A.95) can also be studied *analytically* by examining how the *probability density function, p(x,t),* changes as the noise level changes. We can do this using the *Fokker–Planck* equation. The Ito interpretation of this equation yields, for the time-rate of change of the probability of the (x,t) associated with equations (7A.93–95),[17]

$$\frac{\partial p(x,t)}{\partial t} = \frac{\partial}{\partial x}\,[\gamma x_t^3 - \beta - \alpha x_t]\,p(x,t) + \frac{\sigma^2}{2}\frac{\partial^2}{\partial x^2}\,[g^2(x_t)p(x,t)]. \tag{7A.97}$$

By setting $\partial p(x,t)/\partial t$ equal to zero (appropriate given the Markov process assumption) and solving the Fokker–Planck equation, we obtain the *stationary density function,* $p_s(x)$.[18] The solution for the stationary density function has the form:

$$p_s(x) = Ng(x)^{-\nu} \exp\left[\frac{2}{\sigma^2}\int\frac{\alpha x + \beta - \gamma x^3}{g^2(x)}dx\right] \qquad (7A.98)$$

where $\nu = 1,2$ depending on whether we assume multiplicative or additive noise, respectively, and where N is an integration constant.[19] The behavioral effects of noise can be seen by first solving the equation for the

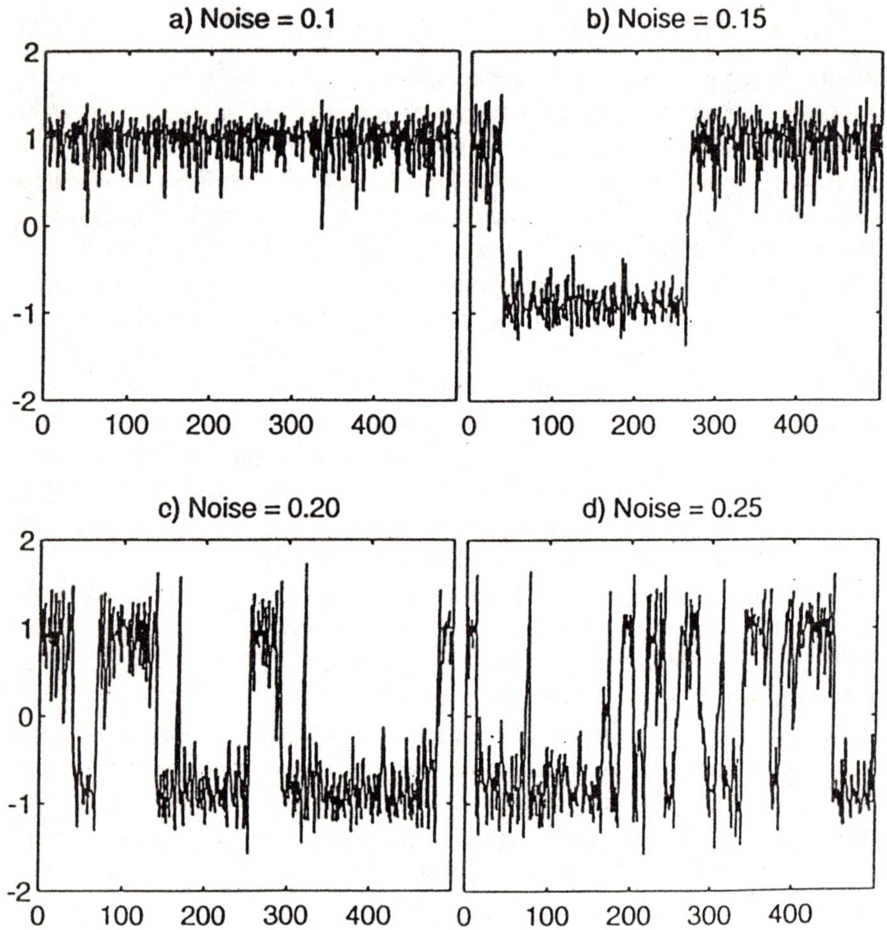

Figure 7A.6 **Time series simulations: impacts of different magnitudes of noise** (The simulations were performed for 500 units of time, each unit being sufficiently small so as to be able to utilize the assumption that there was no effect upon the external system and the environment.) *Source: Isard and Lobo, 1996*

stationary density function, and then depicting the resulting stationary density function for varying noise amplitudes. This we do in Figure (7A.7) for *additive noise*.

Figure 7A.7 shows the stationary density functions for $\alpha = 1$, $\beta = 0.1$, $\gamma = 1$, $g(x) = 1$ and varying magnitudes of σ^2. At a small noise level, (a) in Figure (7A.7), the probability density has two peaks, centered around its most probable values. The two most probable values correspond to the two stable roots of the deterministic equation corresponding to equation (7A.96) and the values of the parameters cited. These two are the roots: 1.04668 and –0.945646. The meaning of the sharpest, highest peak in the stationary density is that most of the time the system is at or near this point with the system experiencing small fluctuations about this point. Thus for very small noise, the system will still be drawn to the vicinity of its

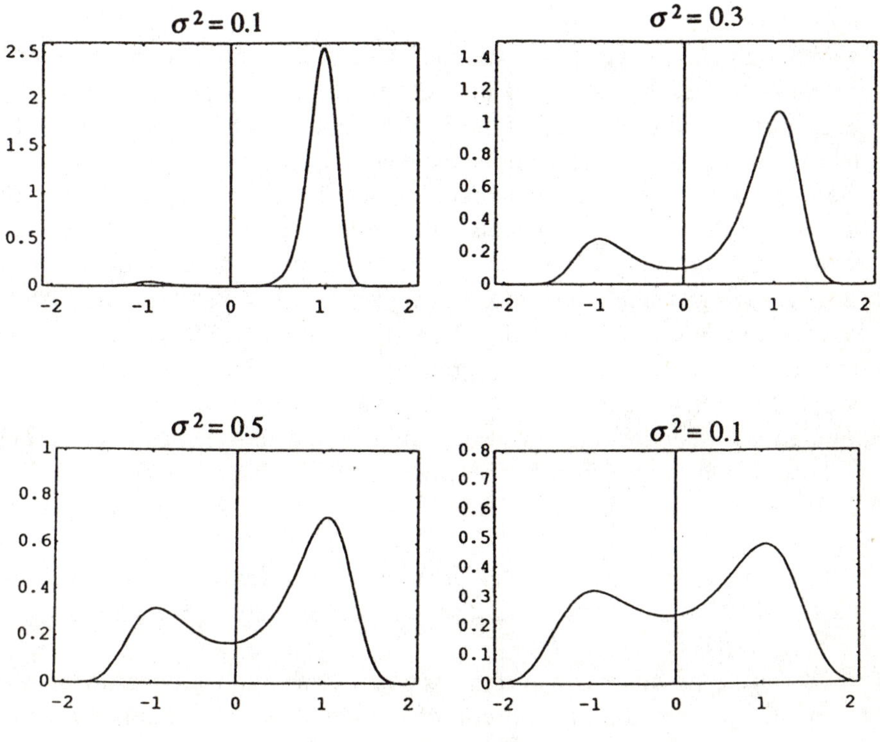

Figure 7A.7 **Some stationary probability density functions for additive noise.** *Source: Isard and Lobo, 1996*

deterministic steady state. The trough between the two peaks represents the boundary between the two wells in the potential function corresponding to equation (7A.89). As the amplitude of the noise is increased, Figures (7A.7b) and (7A.7c), the probability density function becomes more spread out; the boundary between the two wells becomes less pronounced and the probability that the state variable assumes a zero value increases. For a noise level of $\sigma^2 = 1$, as in (7A.87d), the two wells are almost equiprobable.

Figure (7A.8) depicts the impact of *multiplicative noise* upon the probability densities when only the positive quadrant is considered. (There are problems of interpretation and computation when a probability density function is considered for both positive and negative values of x_t.) As is clearly evident, increase in noise leads to decrease in the maximum value of the probability density function, which value for very low magnitudes of noise lies very close to the positive deterministic steady state. As noise increases more and more, the maximum point comes to be closer and closer to the zero value for x_t. However, it never reaches it for finite values of σ^2.

Other models have been examined analytically for major transitions (jumps) induced by noise. In contrast to the model just explored, the probability density function of the Verhulst model (originally designed to investigate population growth phenomena and later employed to examine phenomena in many fields) undergoes an abrupt transition when with increase in the magnitude of noise the natural rate of growth λ comes to equal σ^2. With continuing increase in noise, population extinction can become the most likely phenomenon. Figure (7A.9a) illustrates this. Also, as with the model of equation (7A.95) under multiplicative noise, the predator–prey model shows a gradual shift of the maximum of the probability density function away from the positive value of x (the prey population) when noise is low to one approaching zero (extinction of prey) when noise is large. The former maximum gradually disappears with increase in noise, while the latter increasingly dominates. See Figure (7A.9b).

Still other models, like the genetic, which lead to noise-induced transitions of major impact, are discussed extensively in Horsthemke and Lefever (1984) and Moon (1992).

Of course, all the above analysis is generalizable to many variables and to situations where the transition probabilities depend explicitly on time. But even in the above single variable form it does provide a deeper insight into the stochastic nature of actual transitions. While it does not stem from the dynamic laws at the microscopic level—laws that are yet to be identified

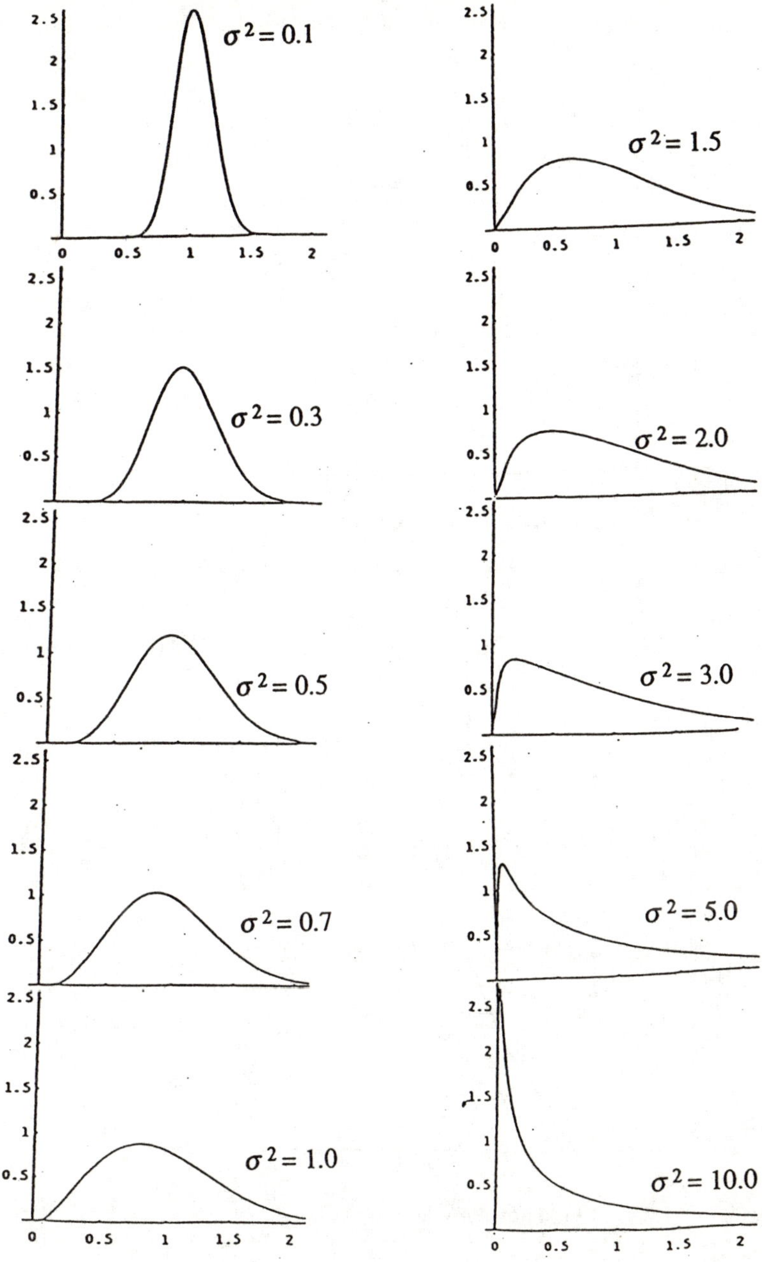

Figure 7A.8 **Some stationary probability density functions for multiplicative noise**

Source: *Isard and Lobo, 1996*

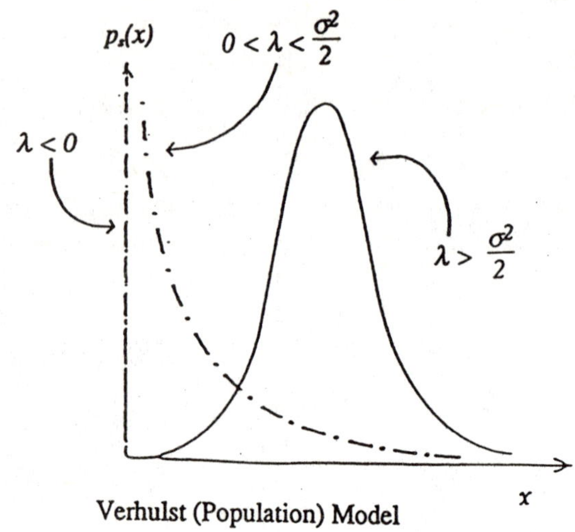

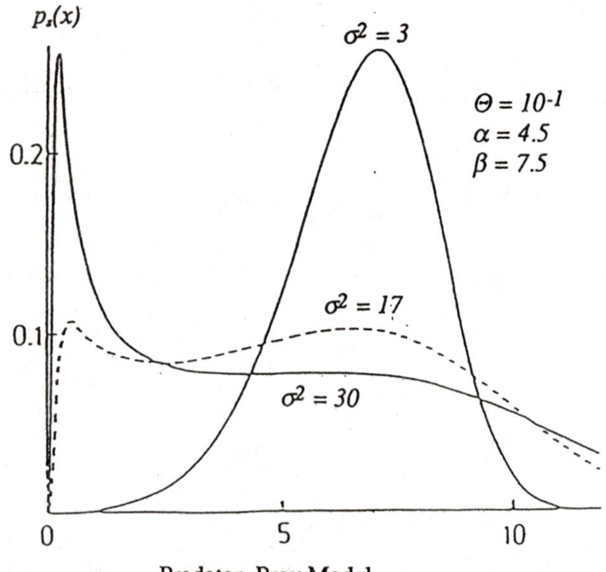

Figure 7A.9 **Stationary probability densities of population numbers**

Source: *Horsthemke and Lefever, 1984, pp. 125, 187*

—it does provide a framework that is not too sensitive to the details of microscopic dynamics and thus permits some generality. Serious shortcomings to its application to reality reside in the many strong assumptions, such as a Markovian process, upon which it relies.

Hence, many analysts choose to adopt the very simple assumption of perfect foresight in a non-stochastic framework, recognizing in full its many limitations. For example, see the dynamical systems analysis in Isard and Liossatos (1979), chaps. 7–9 where a series of stages of development over space-time is presented, starting from self-sufficient farming to, in order: (a) a village trading point, (b) a one-city economy with manufacturing, (c) a one-city economy with manufacturing and services, (d) a many city economy with a hierarchical ordering of cities, and (e) a system involving changing land-use patterns governed by rent, transport costs and other economic factors.

c. *Introduction of a new variable as a perturbation: the case of a multi-nation situation*

Consider first a nonspatial model of n variables $x_1, x_2, \ldots, x_n$ in R^n. Equilibrium solutions, if they exist, are solutions of

$$f_i(x_1, \ldots, x_n) = 0 \qquad i = 1, \ldots, n \tag{7A.99}$$

Assume that the functions $f = (f_1 f_2, \ldots f_n)$ are defined for values of $x = (x_1, x_2, \ldots, x_n)$ and $v = v_1, v_2, \ldots, v_n)$ a set of time-independent parameters of interest, and are smooth (well-behaved) to the extent that the problem requires. Suppose the equations of motion of the system are:

$$\dot{x} = f(x,v). \tag{7A.100}$$

Then the set of equilibrium solutions of (7A.100) satisfy the equation

$$f(x^*,v) = 0. \tag{7A.101}$$

At this point we could consider a perturbation of a state variable within this model. Such is examined in Isard (1976, 1978), Isard and Liossatos (1978), Allen et al. (1981) and Kahn (1981). However, it is equally instructive and interesting to suppose that at some time an error copy, or a new commodity, or a new decision-making unit randomly appears, describable by a new state variable x_{n+1}. (The probability of this appearance might in some models increase with time. For certain situations this increase in probability over time might be inferred or empirically estimated from historical materials.)

Posit that one attractor $x^*(v) \equiv (x_1^*(v), x_2^*(v), \ldots x_n^*(v))$ exists for each $v = (v_1, v_2, \ldots, v_q)$ in some region M of R^q. Also suppose the system (7A.100) changes as follows upon the emergence of the new variable $n + 1$:

$$\dot{x} = g(x, v, x_{n+1}, \varepsilon) \tag{7A.102}$$

$$\varepsilon \dot{x}_{n+1} = G_{n+1}(x, v, x_{n+1}, \varepsilon) \tag{7A.103}$$

where ε is a small parameter and where the vector function g and the real function G_{n+1} have the following properties:

$$G_{n+1}(x, v, x_{n+1}, 0) = 0 \tag{7A.104}$$

which determines

$$x_{n+1} = \Phi(x, v) \tag{7A.105}$$

so that

$$g(x, v, \Phi(x, v), 0) = f(x). \tag{7A.106}$$

In words, the system (7A.102), (7A.103) reduces to (7A.101) at $\varepsilon = 0$. We take R^{n+1} to be the state space of the system (7A.102), (7A.103), and we denote a typical state, a point in R^{n+1}, by (x, x_{n+1}) with $x \in R^n$ and $x_{n+1} \in R$.

We further assume that the system (7A.102), (7A.103) has an equilibrium state $\{\hat{x}^*(v,\varepsilon), \hat{x}_{n+1}^*(v,\varepsilon)\}$, where of course $\hat{x}^*(v,0) = x^*(v)$ and $\hat{x}_{n+1}^*(v,0) = 0$ for all $v \in M$. On the one hand it might be that for v in some subset M^S of M and at least for small values of ε, the equilibrium state $\{\hat{x}^*(v,\varepsilon), \hat{x}_{n+1}^*(v,\varepsilon)\}$ is asymptotically stable. Then we would observe a smooth and very, very slow creation of x_{n+1}. For small ε, the new system would differ negligibly from the old. On the other hand, if v changes and shifts to $M - M^S$, it might be that the new equilibrium state $\{\hat{x}^*(v,\varepsilon), \hat{x}_{n+1}^*(v,\varepsilon)\}$ is unstable for any $\varepsilon \neq 0$, however small. In this case, we say that the system (7A.101) is *structurally unstable*. This means that the stability properties of (7A.101) are radically altered even for infinitesimal perturbations of the form (7A.102) and (7A.103). Typically, the system enters a transition period of turbulence. However it might be that (7A.102) and (7A.103) have another equilibrium state $\{x^{**}(v,\varepsilon), (x_{n+1}^{**}(v,\varepsilon)\}$ that for $v \in M - M^S$ is asymptotically stable, with a nonnegligible magnitude for $(x_{n+1}^{**}(v,\varepsilon)$. Then under finite fluctuations the system *may* be brought to this state, if no other structural changes occur. So the influence of fluctuations, even small parameters like ε, becomes significant via multiplier and accelerator effects. Of course, the new system may also be structurally unstable, which means that at some time in the future a random perturbation might generate another period of turbulence. And so forth.

To illustrate the effect of a new variable, suppose we have n interacting nations (bodies). For simplicity, let the behavior of nation i, $i = 1,2, \ldots ,n$, be characterized by one parameter x_i, $x_i \in R$, and let x_i also denote its state. Then the state of the whole system can be characterized by $x = (x_1,x_2, \ldots ,x_n)$, a point in R^n.

Let each nation i have a payoff function

$$(x_1,x_2, \ldots ,x_n) \rightarrow W_i(x_1,x_2, \ldots ,x_n,v), \qquad i = 1,2, \ldots ,n, \tag{7A.107}$$

a function that depends on v. So to each $v \in M$ there corresponds a family of payoff functions. For any given v, each nation i determines its 'position' x_i by maximizing its payoff as a function of x_i, taking the 'positions' of the other $n-1$ nations as given. To be more specific, let $A_i \subset R$ be the *action space* of nation i. Then

$$A \equiv A_1 \times A_2 \times \ldots \times A_n \tag{7A.108}$$

is the domain of each function (7A.107), or the *system action space*. We also define

$$\bar{A}(A_i) = A_1 \times A_2 \times \ldots \times A_{i-1} \times A_{i+1} \times \ldots \times A_n \tag{7A.109}$$

as the *joint action space excluding i*.

The first-order condition for i to maximize its payoff is

$$\frac{\partial W_i}{\partial x_i}(x_1,x_2, \ldots ,x_n,v) = 0 \qquad i = 1,2, \ldots ,n. \tag{7A.110}$$

In essence, this equation defines the *response function*[20]

$$\hat{x}_i = \hat{x}_i(x_1, \ldots ,x_{i-1},x_{i+1}, \ldots ,x_n,v), \qquad i = 1,2, \ldots ,n. \tag{7A.111}$$

Let $A_i^* \subset A_i$ be the *range* of (7A.111)—the *set of optimum selections* for the ith variable, given all possible values of the other variables. Then we may define the *system optimum space as perceived by i*

$${}^{(i)}A^* = A_1 \times A_2 \times \ldots \times A_{i-1} \times A_i^* \times A_{i+1} \times \ldots \times A_n, \qquad i = 1,2, \ldots ,n. \tag{7A.112}$$

Thus a unique equilibrium point $x^*(v)$ for the n-nation system exists if

$$\bigcap_{i=1}^{n} {}^{(i)}A^* = \{x^*(v)\} \tag{7A.113}$$

for each $v \in M$. Figure 7A.10 illustrates the case for two nations.

The existence of $x^*(v)$ as given by (7A.113) *does not* ensure a round-by-round procedure that converges to $x^*(v)$. That is, it does not ensure a situation wherein on each of several rounds each nation announces a 'position' that maximizes its payoff *assuming* certain values for the 'positions' of the others (these values are initial values on the first round, and previous-round values on subsequent rounds), and wherein the series of rounds converge to $x^*(v)$. We assume that the attainment of equilibrium is a dynamic process, wherein nation i tends to increase x_i whenever $\partial W_i/\partial x_i > 0$ and to decrease x_i whenever $\partial W_i/\partial x_i < 0$. Accordingly, if we use the dynamical systems approach in an *auxiliary* manner to depict the round-by-round process as a state trajectory, then each nation in seeking to maximize its payoff chooses $\hat{x}_i$, thus defining its change of position $\dot{x}_i$ over the unit of time from one round to the next as proportional to the rate of change $\partial W_i/\partial x_i\,(x,v)$. That is,

$$\dot{x}_i = k\frac{\partial W_i}{\partial x_i}(x_1,x_2, \ldots ,x_n,v), \qquad i = 1,2, \ldots ,n, \tag{7A.114}$$

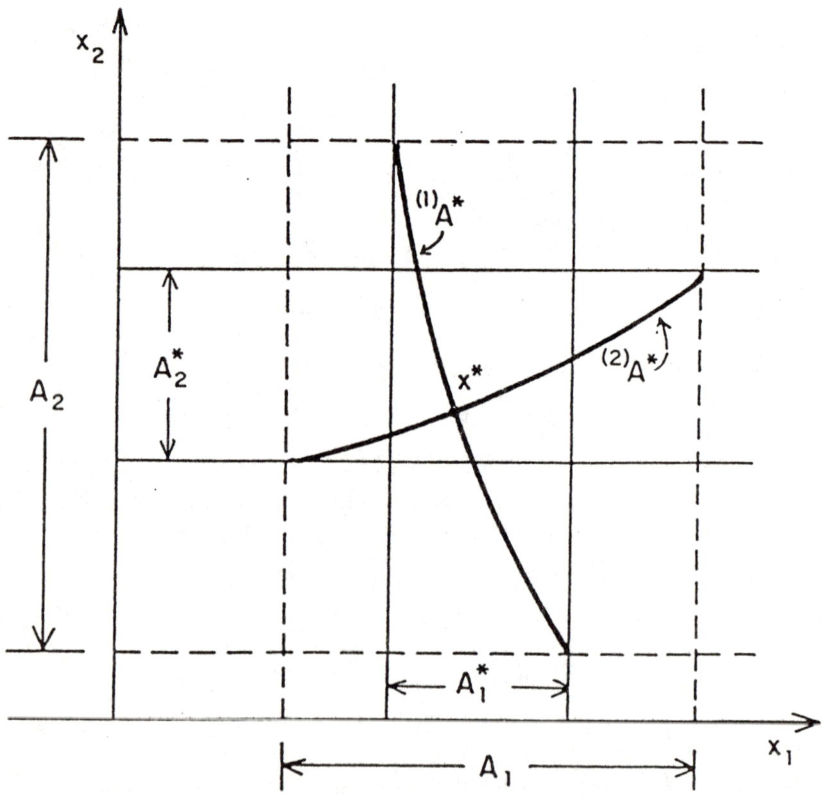

Figure 7A.10 The response functions for two nations and the equilibrium point

k being a proportionality constant with appropriate units to have dimensional consistency. This equation is of the form (7A.100). The existence of $x^*(v)$ as posited by (7A.113) amounts to saying that $x^*(v)$ is a unique equilibrium state for (7A.114) for each v.[21] The convergence of the round-by-round procedure is ensured if $x^*(v)$, as an equilibrium state of (7A.114), is asymptotically stable.[22] If, for some v, $x^*(v)$ is unstable, this implies that the system does not converge, but enters a stage of turbulence, although each nation may continue to maximize its payoff in the sense depicted above.

So far we have assumed that for each v there exists a unique equilibrium state, which is an attractor. However, from our previous discussion, it is very likely that more than one attractor will exist for each v in a subset of M. When such a situation obtains, we may have indeterminacy with regard to equilibrium states. For example, if there are two attractors, all nations might obtain a higher maximum payoff at the same attractor for some values of v. On the other hand, for other values of v, some nations might obtain a higher payoff at the first attractor, while others might do so at the second. In the latter case, a *conflict* exists. We can, for example, use Figure 7A.4 to make these points more explicit when we symbolically replace β by $v = (v_1,...,v_q)$ and take $x = (x_1,...,x_n)$. In Figure 7A.4 we may imagine that the system of nations is moving in macro time T along the lower branch from b' to d'. As it crosses d', a second attractor appears just after a. If all nations have the same welfare function, that is,

$$W_1(x,v) = W_2(x,v) = \ldots = W_n(x,v) \tag{7A.115}$$

and if each is concerned with payoff maximization and behaves logically (rationally), we can imagine that each independently chooses to stay on the lower branch until c' is reached. Then simultaneously all make the same decision and jump to c. Also, if all are alike, but each does not and cannot perceive changes of action beyond the immediate neighborhood, they all stay on the lower branch until a' is reached, then simultaneously all make the same decision to leave the lower branch, and ultimately they arrive at d through a series of moves. Under these circumstances, we have no need to identify and employ a system payoff function.

However, when the nations are not alike and have different welfare functions, there is no general criterion by which all simultaneously choose to be on one branch. All we can say is that at a' no nation maximizes; so when a' is reached (if it ever is), they all jump simultaneously. Depending on how different their welfare functions are, and on the extent to which they can perceive big as well as little jumps, we may imagine that once d' is crossed, sooner or later one and then more than one nation will, out of self-interest, want to jump to the upper branch, while others will want to remain on the lower branch. Since they all must be on the same branch—*assuming there can be only one multinational organization*—conflict arises. In this case we must have recourse to some technique of conflict resolution. As an example, we might search for a feasible system welfare function. Suppose the functions W_i are such that

$$\frac{\partial^2 W_i}{\partial x_j \partial x_i}(x,v) = \frac{\partial^2 W_j}{\partial x_i \partial x_j}(x,v). \tag{7A.116}$$

Then there exists a function

$$x \rightarrow W(x,v) \tag{7A.117}$$

such that

$$\frac{\partial W}{\partial x_i}(x,v) = \frac{\partial W_i}{\partial x_i}(x,v)\,. \tag{7A.118}$$

In this case W attains a maximum if and only if each W_i attains a maximum. This implies that whenever it is desirable from the system standpoint to shift from the lower to the upper branch, as discussed in the previous section (for example, to jump from c' to c), then it will be desirable simultaneously for all nations to do so, provided that they have all accepted the welfare function (7A.117). However, conflict still remains among the nations in that not all nations need become better off when the jump is made from c' to c. Some may be worse off in the new state, but if they want to live within the system they can only choose x_i to maximize W_i given $(x_1, \ldots, x_{i-1}, x_{i+1}, \ldots, x_n)$, and this automatically maximizes W.

By and large, n-nation situations satisfying (7A.116) are rare. Typically, nonidentical welfare functions will exist, and asymmetrical relations will obtain among at least one pair of nations. In these circumstances, it becomes difficult to specify the point of transition.

To consider one more important aspect in the analysis of the transition problem, introduce another behaving unit, the $(n+1)$th body. Imagine it to be a worldwide behaving unit, which we now designate the *world unit.* Suppose the payoff functions are modified to read

$$(x_1, x_2, \ldots, x_n, x_{n+1}) \rightarrow$$

$$\bar{W}_i(x_1, x_2, \ldots, x_n, x_{n+1}, v, \varepsilon), \quad i = 1, 2, \ldots, n, \tag{7A.119}$$

with the payoff function of the world unit being

$$(x_1, x_2, \ldots, x_n, x_{n+1}) \rightarrow \bar{W}_{n+1}(x_1, x_2, \ldots, x_n, x_{n+1}, v, \varepsilon). \tag{7A.120}$$

Consequently, setting

$$g_i = \frac{\partial \overline{W}_i}{\partial x_i} \tag{7A.121}$$

$$\frac{1}{\varepsilon} G_{n+1} = \frac{\partial \overline{W}_{n+1}}{\partial x_{n+1}}, \tag{7A.122}$$

the system described by (7A.102) and (7A.103) pertains, and once again we take the conditions (7A.104) to (7A.106) to hold. Then, for small ε, we might posit that the curve of Figure 7A.4 is modified, as depicted by the dashed curve in Figure 7A.11.

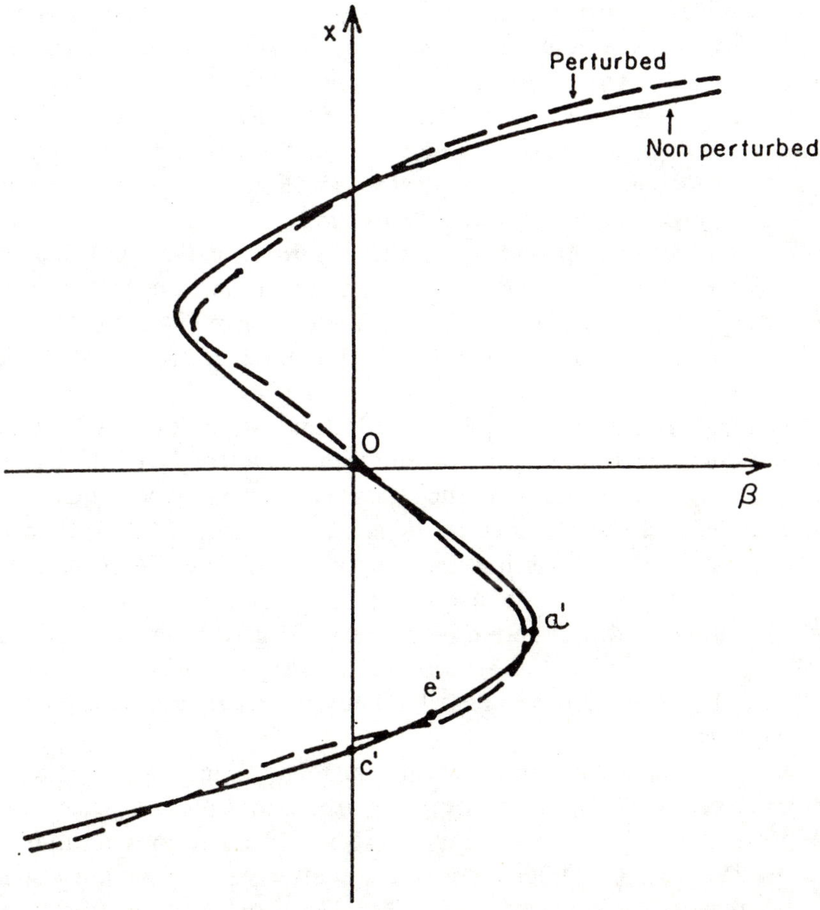

Figure 7A.11 Comparison of the variation of equilibrium states with β for a perturbed and nonperturbed system

If, at least for small values of ε, this is the only effect, and if the stability nature of the displaced branch is like that of the old, then the n-nation system is *structurally stable* under sufficiently weak coupling of the new unit. It might be, however, that for any $\varepsilon > 0$, however small, there exist critical values of v'—for instance, beyond e' in Figure 7A.11—where all points on the dashed curve become unstable and one or more branches of stable equilibrium states come to appear. A typical characteristic of these branches is that for any given v, the levels of $x_i^{**}(v,\varepsilon)$ $(i = 1,2, \ldots ,n)$ are significantly different from those on the dashed line, and in addition the level of $x_{n+1}^{**}(v,\varepsilon)$ is sizable [whereas the level of $\hat{x}_{n+1}^{*}(v,\varepsilon)$ is infinitesimal *along the dashed* line for small ε]. In this case the n-nation system is structurally unstable even under a weak coupling of the $(n+1)$th unit—a small parameter ε has large effects. So as the system crosses e', it needs to jump to a stable branch if it is to survive. If there is more than one stable branch, it will be necessary to select one of these branches to jump to, perhaps with the help of some cooperative procedure. If there is none when the system crosses e', it is driven to destruction.

Now consider $v \in M - M^S$. The phase portrait of (7A.100) does not change. However, for $v \in M - M^S$ it is no longer the case that any small structural perturbation of (7A.100) such as the system (7A.102), (7A.103) yields a phase portrait that looks like the phase portrait of (7A.100). Rather, the phase portrait of the system (7A.102), (7A.103) is radically different from that of (7A.100). Equilibrium states of (7A.102) and (7A.103) that are close to stable equilibrium states of (7A.100) become unstable, and new equilibrium states (branches) appear. Consequently, we can no longer use the structural relations of the system (7A.100) to model or approximate the equilibrium states of (7A.102) and (7A.103), and to provide insights or other information regarding that system. Moreover, the data pertaining to that system no longer can be used to estimate (fit) the structural relations of the 'ideal' system (7A.100). Of course, for (7A.102) and (7A.103) to be useful as a model, it must by definition have at least one stable branch.

Returning to our n-nation world-unit situation, we may speculate that in this more complex case, where new key interactions do take place due to the presence of the $(n+1)$th unit, there may be new possibilities for destabilization via multiplier and accelerator effects even though ε is small. For instance, if v just crosses d' in Figure 7A.4 (when v replaces β), we might conceive of some atypical nation wishing that the system were on the upper branch rather than the lower, because it would be better off on the upper (although the system would be worse off). If a perturbation were to occur in some other nation content with the existing situation, we may imagine that no multiplier-accelerator effects would occur, and so the perturbation would regress. On the other hand, if it occurred in a dissident nation, we may imagine there would be local multiplier and accelerator effects. The probability, however, is extremely low that there will be much, if any, spread of these effects to neighboring nations—though it is nonzero.

Next, imagine that the value of v changes further and gradually approaches that at c'. In general, we may speculate that the basins of the attractors along the lower branch steadily shrink as v changes—and much more rapidly than in the previous model, where the $(n+1)$th unit was not present. That is, as already indicated, at some point e' beyond c', but closer to c' than to a', the basin shrinks to zero as the system crosses e'. Put otherwise, the addition of the $(n+1)$th unit for any ε, however small, changes the nature of the equilibrium states so that everything beyond e' becomes completely unstable, even under the slightest fluctuation. In terms of the n-nation world-unit system, this suggests that as the system crosses e', a perturbation has a probability of almost unity of being located in a dissident nation, and further, there is a probability of almost unity that the local multiplier and accelerator effects will spread to the many surrounding dissident nations.

At e', the system might conceivably jump to the upper branch if that branch is still a set of attractors. But it also may be the case that in addition to the upper branch there are one or more other sets of attractors, so that the system may jump to any one of them—or collapse because of the absence of attractors for some location that the system attains. That is, after e' is reached, all the equilibrium states on the lower branch to the right of e' in the old system become irrelevant. They are unstable even under the slightest perturbation. Also, the upper-branch equilibrium states may be unstable or come to have reduced basins. At the same time, one or more new stable branches may come to exist and á la Prigogine it would be chance that would determine which one, if any, would be reached. Or it may turn out that no stable branches exist, in which case the system collapses.

Conceptually, the system may be still more generalized, as in Isard and Liossatos (1979), where in a sequence of stages, a new behaving unit is added at each stage.

Before closing this section, we note that there may be much more interaction among the nations than via the system welfare function just examined. The variable x_i, say i's level of military expenditures, specifying the position of nation i and taken to be its state, represents only one dimension of a possible n-dimensional policy space. Further, it does not consider i's geographic position in physical space. Nations can interact directly, as is well known in arms races where the level of i's military expenditure or other policy is directly related to the levels (positions, states) of other nations, especially its neighbors. At least in one important aspect, namely economic, a nation's welfare is affected by its geographic position through import and export of commodities, movement of capital, population migration and diffusion of ideas and technology. Such phenomena require a much more complicated welfare function. In a continuous but bounded geographic space the function might have a term such as $\gamma_i(\partial\Phi_i^2(z,t)/\partial z^2)$—where $\Phi_i(z,t)$ represents a stock, say of capital, at a location along a z axis in geographic space, where $\partial\Phi_i^2(z,t)/\partial z^2$ is a diffusion term representing the difference between inflow and outflow of capital or population or commodities at z, and where γ_i is a diffusivity constant. Incorporation of such a term in any nation i's welfare function could lead to all kinds of amplifications of perturbations, and unpredictable space-time patterns of the diverse nations' stocks being examined—patterns most likely to be inhomogeneous. Thus, too, the pattern of welfare among the nations and the system's welfare would be subject to all kinds of amplifications and unpredictabilities. For further discussion see Isard and Liossatos (1979), pp. 341–4. These observations hold also for systems composed of other types of interacting hierarchical subsystems having still more complicated interconnections and patterns of information flow and interdependent decision making.

7A.6 Summary

What have we accomplished in this Appendix? First we have presented a series of models which relate to the analysis of Chapter 7, each roughly sketched whose full statement appears elsewhere. Also, we have referred to a number of models which have not been sketched because of their highly technical nature and limitations of space.

We began with a dynamical system model of a single point (nonspatial) region. It presented a single rate-dependent process whose time-path from any initial point of time t_0 to any future terminal point t_l was fully determined once the initial state and a scrap value at the terminal point were given. It contained no element of control embodying any reflective, cognitive-based activity or interaction of any units involved. Conceivably, however, we might claim that setting a desirable and feasible scrap value involves such activity or interaction and implicit control.

Next, we did introduce implicit control by including an objective (performance) function in the model—one that could be set by a central planner or behaving unit. Such control, however, is not the rate-independent measurement-control activity of reality that we would like to incorporate into a model.

The third model embodied more of a measurement-control process. It introduced a coordinator, often designated an ombudsperson, who at his/her level of a hierarchy was concerned with system welfare. He/she attained such through imposing coordination inputs (in effect constraints representable by subsidies and penalties) upon two myopic optimizing regions (subsystems) at a lower level of the hierarchy which together with the coordinator comprised the whole. This model was a static one which, however, did illustrate the principle of optimum loss of detail. However, it did not really reflect the operation of a rate-independent control process.

Our fourth model added a dynamical systems framework to the previous one. But once again the control exercised by each of the two regions and also by the ombudsperson is a time path of instantaneous actions (decisions) where the coordinator establishes a time path of control inputs that is the same as the time path for these inputs which each of these regions most desires. No time is allowed for actual measurement, comparison and other cognitive activity of actuality.

The next set of models covered the reality of space, thus permitting analysis of the breaking of spatial symmetry as well as temporal. We began with a system of many point regions, a single decision-making unit, a central planner, and a single (composite) good which could be shipped between any pair of regions. There we also referred to a model with an optimizing central planner with many behaving units (consumers, producers and traders) in each region, each unit myopically optimizing, with many commodities, with investment and with limited government production and distribution. However, an ombudsperson is required to impose constraints on the central planner to insure that the distribution of goods to each consumer is in accord with the bundle of goods that the consumer would demand as a utility maximizer.

To be able to employ more powerful tools of analysis, we then sketched a continuous space model, one with many shipments of different commodities among the many pairs of points (locations). Next we consolidated the diverse flows into a single aggregate local variable, $U(x,t)$. This last step permitted us to develop a field description of spatial interaction to parallel the field descriptions that exist in physical models—for example electric current flowing through a transmission line, or a force acting on a spring—to make possible exploitation of some of their relationships. Such field models capture interaction of all the units in a system and are important in suggesting how the interactions of behaving units in a social system can be incorporated as a macro variable into social science models. In particular, corresponding to the local macro variable $U(x,t)$ [in an electric transmission line, $I(x,t)$], we can conceive of a macro variable ${}_iV(x,t)$ to indicate two types of participation by ombudspersons: (1) direct participation of all ombudspersons in the system in a decision or action to be taken at (x,t); and (2) direct participation of the ombudsperson at location (x,t) in the decisions (action) to be taken at all locations. However, these two types of participation are envisaged to have most significance for measurement-control processes, and in particular those processes which are reflective and cognitive-based. These types involve rate-independent processes that take place in ordinal time, and thus far we have not been able to develop a model which combines them with the rate-dependent processes and instantaneous decision making that we have covered.

With a space-time model of many interacting units and one or more central planners and ombudspersons, which is consistent with the breaking of both temporal and spatial symmetry, and thus with a non-hierarchical system, or one with two or more interacting levels in a hierarchical system, we attempted to introduce Prigogine leaps.

We set up a model of the growth of a population of a city having a welfare function reflecting the basic forces behind a city's population growth. Changes in the values of the parameters lead to a pattern of growth that corresponds to a Thom cusp phenomenon, as in Figure 7A.4. We ask: why, where and when does the jump occur from a lower to an upper branch? If we focus strictly on local stability conditions, the jump does not occur until the end of the lower branch is reached. However, if we allow noise—as characterizes reality from microscopic interactions—the jump may occur at all points when a corresponding position on an upper branch exists. We set forth reasons—in terms of the probability or nonprobability of a large enough perturbation and subsequent amplification at every point on the lower branch. (However we cannot identify a critical value as in the Bénard example when the situation is well defined—perhaps because as yet we are not able sufficiently to define a corresponding social system situation.) We may, for example, imagine that the probability of the amplification of a small or even large perturbation remains very small until some point as c' is closely approached, a point where the probability increases sharply to reach a maximum, and then falls off sharply beyond this critical point. In effect, within a small area around c' the probability of a jump to the upper branch may be taken to approach 100%—the corresponding probability measures at the relevant points being perhaps representable by a δ-function.

To handle a system under continual bombardment by noise (perturbation) at all (x,t), one can consider employing the Master Equation approach. This allows the analyst to set up a deterministic model for the expected or mean value of say a population. But this approach requires the assumptions of a Markov process (that the future is dependent only on the present and not the past) and that the mean square deviation of population from its equilibrium state be bounded with time and tends to a steady value. In general, these assumptions viciously violate reality and hence this Master Equation approach is not acceptable. Nor is the approach that additionally employs the use of a Fokker–Planck equation—an approach which examines the impact of rapidly fluctuating external noise (Brownian–type movement) from the environment. Nonetheless, whether we consider the noise additive or multiplicative, valuable insights are obtainable regarding the probability of major upswings or downswings, even extinction, of systems, gradual or sudden.

Lastly, we considered perturbations that introduced a new element (variable) into the system—a new commodity, a new technology, a new species, a new nation. Once again, we were able to speculate on conditions that might cause the system to take a Prigogine leap—and even over time to a sequence of Prigogine leaps leading to more and more levels (complexity) of a hierarchical structure—as in a multination system considered in Chapter 10. But such speculation did not produce new light upon the question of why, where and when the system takes such a leap (or degenerates into nondeterministic chaos or disintegration).

In sum, this Appendix adds precision to the insights of Chapter 7, sharpens some of the basic questions raised, suggests new research thrusts but does not provide answers.

Notes

1 In what follows we restrict u_1, u_2, y_1, and y_2 to positive values.

2 To simplify the mathematical statements, we assume a depreciation rate of zero. In view of previous discussion, the incorporation of depreciation of capital into the model is straightforward.

3 In view of the discussion in the preceding section, the introduction of an investment vector as a second control variable is straightforward.

4 If $\partial f/\partial C^J \to +\infty$ as $C^J \to 0+$ (for all $C^L > 0, L \neq J$), then (7A.69) is satisfied. If in addition the Hessian matrix is negative definite (implying strict concavity), then the equations (7A.69) define, for any given $p(t)$, a unique vector $C(t)$ that maximizes H.

5 Strictly speaking, the points (regions) in the sequence to be considered do not need to be on a straight line. We only require that there be a definite ordering of them and that the distance between any two points be the sum of the distances between the intermediate pairs of neighboring points.

6 In what follows, we often omit the word density when it is clear from the text that the magnitude referred to is a density.

7 By so doing we can obtain

$$\frac{\dot{K}^J(t)}{\Delta x^J} = \frac{Y^J(t)}{\Delta x^J} - \frac{C^J(t)}{\Delta x^J} + \sum_{K \neq J} \frac{s^{K \to J}(t)}{\Delta x^J \Delta x^K} \Delta x^K$$

$$- \sum_{L \neq J} (1 + \sigma^{J \to L}) \frac{s^{J \to L}}{\Delta x^J \Delta x^L} \Delta x^L.$$

8 Hereafter, whenever we have a function $(x,t) \to U(x,t)$, we shall adopt the convention that

$$\overset{x}{U}(x,t) \equiv \frac{\partial U}{\partial x}(x,t).$$

9 With our conventions, exports from x become $\int_x^B s(x,x',t)dx'$ and imports into x become $\int_0^x s(x',x,t)dx'$, so that we may rewrite (7A.82) as

$$\dot{K}(x,t) = F(K(x,t),x) - C(x,t) + \int_0^x s(x',x,t)dx' - \int_x^B s(x,x',t)dx'$$

$$- \sigma \int_{x'=0}^{x} \int_{x''=x}^{B} s(x',x'',t)dx'dx'')).$$

10 It should be noted that this model—a type of field description model—parallels a number of models of physical systems (see Isard and Liossatos, 1975). In particular, if we examine the parallel with the

system in an electric transmission line: $K(x,t)$ corresponds to charge density $\rho(x,t)$; $U(x,t)$ corresponds to $I(x,t)$ the electric current (a flow); l corresponds to L an inertial generating element; ξ corresponds to $1/C$ an elastic constant; v corresponds to R the coefficient of friction of the wires (electric resistance); and μ, the rate of capital depreciation (in eq. 7.A.8), corresponds to the loss coefficient G/C (charge density being lost to the insulator between the wires since no insulator is perfect).

11 Observe that still other welfare functions may be specified, involving perhaps higher-order terms, where each term has a rationale. For example, a welfare function may be constructed wherein agglomeration and deglomeration economies are disaggregated to correspond to different types of specific forces. One force may correspond to localization economies, a second to urbanization economies, a third to congestion diseconomies, a fourth to pollution diseconomies, and so forth. For each welfare function considered average and marginal contributions of population can be developed. This possibility of constructing welfare (or payoff) functions composed of many terms, one for each type of relevant force, allows us to generalize the analysis to cover all kinds of organizations (business, religious, social, political, multinational)—and even to cover individual organisms, including human beings, whenever a general welfare (potential) function can be hypothesized.

12 At the boundaries of the interval, two of the three roots obviously become identical.

13 Where $|x| = \sqrt{\alpha/3}$, we have $d^2W/dx^2 = 0$ but $d^3W/dx^3 \neq 0$; hence this state is unstable.

14 In the construction of Figure 7A.5 we have used the specific function

$$W = \frac{450{,}000}{2}(\bar{x} - 2000)^2 - \frac{1}{4}(\bar{x} - 2000)^4 + \beta(\bar{x} - 2000)$$

where $\bar{x} = x + 2000$ and represents the actual population. (Here we take the unit of x and $\bar{x}$ as 10,000 individuals.) Hence by comparison with (7A.89), $\alpha = 450{,}000$.

15 In this quotation, an increasing Gibbs function corresponds to a decreasing welfare maximum, and the words in italics 'equally attractive state' replace his 'even more attractive,' which we believe is more appropriate in the context of our discussion. Also, we have replaced his notation with that of our Figure 7A.4. All italics are our own.

16 A Wiener process is a discrete approximation to Gaussian White Noise.
17 For a formal and detailed exposition of the analytical framework, see Horsthemke and Lefever (1984).
18 See Isard and Lobo (1996).
19 Where $p_s(x) = \lim_{t\to\infty} p(x,t)$.
20 Equation (7A.111) is the analytical expression for the mapping

$$\hat{x}_i : \bar{A}(A_i) \to A_i.$$

The existence of (7A.111) is locally ensured if W_i is twice continuously differentiable on $A \times M$ and if

$$\frac{\partial^2 W_i}{\partial x_i^2}(x_1, x_2, \ldots, x_n, v) < 0$$

for all $x = (x_1, x_2, \ldots, x_n) \in A$ and any $v \in M$.

21 This means that the system (7A.110) must have a unique solution. That is ensured locally if the Jacobian determinant $(\partial^2 W_i / \partial x_i \partial x_j) \neq 0$.
22 This requires that the Jacobian matrix $(\partial^2 W_i / \partial x_i \partial x_j)$ be negative definite at $(x^*(v), v)$.

References

Allen, P.M. et al. (1981), *Models of Urban Settlement and Structure as Dynamic Self-Organizing Systems,* Report to the U.S. Department of Transportation, Washington, D.C., mimeographed, February.

Callen, H.B. (1960), *Thermodynamics,* New York: Wiley.

Doering, C. R. (1991), 'Modelling Complex Systems: Stochastic Processes, Stochastic Differential Equations, and Fokker–Planck Equations,' in Nadel, L., and Stein, D. L. (eds.), *1990 Lectures in Complex Systems,* Redwood City, CA: Addison-Wesley Publishing Company.

Granger, C. W. J. and Teräsvirta, T. (1993), *Modelling Nonlinear Economic Relationships,* New York: Oxford University Press.

Horsthemke, W. and Lefever, R. (1984), *Noise-Induced Transitions: Theory and Applications in Physics, Chemistry and Biology,* New York: Springer-Verlag.

Isard, W. (1976), 'On Regional Science Models: Parallels from Biological Science,' *Papers,* Regional Science Association, vol. 36, pp. 77–85.

Isard, W. (1978), 'Strategic Elements of a Theory of Major Structural Change,' *Papers,* Regional Science Association, vol. 38, pp. 1–14.

Isard, W. and Smith, T., et al. (1969), *General Theory: Social, Political, Economic and Regional*, Cambridge, MA: MIT Press.

Isard, W. and Liossatos, P. (1975), 'Parallels from Physics for Space-Time Development Models: Part I,' *Regional Science and Urban Economics*, vol. 5, no. 1.

Isard, W. and Liossatos, P. (1977), 'Models of Transition Processes,' *Papers*, Regional Science Association, vol. 39.

Isard, W. and Liossatos, P. (1978), 'A Formal Model of Big Step Disarmament and Domino Effects,' *Journal of Peace Science* (now *Conflict Management and Peace Science*), vol. 3, no. 2, pp. 131–45.

Isard, W. and Liossatos, P. (1979), *Spatial Dynamics and Optimal Space-Time Development*, New York: North Holland.

Isard, W. and Lobo, J. (1996), 'Impact of Noise on Regional Development,' forthcoming.

Kahn, D. (ed.) (1981), *Essays in Societal Systems Dynamics and Transportation*, Report of the Third Annual Workshop in Urban and Regional Systems Analysis, U.S. Dept. of Transportation, Washington, D.C.

Kaplan, D. (1995), *Understanding Nonlinear Dynamics*, New York: Springer-Verlag.

Kitahara, K. (1975), 'The Hamilton–Jacobi–Equation Approach to Fluctuation Phenomena,' in G. Nicolis and R. Lefever (eds.), *Advances in Chemical Physics*, vol. 29, pp. 85–111, New York: Wiley.

Lasota, A. and Mackey, M. C. (1994), *Chaos, Fractals and Noise*, New York, Springer-Verlag.

Levins, R. (1973), 'The Limits of Complexity,' in H. H. Pattee (ed.), *Hierarchy Theory*, New York: Braziller.

Malek-Mansour, M. and Nicolis, G. (1975), 'A Master Equation Description of Local Fluctuations,' *Journal of Statistical Physics*, vol. 13, no. 3.

Mesarovich, M. D., Macko, D. and Takahara, Y. (1970), *Theory of Hierarchical, Multilevel Systems*, New York: Academic.

Moon, F. C. (1992), *Chaotic and Fractal Dynamics, An Introduction*, New York: Wiley.

Nicolis, G. and Prigogine, I. (1989), *Exploring Complexity*, New York: W. H. Freeman.

Osaki, T. (1985), 'Non-Linear Time Series Models and Dynamical Systems,' in Hannan, E. J., Krishnaiah, P. R., and Rao, M. M. (eds.), *Handbook of Statistics, Vol. 5*, New York: Elsevier Science Publishers B.V.

Pattee, H. H. (1973), 'The Physical Basis and Origin of Hierarchical Control,' in Pattee, H. H. (ed.), *Hierarchy Theory*, New York: Braziller.

Prigogine, I. and Stengers, I. (1984), *Order Out of Chaos*, London: New Science Library.

Risken, H. (1989), *The Fokker–Planck Equation*, Berlin: Springer Verlag.

Riste, T. (ed.) (1981), *Nonlinear Phenomena at Phase Transitions and Instabilities*, New York: Plenum Press and NATO Scientific Affairs Division.

Simon, H. (1962), 'The Architecture of Complexity,' *Proceedings of the American Philosophical Society*, vol. 106, no. 6, pp. 462–82.

Turner, J. S. (1975), 'Finite Fluctuations, Nonlinear Thermodynamics and Far-From-Equilibrium Transitions Between Multiple Steady States,' in Nicolis, G., and LeFever, R. (eds.), *Advances in Chemical Physics*, vol. 29, New York: Wiley.

8 Evolution of Western painting

8.1 Introduction

Having concluded our examination of areas in the physical, biological and social sciences with a number of unanswered questions and directions for further research, we now proceed in the next set of chapters to study the evolution of selective aspects of the social system. Is there insight obtainable from such study that can help answer some of these unanswered questions and identify more precisely and even move along one or more desirable research directions?

In this chapter we will concentrate for the most part on the evolution of painting in Western society. We do so because the historical record of painting in this society is rich in stages of development and is as complete as that in any other society. While the evolution of art must also consider the record of architecture, sculpture, music and other activity that can be characterized as art, we confine ourselves in this chapter to painting to avoid the difficulties of generalization over different media and modes. However, in Chapter 10, we do present supplementary material on architecture, sculpture and music.

By and large, the evolution of painting in Western society has had its ups and downs and sidetracks, and even at times has suffered retrogression, at least in certain aspects. Yet over a broad enough span of time, it may be said to have been cumulative. The ideas, aims, media, techniques, shapes and forms of the past may be found in one way or another in the present. Art of course always evolves in a political, social, economic and cultural environment consisting of beliefs, rituals, magic, myths, science and so-called objective truths, ethics and moral principles, ways of thinking, and traditions. At most any stage of history it conditions this environment, often by looking into the future. In particular, it conditions through communicating messages—messages concerned with eliminating fears and anxieties, with recording truth and falsehood and good and evil, with decrying social conditions and establishing moral codes, with depicting both beauty and ugliness, and with portraying both tragic implications of man's actions and man's potentialities. But at the same time art is conditioned by the environment. For example, it is restricted by the availability of materials and tools and traditions and social customs—and this is so whether we consider cave-man art at one extreme, or the highly abstract paintings of today at the other.

We begin with cave-man art. Then we trace the subsequent history of Western painting, through Egyptian art, the Greco-Roman Classical-Hellenistic period, the Medieval Age, the Renaissance era and the flowering of painting therein, the post-Renaissance relative stasis, and finally the revolution which has been dubbed surging individualism unbounded. We pay particular attention to developments leading up to high points—which à la Prigogine we might designate far-from-equilibrium situations. Closing remarks are contained in a summary section.

8.2 Cave-man and other prehistoric art

As all art, primitive art may be said to be the outcome of mutual interaction between human beings and their environment, and to involve explicit messages and communications. This art has been considered by some to be of extraordinary caliber, many of the images being skillfully and beautifully executed, particularly in their naturalistic representation. In referring to the painting of a bison, de la Croix et al. (1991, p.34) state: 'not only do the outlines have the elastic strength and fluency that we find in Chinese calligraphy and brushwork, but also the tone is so managed as to suggest both the turning under of the belly of the pregnant animal and the change in the color of the coat.'

Among scholars there is little agreement on the purpose or message of this art—what it aimed to communicate to behaving units, including the artist himself, or some unknown godly-type power. It might have been to: work magic or serve as a religious rite or spiritual action; seek good luck or appeal to some unknown power for support in hunting; express a wish (for example, to kill a prey when a spear is shown having pierced a bison); or provide a feeling of security; or explore the medium of painting to see what vision of the vitally important world one could reproduce; and/or enjoy one's creation. Or the painting may have served to satisfy the artist's curiosity to see whether he could set down on a wall what he had seen, the characteristic of curiosity, the exploration into the unknown, having always been an attribute of a 'brain-endowed'man.

Despite all these possible purposes, in general prehistoric art, as most pre-Hellenic, is alive and oriented to action. Dorner (1958) maintains that the art was 'pragmatically effectual' and as little concerned with beauty, calmness and other attributes that take one beyond the vicissitudes of action. He considers the mental processes of the prehistoric and pre-

Hellenic mind, desires and creations to be directed by and large by sensory experiences. The smell of the bison, its taste, its movement provoke the painting. Nonetheless, the art itself reflected the state of the society, the tools at its command, and the knowledge it possessed. In the early years of this society's painting, a single line or set of lines depicting a bison or other animal or object was sufficient to convey the message. Though the art in a formal Euclidean sense had two dimensions (x and y), it was two dimensional only minimally. Often it was primarily the movement of a single brush or other tool in its early stages—to produce the vision of an animal or other object. Sooner or later, of course, primitive man must have become aware of processes—how hunters in a team could cooperate to pursue a prey (as depicted in the painting scene at El Cerro Felio). He thus commenced to depict aggregates of sensual images—aggregates involving a schemization, a way of putting together symbols which was necessarily a manifestation of mental activity beyond pure sensual reaction.

8.3 Egyptian advances

During the period extending from the era of cave man (the hunting society) to that of the agricultural river valley civilization, the record of painting is very incomplete. The transition that occurred may have been gradual or abrupt. One may be inclined to accept the point of view of the historian Artigani (1981), who employs notions of the Prigogine school and considers the transition as a leap to a new, stable, relatively far-from-equilibrium situation. Details on this transition will undoubtedly be filled in by future archaeological finds. At this point, all one can do is to present the characteristics of painting in river valley civilizations and then examine the continuous record thereafter. Unfortunately, the only significant store of painting that has survived from this period is that of the Egyptians—buried in their tombs. Little remains of Mesopatamian and other painting.

To understand Egyptian art, one needs to have in mind the transition from a nomadic to a communal life. By and large food gathering replaced food hunting; animal breeding took the place of animal chasing.

In comparing Egyptian art with that of cave man, we observe a shift

> from naturalistic or direct imitation of nature to a more geometric art based on formal principles and traditional conventions. Concrete images yielded to abstract forms, naturalism to stylization, imitation to idealization, the actual to the metaphorical. In short, the rendering of reality was replaced by accepted conventions. These are the polar extremes in relation to which the art of all following periods tended to express itself—the faithful portrayal of the natural appearance and 'speaking' likeness on the one hand, and the conceptual, geometric, stylized conventions of the established formulas on the other (Fleming, 1986, p.10).

Egyptian art from early on was extremely inflexible—as if by formal God-given law novelties were precluded—achieving a consistency and permanence rarely matched in Western art. More concretely, the Egyptian artist, as seen in tomb paintings, limited himself to the picture plane, with no attempt at capturing depth, or striving for a three dimensional presentation of a figure or object, or even a two-dimensional one against a background. Moreover, there is no evidence of his seeing time as a dynamic force. His painting aimed not at prettiness but at presenting in a clear, permanent and as complete as manner as possible, the knowledge he possessed. Thus heads are drawn in profile, while eyes had to be represented front face if knowledge about a person's eyes was to be shown.

> The torsos are frontal, but the arms and legs offer a side view. And though the figures are shown from the right side, they have two left feet so that both big toes are toward the front. If a pool or river is included in a landscape, the view of it is from above, but fish, ducks, plants, and trees in and around it are shown sideways. Important personages always appear larger than their families, followers, or servants. Once these conventions are taken for granted, the scenes appear remarkably lifelike. Naturalistic detail is rendered so accurately that botanists and zoologists can recognize each separate species of plant and animal life. The Egyptian artist also knew how to portray the fur and feathers of animals and birds by breaking up the color surfaces with minute brush strokes of various hues (Fleming, 1986, p.15).

Egyptian art of course reflected a much more advanced level of thinking and analysis based on many more years of experience in conquering a harsh environment and technological development. The aggregates of elements used by prehistoric man to depict a process now gave way to conglomerates of them to portray the much more complex interrelations of elements in a society—and in ways that were more complete and often more precise. Belief in many gods and the hierarchical structure of power played dominant roles in the painting which was still strongly action-oriented and colored by mysticism.

The iron bar of convention was, however, shaken during the reign of the pharaoh Akheniton. He set in motion events that might be viewed as a Prigogine-type perturbation with some amplification, but not enough to lead to a lasting major change in the society and art style. (Akheniton also adopted monotheism, rejecting the many-god mysticism.) The unfinished bust of Nefertiti, his queen, presents the living likeness of a real personality exuding genuine human warmth. There also has come down to us a portrayal of Akheniton himself, informally fondling his baby daughter, and offering Nefertiti a flower while she lovingly holds on her lap two infant princesses. This humanistic informal development carried over only briefly into the following reign of Tutankhamen.

8.4 Art of the Greco-Roman, Classical-Hellenistic period

Pre-hellenistic Greek art was essentially an outgrowth of influences emanating from Mesopatamia, Egypt and other Mediterranean cultures. Changes that took place were of a gradual form. However, starting with a few years just before the 5th century BC and continuing through that century, one observes a sudden major transformation in Greek Art—a Prigogine jump to another far-from-equilibrium state. The repulsion of Xerxes and the Persians (479 BC) was one of the set of perturbations (but just one) that was amplified in the Greek society—an amplification that constituted the Golden Age and the flowering of Athenian democracy. One of the most rapid, if not the most rapid change occurred in Western art. The classical style that evolved was to dominate Western art for a thousand years or so. There were three crucial elements in this revolutionary development. One was the Greek idea of freedom that led to autonomy in the depiction of the human form, and raised questions regarding free will, rationality, the nature of civilization and barbarism, the realm of reasoning and science, and the role of politics, government, and science.

A second crucial element of the new, classical style was the mythic dimension in which gods and godesses as well as humans were involved. Emanating from this dimension were themes which were to continue for a long time to be a rich source of meaning. This dimension, though undoubtedly present in the lost painting of this era, is best seen in Greek architecture and sculpture, in the Parthenon and the statues of Gods and Goddesses 'with their beautiful bodies, richly clothed and finely coiffured, their heads adorned with splendid diadems, their mythical personalities symbolized in their poses and attributes' (Cole and Gealt, 1989, p.x).

The third crucial element was its humanism, in particular its tragic vision of life . . . 'the portrayal of the inevitability of human suffering . . . unusual in the public art of other cultures. But in Western art, pagan and Christian, the concept that suffering brings wisdom is one of the great themes and it is derived from the Greeks' (Cole and Gealt, 1989, p. x).

While the Greeks excelled in painting as well as architecture and sculpture—and perhaps their painting was even of superior quality—little has survived. No large scale works in fresco have come down to us, although remains of later Roman wall painting provide some indication of what Greek painting must have looked like. Fortunately, numerous important paintings are on vases and other ceramics found in archaeological digs. The many splendid objects recovered portray in living form images of men and gods, their mortal and divine tragedies, their humanism, spirit, feeling and emotions. An outstanding example is the

KRATER c.515 BC by Euphronios, a painting on a vessel used for mixing wine and water. It 'is an example of the monumental nature of the subject matching the grandeur of the vessel's shape' (Cole and Gealt, 1989, p. 11). Somber action involving animated figures is depicted in a scene 'raised to a level of seriousness and pathos seldom equalled in Western art' (Cole and Gealt, 1989, p. 11).

Further, there is evidence that the legendary Athenian painters Polygnotus and Appollodorus had respectively: (1) started to work out the principles of perspective painting (a non-amplified perturbation which preceded that of Giotto in the 14th century); and (2) began to successfully exploit the use of light and shade and the finer gradations of color (another perturbation that died out).

At this point, it is instructive to consider the views of Dorner regarding the general contribution of Greek painting to the evolution of this art. Dorner claims that up to the classical age, art was concerned primarily, if not entirely, with sensuous changeability, becoming increasingly sensuously aware of the growing complexity of life—a result of accumulated knowledge and technological advance. When during the classical 5th century BC the Greeks introduced into their art a combination of rationality and humanism (so evident in the seriousness, somberness, and pathos of the animation, and the quality of expression noted above), they captured a third dimension. This dimension served to supplement the x and y (2-dimension) scheme of archaic art which is so explicit in the rigid traditional Egyptian painting. As Dorner states, this third "dimensionality is only possible through the polaric antithesis of spiritual Being and sensuous Becoming, and this again is the essence of rational thinking which relates all sensory changes to one idea as their cause. So, the birth of rational thinking asked for a new x-ray-like vision that saw form through changeable images, i.e. that saw 'beauty' and created 'works of art'" (Dorner, 1958, p. 56).

While in the discussion to follow we will reserve the notion of a third dimension to be used to designate depiction in the physical material sense, the forceful statement of Dorner represents another perspective on the dramatic Prigogine jump that occurred in the classical period of Greek art and was preserved during the Hellenistic era that followed.

8.5 Medieval painting

The Hellenistic post-classical period that saw the diffusion and spread of Greek art to all parts of the empires of Alexander the Great and Rome did not involve further significant development in painting. Though the Romans achieved outstanding engineering and architectural feats, and though the surviving Roman frescoes do have some non-ideal real features, the Roman ruling class, conservative in character, did not encourage innovation and exploration with new stylistic forms.

With the barbaric invasions and disintegration of the Roman empire, the conviction that ultimate reality was accessible to the rational intelligence of man was overthrown. Moreover, the notion of ideal (perfect and beautiful) forms of the gods and goddesses, athletes and other human figures and objects—which perhaps were initially designed to appease the gods and goddesses, to seek their assistance—confronted reality. Out of reality—its ugliness, its lack of grace, the omnipresence of imperfections, and its vicissitudes—came the forces that shaped medieval art, the one aspect of Greek art that survived being the limited religious spirituality that it possessed. These forces were strongly influenced by the emergence of Christianity. During the later years of the Roman empire, Christianity had grown significantly in acceptance, and by 313 AD it was legalized. Thus after the collapse of the Roman empire, and with the continuing crumbling and disintegration of the imperial administrative structure, Christianity, which had survived, came to fill as best it could the administrative power function. It came to restore order and impose a new code of law. With its spiritual message it provided the bonds that held together the former lands of the Western Roman empire while dominating the Byzantine (Eastern Roman) empire until the fall of Constantinople in 1453. Christianity became the force that provided hope and security for the individual in the world of great uncertainty, soothed his anxieties, and enunciated simple principles for escaping from the greed and evils of the day and for marking out a path to security and the good life.

The dominance of the Christian message is clearly depicted in the painting of the period. While bits of the classical style remained and were fused with the visualized, abstract concepts of the so-called barbaric Celts and Germanic tribes of the North, the message was antimaterialistic and highly ritualistic. In early Christianity 'the material world, including man, had no meaningful place in the system of representation that was intended to make the elements of faith visible to the faithful. Indeed, humanity's place was seen to be so low on the hierarchical ladder that representations of actual people ceased to matter' (Cole and Gealt, 1989, p. 38). Man became inconsequential for depiction. Gone was the beauty, the ideal form and rationality of the Classical Age. However, expression and humanism did remain, albeit particularistic, in the humility, compassion, suffering and hope of the new, religious message. While the representation of scenes with the presence of people could not be discarded, all knowledge about physical form and spatial arrangement could be and were in the overwhelming concentration on the message. See for example, the mosaic the *Miracle of the Loaves and Fishes,* c.504 AD, which lacked the precision present in previous Roman mosaics.

In effect, the painting may be said to have become one-dimensional in message à la Dorner. Outlines of most everything were crude and just enough to convey the message. To employ a concept of biological evolution theory, a process of paedomorphism—a term used to indicate retreat of a species from overspecialization—may be said to have taken place. Especially during the early Christian era the escalating administrative incompetence of the Roman empire and the steady dissolution of its social and political bonds led to major deterioration in the environment for artistic performance. One might contend that art had to take a step backward in order to evolve anew in a more fruitful and energetic direction. It had to fall back upon the spiritual, mystic dimension that had been, according to many scholars, one of the basic props in the development of cave man art and had always been present—a dimension that Christianity as it emerged most abundantly provided, more explicitly and forcefully than ever before.

In time, it may be said that the art of painting gradually progressed again. Forward movement in a varied, multi-dimensional form took place. Facial expressions, harmonious arrangements of people and objects reappeared to make the basic religious message more clear and effective. During the late monastic Romanesque period (11th to 12th century) knowledge and objectivity became increasingly incorporated, leading into some of the great Renaissance works, especially after spatial perspective was added. During these later Middle Age years, one finds reconciled the classical (Greek) and nonclassical (Celtic and Germanic) roots of Western art—art, however, still rigidly oriented to the hereafter of heaven, although man *per se* starts to take a more significant role.

8.6 The Renaissance era: the flowering of painting

With the growing economic development and wealth from trade of the Italian city states, the way was opened for a tremendous flowering in painting. Already in the 13th and 14th centuries, innovations were being introduced into religious art. Perhaps the greatest innovation was the incorporation of a third (fictitious) dimension (the dimension to accompany the x and y) in painting—a dimension that yields the illusion of depth on a flat surface, a dimension that is associated with spatial perspective. At first, it was a perturbation, as seen in the classic work *Madonna Enthroned,* c. 1310 by Giotto. This perturbation, however, was to be amplified. Brunelleschi discovered the mathematical laws of perspective, and in Masaccio's *The Holy Trinity, The Virgin, St. John and Donors,* 1427, we have the first painting with accurate perspective, a painting that conveys the direct impression of a hole in the wall.

In one sense, the achievement of scientific perspective might be viewed as the attainment of a new level of stratified stability à la Bronowski. Once achieved it remained in the tool kit (the stock of knowledge) of the painter—not necessarily to be explicit in an artist's work, but always available for use, to be discarded when required to realize the artist's desired effects. Like many innovations, this perturbation spread rapidly through the surrounding environment, became amplified and lifted the world of the artist to a new far-from-equilibrium stratum.

Along with perspective there emerged in the glorious days of the 15th century Italian Renaissance: a reinvigoration, magnification and reinvention of classical tradition including its mythology; a return to naturalism and the worship of the ideal form; and a new dignified humanism which involved an even finer meshing of the classical and nonclassical strains in Western civilization.

The earliest indication of an emergence of realism and humanism was perhaps in the works of Cimabue (his *Crucifex*, c.1290) followed by Duccio Di Buoninsegna (his *Maestà Altarpiece,* c. 1311) and others. Their works together with the early sculpturing of Ghiberti and Brunelleschi set the stage for the scintillating Italian Renaissance epoch. In the paintings of Masaccio (*Tribute Money,* c.1425), Fra Filippo Lippi (*Coronation of the Virgin,* c.1440), Fra Angelico (*Descent from the Cross,* c.1434), and other Florentine artists during the 15th century, man along with sacred figures was ennobled. In this art no longer did God alone control man's destiny as in medieval thought. Man, too, could plan and with effort nurture his own individuality and progress. The future can be determined, at least in part, by his own powers and can be bright and glorious. Man need not be pessimistic; misery and suffering can be avoided. Portraits (*The Duke and Duchess of Urbino*, after 1474, by Piero Della Francesca) came to express his individualism. In Venice and North Italy, too, the joint interplay of the sacred and secular was depicted, but there, it was by Bellini (*Saint Francis in Ecstasy*, c.1485) and his followers in a less mystical and more poetic fashion and with more roots in nature, landscape and the material environment.

While the Early Renaissance in Italy was gradually blossoming, up North in the early 15th century great achievements were realized in the work of the two Flemish masters, Jan van Eyck and Rogier van der Weyden. Nature, people and light were being depicted in an amazing detailed manner—meticulous, precise and comprehensive. To them, the material world was a manifestation of God's work. In the former's *Giovanni Arnolfini and his Bride*, 1434, one finds an honest, frank, and most humane depiction of a real life event, implicitly overflowing with emotion, and undertaken with infinite care and patience—an illusion of nature being achieved by the addition of detail upon detail. In van der Weyden's masterpieces *Descent from the Cross,* c. 1435, and *Entombment*, c.1450, manifestations of devotional art that was also present in the North, we see how the intense realism of van Eyck is employed to achieve dramatically great emotional depth—painting so effective in the use of color and light. Also important is the *Portinari Altarpiece*, c.1476, by Hugo van der Goes, which was admired by the Florentines and which increased the extent and intensity of interchange of innovations between Italy and the North.

The last years of the 15th and the first half of the 16th centuries saw

Western art reach dazzling heights. Dürer and Grünewald in the North together with High Renaissance titans, Leonardo, Michelangelo, Raphael and Titian among others, constituted an assemblage of genuises, perhaps never before nor subsequently equalled. Perhaps Leonardo's *Last Supper,* c.1495-98 (upon which we comment on p. 239) and Michelangelo's *Sistine Chapel,* 1508-12 are the most distinguished of the resulting masterpieces. Of the heights of harmony attained by Michelangelo, Cole and Gealt (1989, p. 146) write: The ceiling of Michelangelo's Sistine Chapel

> is essentially an unmatched series of figure studies. Organized by the simulated architecture that divides one area from another, the figures embody both grand and simple ideas, acting in some areas as merely decorative devices and in others as the actors in a great biblical drama. The prophets and sibyls, for example, express their serious purpose through their dynamic, monumental presence and vivid expressions. Whether seated or standing, Michelangelo's figures are bursting with energy and movement. Seemingly carved out of paint, they command the space around them . . . Though tracing the Fall of Man from Genesis, the ceiling in fact celebrates the rise of man, whose beauty and perfection are gifts of God. That idea is best expressed in the *Creation of Adam.*

There we see as Gombrich (1989, p.235) writes:

> Adam is lying on the ground in all the vigour and beauty that befits the first man; from the other side God the Father is approaching, carried and supported by His angels, wrapped in a wide and majestic mantle blown out by the wind like a sail, and suggesting the ease and speed with which He floats through the void. As He stretches out His hand, not even touching Adam's finger, we almost see the first man waking, as from a profound sleep, and gazing into the fatherly face of his Maker. It is one of the greatest miracles in art how Michelangelo has contrived thus to make the touch of the Divine hand the centre and focus of the picture, and how he has made us see the idea of omnipotence by the ease and power of this gesture of creation.

Though Leonardo and Michelangelo may be considered the greatest of the great, major achievements of others were realized. Dürer, considered by some to be the German counterpart of Leonardo, with loving patience and meticulous care brought to perfection, in the tradition of van Eyck, the turn of Gothic art towards the imitation of nature, yet not without a search for ideals of proportion. See his *Hare* (1502) and *Adam and Eve* engraving (1504) wherein he struggles for a convincing vision of a sacred story. His contemporary and equal, Grünewald, applied remarkable psychological insight and expressive power in depicting the human condition; see his religious masterpiece *Crucifixion*, c.1510-1515. And perhaps within the category of titans one might include Brueghel the Elder who somewhat later established the genre scene as a basic category of painting; his *A Country Wedding,* c.1565 depicts insightfully the unvarnished reality of people feasting, dancing, suffering and experiencing grief and other emotions as part of nature.

This reality of the North was in great contrast with the idealistic, mythic proportions, beauty and perfection of the South. In beauty there has been and was no equal to Raphael. What can match his Madonnas (see the *Madonna del Granduca,*c.1505 for classical perfection); and in his *The Nymph Galatea,* c. 1514, there is attained an amazingly balanced arrangement of a diversity of freely, moving figures—a balance that previous generations had never been able to achieve.

Titans, too, were in the North of Italy. Titian boldly departed from established conventions; and he was able to do so and still attain harmony through welding things together by the way he contrived light, air and color. See the sumptuousness of his coloring in his altarpiece *Assumption of the Virgin,* 1516-18. As Honour and Fleming (1992, p.142) remark:

> he was perhaps the greatest Renaissance painter of all. He had none of Leonardo's scientific interests or Michelangelo's religious and poetic ones. Nor was he, like Raphael, an architect as well. It was Titian, however, who became in a real sense the founder of modern painting, for he made oil colour on canvas the main medium of later Western art. He exploited all its possibilities for the first time, from the animation of the picture surface with vigorous brush strokes on the rough texture of the canvas to the contrast between rich, creamy highlights thickly laid on and deep, dark hues delicately modulated with glazes.

Among the titans, one might also be inclined to include Tintoretto, although he may be said to belong to the following period. He combined the coloring innovations of his master Titian with the draftsmanship of Michelangelo to create paintings of striking originality in their sharpness, irrational perspective, and drama. See his *Removal of the Body of Saint Mark,* 1562.

8.7 Post-renaissance stasis

Once the heights of the Renaissance were attained—to which Cranach, Bosch, Bellini, Giorgione, and others contributed—a period of relative stasis set in. It was not that there were no great painters. Rather it was a plateau period, relatively speaking—a period of a number of noteworthy achievements (refinements as perturbations)—but not one of ushering in an amplification of a significant perturbation—a period when the extremely impressive achievements of the Renaissance were by and large taken as the way of art until late 19th century.

There was the time of Mannerisms during the middle and late 16th century—a period characterized by attempts to outdo the past masters, to seek excessive refinements, to do so by experimenting in new, startling strange and sophisticated ways. The works of Tintoretto are often assigned to this period, and certainly those of Correggio (*Jupiter and Io*, c.1532), Parmigianino (*Madonna with the Long Neck*, c.1535), Bronzino (*Portrait of a Young Man*, c.1550), El Greco (*The Opening of the Fifth Seal*, c.1610) and other intense, deeply emotional religious art with figures elongated to achieve maximum tension.

No significant amplification of perturbations were realized in the 17th century, although this century ushers in many artists of high standing producing paintings of signal character. There are: the religious works of Caracci, Caravaggio, Reni, Bernini and Gaulli in Italy; the landscape paintings of Poussin and Lorrain in France; the works of Velasquez and Murillo in Spain; and most importantly the varied compositions of the Flemish and Dutch, which Cole and Gealt (1989) consider to constitute a mini-Golden Age. In the last group there were the glorious works of Rubens, about whose many improvements over the High Renaissance Fleming (1986, pp. 304-5) writes:

> Rubens succeeded in combining the rich color of Titian and the dramatic tension of Tintoretto with an unbounded energy and physical power of his own. His conceptions have something of the heroic sweep of Michelangelo, though they lack the latter's thoughtfulness and restraint. His complex organization of space and freedom of movement recall El Greco, but his figures are as round and robust as the latter's were tall and thin. His success in religious pictures, hunting scenes, and landscapes, as well as the mythological paintings that suited his temperament so well, shows the enormous sweep of his pictorial powers. For sheer imaginative invention and brilliant handling of the brush he has rarely, if ever, been equaled.

The unrivalled giant of this Flemish/Dutch Golden Age was Rembrandt. He admired and collected the works of other artists and often borrowed from them, but went well beyond them. As Fleming (1986, p.315) states:

> Towering above all other Dutch painters because of the breadth of his vision, the power of his characterizations and the uncompromising integrity of his ideas, stands the figure of Rembrandt van Rijn in lonely eminence. Like his contemporaries, Rembrandt painted portraits, genre scenes, historical subjects, and landscapes, but unlike them he refused to specialize and succeeded magnificently in all. Furthermore, he brought a new psychological depth to his portraiture,

> an unaccustomed liveliness to his genre scenes, a greater dramatic intensity to his religious pictures, and a broader sweep to his landscapes than had been achieved before in the Northern tradition.

Above all, Rembrandt went beyond Titian by his

> discoveries of the power of light, in all its varying degrees, to illuminate character both from without and from within, to define space by varying degrees of brightness, and to give life to that space by the flowing movement of shadows (Fleming, 1989, p. 315).

In brief, although Rembrandt's work brought painting to heights greater than ever before, his works like Rubens' did not constitute a qualitative leap nor did they when viewed together with the works of his 17th century predecessors and contemporaries: Hals, Vlieger, van Goyen, van Dyck, Steen, Kalf, Ruisdael and Vermeer.

The slowly ascending level of painting with small outbursts of achievement that might be said to characterize painting during the last half of the 16th and the 17th centuries continued, perhaps with slower forward movement, during the 18th century. There was the especially graceful, delicate and exquisite work of Watteau, and fine pieces by Boucher, Fragonard, and Clodion, other rococo painters. In England, Hogarth, Reynolds and Gainsborough produced works much more oriented to ordinary mortals and in keeping with reason as did Chardin in France; and in Italy, Tiepolo and Guardi excelled in frescoes and presentations of views. Cole and Gealt sum it up: 'By turns logical, intuitive, polemical, impulsive, brilliant or pedestrian, the art of the eighteenth century built on and exaggerated the tendencies of earlier periods. It became more refined, more delicate, more sensuous, more intellectual, more emotional, and more secular' (Cole and Gealt, 1989, p. 193). No major breakthrough or amplification occurred.

8.8 Revolution: surging individualism unbounded

It was during the late 18th and 19th centuries that a series of breaks in tradition took place—reflecting the accumulation of a number of social, political, economic and intellectual developments. The onset of the French Revolution in 1789, exhibiting major discontent with traditional ways, emancipated the artist from the social, political and other pressures to pursue accepted practices. Later, the Industrial Revolution disrupted the economic position of the artist. No longer did the economic order place highest value on solid pieces of craftsmanship; rather mass production of standard products by a factory system came to dominate. These forces and others disrupted the traditional craftsman-to-client relationship through which an artist could find means of livelihood and/or gainful employment—and in particular, the tie of the artist to a patron. This introduced considerable insecurity into the life of an artist. He was on his own and had to find a market for his product. At the same time he was released from the demands and whims of a patron or client. The result was that the artist became largely unrestricted in the pursuit of his own directions. He could go beyond logic and allow his subconscious state of mind and his emotions to guide his work. He could criticize conventions fearlessly and discard them at will. He could explore wild, fanatic and even fantastic depictions. In brief, he could express him/herself as a fully unconstrained, creative individual. As a consequence of this greatly enhanced potential for individuality and the subsequent breaks in tradition that occurred, a great diversity of styles emerged in the ensuing decades of the 19th and 20th centuries.

A first clear break was in the paintings of David, the so-called official artist of the French Revolutionary Government, although some art historians would begin with the unconventional works of Hogarth and his deliberate attempt to create a new style for a new public. However, one can claim that Hogarth belonged to the Age of Enlightment preceding the French Revolution.

Following his own instincts and emotional drives, David chose to paint heroic-type scenes and in doing so, draw upon Greek and Roman classical works on heroes (e.g. in his *Marat*, 1793). He instituted a neo-classical style with a strong historical strain which was pursued by Benoist, Gros, Ingres and other contemporaries and followers. In a much more vivid, imaginative and independent manner, Goya revealed the vanity, ugliness, greed, emptiness and suffering of the Revolutionary era (see his *Executions of the Third of May*, 1808, 1814-15). So did Gericault (see his *The Raft of the Medusa*, 1819). Blake was even more fantastic, grandiose and removed from convention (see his *The Ancient of Days,* 1794). Another break, but like a ripple, were the back-to-nature advances in landscape painting which came at this time to be accorded much more dignity. Friedrich (*The Wanderer above the Mists,* c.1817-18) produced landscapes in a lyrical and poetic manner; Constable (*The Haywain,* 1821) in a restrained, unpretentious way with simplicity and honesty); Turner (*Steamer in a Snowstorm,* 1842) in a more daring, visionary and dramatic manner; and Corot (*Volterra,* 1834) in a conscientious effort to reproduce nature without loss of the emotion it conveys.

It is difficult to establish the date when all the ripples, perturbations and spiked breaks in tradition synergistically worked to establish definitively, via amplification, the Permanent Revolution that transpired. Perhaps it may be said to have occurred with appearance of the art of Delacroix (see his *Arabic Fantasy,* 1834)—whose revolutionary depictions became the rallying point for those who opposed the conservative and smooth perfectionism of the times' neoclassicism). Gombrich (1989) considers the ascendancy of the strong romanticism of Delacroix and his followers (in the competitive arena of styles) to be one of the first of three ingredients of the Permanent Revolution. A second ingredient was the drastically new emphasis on subject matter. Although genre scenes had been found acceptable along with works depicting dignified personages, painters began to emphasize portrayals of plain, ordinary figures *per se* and their activities. Nullet (see *The Gleaners,* 1857) dignified typical peasants with a true representation of all their non-idealistic features. Courbet (see his *Bonjour, Monsieur Courbet,* 1854) injected a still more revolutionary 'realism' into his works—sincere, non-artificial depictions of people and things, as he saw them, as part of nature and not as the imagined and/or envisaged beauty, perfectionism and idealism of conventional art. See also Daumier (*Third-Class Carriage,* c.1862). [The *Pre-Raphaelite Brotherhood* movement in England led by Rossetti (see his *Ecce Ancilla Domini,* 1849-50) was motivated by the same objective as Courbet and members of the *Realistic* school. However, it failed from inability to portray anew and with forceful effectiveness the 'primitive' devout religious messages and related statements of those who preceded Raphael. It was a perturbation that died out.]

The third ingredient, the most significant, was the introduction of a totally new dimension into painting. As did the Egyptian, one can paint what he knows from science and/or long and careful observation of an object, person, or scene involving elements, either as an aggregate of independent units or more often as interconnected and interacting. And like Leonardo, Michelangelo, Raphael and other High Renaissance artists, one can see and produce fully what science and unending observation reveals when one has the time to do so and can put out the necessary effort. But most behaving organisms do not have the time to observe meticulously a situation, nor have they learned from science and extensive experience the exact features of elements and their interconnections. Rather, one usually obtains an impression, a quick sizing up of a situation, as when he crosses a street or notes the weather. Such is highly subjective, being dependent on the brief attention one gives to a situation and one's limited accumulation of knowledge and experience. And it is just this depiction of impressions, of instantaneous evaluations of a situation that characterizes most of the daily living of humans, that the *Impressionists* captured. They extended Art to embrace permanently another dimension. More particularly, when one sees objects, persons and scenes, one does not see each object with its own color and specific features but rather a set of colors and features that meld, very often vaguely and diffusively to various degrees, with each other.

Manet is usually regarded as the leader of this movement. Like Courbet he sought artistic honesty, without studio tricks and subterfuges. For example, see his *Olympia*, 1863, a forthright, sexually explicit work that outraged society, his *Déjeuner sur l'herbe,* 1863 and his *The Races at Longchamp,* c.1872. His followers include the major figures:

Monet (*Impression: Sunrise,* 1872-73, and *St. Lazare Train Station, the Normandy Train,* 1877);
Whistler (*Arrangement in Grey and Black,* 1871);
Degas (*Ballet Rehearsal,* 1874);
Renoir (*Le Moulin de la Galette,* 1876);
Cassatt (*Woman in Black at the Opera,* 1880);
Pissarro (*The Boulevard des Italiens, Morning, Sunlight,* 1897);

and others, often associated with a Post-Impressionist era:

Seurat (*A Sunday Afternoon on the Island of La Grande Jatte,* 1884-86);
Cezanne (*Woman with Coffee Pot,* 1890-92, and *Still Life with Basket of Apples,* 1890-94);
van Gogh (*Night Café,* 1888); and
Gaugin (*Spirit of the Dead Watching,* 1892).

When we arrive at the 20th century, our perspective on developments in art becomes less clear—we do not have enough time to judge what has been transpiring. Clearly, a permanent revolution had been initiated in the middle and last parts of the 19th century. How far the amplification of the perturbations of the late 18th and 19th centuries carry into the 20th is not clear—whether to the World War I era or beyond. There have been many developments since the early 1920s. They can be viewed as: (1) further developments of the heights achieved in the early decades of the 20th century; or (2) except for superlative creativity of one, two or three giants, just a period of many 'isms', like that of the mannerisms in the times subsequent to the High Renaissance, which saw no truly great general advance.

Expressionists reached a peak of jarring boldness in the early years of the 20th century. The associated movement, dubbed *Expressionism*, contrasted sharply with Impressionism, but was also highly individualistic. It probed the 'withinness' of a world of emotion and psychological states, rather than the 'withoutness' and the fleeting, snapshot reality of Impressionism. Following the post-Impressionists van Gogh and Gaugin, the Expressionists sought emotional and sensuous expression, but they turned away from simplicity and sincerity, exaggerating and distorting forms to convey their message. Moreover, following Cezanne in his efforts to achieve enduring structural solidity, form and depth on a flat canvas, they, too, employed bold colors as one means.

In France there emerged in the first years of the 20th century a school of expressionists dubbed *Fauves* ('wild beasts'). It was designated such because of its use of explosions of bright colors and barbaric harmonies, coupled with open disregard for the forms of nature. Matisse was the leading exponent of this movement, developing works involving masses of brilliant color and unsophisticated decorative patterns which yet exploited color intensity and glitter. See his *Portrait with Green Stripes,* 1905 and *The Red Room,* 1908-9. Other figures were Derain (*Effects of Sunlight on Water,* 1905) and Roault (*Prostitute at her Mirror,* 1906).

In Germany, major works by expressionists were inspired by Munch (see his *The Scream,* lithograph, 1895). Those in *Die Brüche* movement were engulfed in revolutionary social protest (see Nolde, *Dancing Around the Golden Calf,* 1910 and Kirchner, *Berlin Street Scene,* 1913). Those in the *Der Blaue Reiter* movement, which came later, were less motivated to depict the violence and misery of reality, and much more inclined to naturalistic and imaginative portrayals. From this movement, *Abstractionism* emerged. Here, Kandinsky was the outstanding figure. In the choice of colors and lines, he went much further than Matisse. What primarily mattered to him, especially in his later years, was the expression of feelings through such choice, and secondarily (if at all) a recognizable object or other imitation of nature. See his *Improvisation 30 (Cannon),* 1913 and *Composition 238: Bright Circle,* 1921. He aimed with pure color to bring out the 'music' in painting and to rival music in expressiveness. Marc (*Deer in the Woods,* No. I, 1913) was another early abstractionist.

Still, the challenge posed by Cezanne—namely to produce the appearance of solid structure and form on a flat canvas—remained unmet despite the efforts of many who took up this challenge. It fell to Picasso, the outstanding figure in the *Cubism* school, and other Cubists to meet it headlong. Increasingly, Picasso drew upon the artist's freedom to distort reality, to use imaginative design, to exploit intense colors and to extract the most telling effects from primitive and African art whose spontaneity, energy, sincerity and simplified geometric forms appealed to and influenced many in the early 20th century. He increasingly followed Cezanne's advice to use spheres, cones, cylinders and in general basic solid shapes to capture depth. One sees this in the progressive development of his works, from *Absinthe Drinker* (1902) to *Les Demoiselles d'Avignon* (1907), to *Violin and Grapes* (1912), to *Head* (1928), and finally to *Guernica* (1937), the last being a comprehensive vision of a major world debacle, an encyclopedic depiction.

Associated with Picasso was another great cubist, Braque (see his *The Portuguese,* 1911). They had major influence upon other artists and movements: Klee, also influenced by abstractionism (see his *A Tiny Tale of a Tiny Dwarf,* 1925); Duchamp and Leger in their use of the cubist's principle of fragmentation (see, respectively, *The Nude Descending a Staircase, No. 2,* 1912 and *The City,* 1919). Also, there were: (1) the *Futurists* who attacked the past, advocating the destruction of academies, museums and the like as barriers to progress, extolling the glories of industrial achievement, for example, Severine (*Armored Train,* 1915), Balla (*Abstract Speed the Car has Passed,* 1913), Broccioni (*The City*

Rises, 1910-11), and Stella (*Brooklyn Bridge,* 1917-18); and (2) the *Constructivists* who aimed to build up pictures using the simplest geometrical elements with clarity aimed at reflecting the objective laws of the universe—as a logical outcome of cubism and abstractionism—for example, Mondrian (*Composition with Red, Black, Blue, Yellow and Grey,* 1920) and El. Lissitzky (*Proun 99,* 1924-25).

Two significant movements, related and having significant impact largely in the interwar period, were *Dada* and *Surrealism*. These movements were, as the above, highly individualistic, but even more diametrically opposed to the fleeting, visionary subjectivity of the Impressionists. Dada, at its deep internal subjectivity and mad logic, furiously attacked every cultural standard and every form of artistic activity as seen in: Arp's *Mountain Table Anchors Navel,*1925; Duchamp's *The Bride Stripped Bare by her Bachelors, Even (The Large Glass)*, 1915-23; and De Chirico's *The Elephant of the Celebes*, 1921. The Dadaists' manifestations of spontaneous expression, liberating the creative process from logical shackles, paved the way for *Surrealism* with its illogical orientation to subconscious Freudian associations, but which also was anticipated by childhood reveries and folk-type art such as that of Chagall (see his *I and the Village,* 1911). The manna for such associations were dreams, nightmares, obsessions, and wild uncontrolled and uninhibited multi-interpretable portrayals of objects and interconnections. For example, see: Ernst, *Europe After the Rain*, 1940-42; Miro, *Harliquin's Carnival,* 1924-25; Dali (often regarded as the leading exponent), *The Persistence of Memory,* 1931, and *Apparition of Face and Fruit-Bowl on a Beach,* 1938; Matta, *Listen to Living (Écoutez Vivre)*, 1941; and Magritte, *Delusions of Grandeur*, 1961.

As already noted, we lack sufficient perspective to classify, or otherwise assign appellations, tendencies and interpretations to the art of the 20th century, especially that of the last half; it is so close to us. One might consider the entire period from the late 18th century through the 20th as the period of *Permanent Revolution: Surging Individualism Unbounded.* But also, the more restricted period from the Impressionists through World War I and even the interwar years, when Paris was by far the dominating center of art, both directly and indirectly, might be viewed as the era of major amplification. Subsequent art then might be viewed as reflecting all kinds of emphases and narrowly defined directions that developed to absorb the urge of painters for creativity—in effect to be placed in the category of diverse 'isms' (movements) found in the history of painting, being primarily refinements, exaggerations, and excessive sophistications of the basic lines established during the Paris amplification.

In general, in post-World War II art, abstract expressionism continued as a strong underlying movement, as in Pollock (*Blue Poles,* 1953), de Kooning (*The Time of the Fire,* 1956, and *Untitled XVII*, 1984), Rothko (*Green on Blue*, 1956), and Stella, (*Avicenna*, 1960). More specialized movements were present, for example *Pop Art* (Rauschenberg, *Monogram*, 1955-59, and Warhol, *Disaster*, 1964), reflecting particular emphases or mannerisms.

8.9 Commonalities, parallels, and relationships with concepts, processes and analyses in other areas of knowledge

To start the discussion of this section, we consider the phenomenon of symmetry breaking as found in the evolution of the physical systems discussed in Chapter 5.

The first evidence of painting of which we have a record is cave-man art. As the cave-man artist (as well as his fellow humans) looked upon the wall of a cave prior to his artistry, he undoubtedly found no cognitive message or impression there, whether for others or himself. As with the characterization of molecular chaos in Figure 6.9, there was nothing that defined a meaningful space or time in that wall. But once the cave-man painter scratched on that wall an outline of a bison pierced by a spear, perhaps to pass on a message, express a wish or otherwise, one can claim that he broke the spatial symmetry of the wall. Within the limited bounds of his visibility, each observed point on the wall came to have a meaningful spatial orientation. The single or several scratches of a crude tool or strokes of a primitive brush was sufficient to provide a position for any point noted. Either a point was a bit of the bison, or a location near the bison or far from it—as also when later an outline of a hunter was added to the sketch.

There were, of course, a series of advances in this artistry when an aggregate denoting a relatively complex process (e.g. the trapping of prey by an organized team of hunters) was depicted and color based on available natural materials employed.

A second possible symmetry breaking (though questionable) with regard to painting may be said to have occurred when the Egyptians during their very early development as an urban civilization imposed on painting a rigid and fixed form, although at this early stage an artist through his creative effort may have greatly influenced the particular rigidities which were laid down. In effect, painting may be said to have been restricted to a defined set of x, y coordinates in a pervasive rectangular frame which acted to restrict free flowing work. But lacking an adequate historical record we cannot firmly establish whether the transition from cave-man art to Egyptian was gradual or a Prigogine jump.

The next clear break in symmetry, anticipated by a perturbation during pharoah Akheniton's reign involving flexibility and humanism that was later squashed, took place in Greek painting. It followed the gradual diffusion of practices of the Egyptian, Mesopotamian and other cultures in the pre-Hellenistic era. Starting from about the end of the 6th century BC, reaching a zenith during the reign of Pericles and certainly extending until the death of Alexander the Great, there was, as already remarked, a tremendous outburst in the development of Greek art and culture—the classical Golden Age. In general, the development was associated with the emergence of Athenian democracy and the Greek notion of freedom, of man's autonomy and ability to master his environment through reasoning and other intellectual activity—all this against a mythical religious background replete with gods and goddesses, and a tragic vision of life.

In this humanistic painting that comes down to us primarily on vessels, an object or persons had to be visually depicted locally with a third dimension. This was done by the use of foreshortening and other means, even though the painting was restricted to a flat or curved plane. The particular set of points employed to identify the object or person as three-dimensional were distinct from the neighboring and other points that produced only a two-dimensional effect. This shattering of the two-dimensional rigidity of Egyptian painting with the active involvement of a third dimensional effect, albeit only locally, constitutes a third symmetry breaking—a breaking of the non-use of this effect for communicating a message or other purpose.

No truly outstanding development in painting occurred during the subsequent thousand years. After the period of barbaric invasions that destroyed the Roman empire, there followed the Middle Ages, sometimes designated the Dark Ages. While the painting during these ages provided an extremely important message to the people that counteracted their insecurity and anxiety—a humble message of compassion, suffering and hope—gone was the ideal form, rationality, and confidence in man's ability to master the exigencies of his environment, although there was some advance, particularly in harmonious arrangements of people and objects.

Medieval painting, as already noted, was from one standpoint a step backward in that the previous advances through symmetry breaking were hardly exploited—at least in the early years. However, the religious/spiritual message emerged more forcefully than ever before.

In time, the Middle Ages emerged from their depths, and gradual advances came to provide some background conditions for the resurgence of painting: the appearance of facial expressions to complement the strictly religious message; and a beginning at the uplifting of the role of man, and some reconciliation of the classical (Greek) and nonclassical (Celtic and Germanic) themes and concepts.

A major perturbation occurred at the end of the Middle Ages, namely the introduction of spatial perspective, a highly creative innovation. Initiated by Giotto, further developed by Masaccio and given mathematical precision by Brunelleschi, it enabled the complete visual capture of the third dimension—a third dimension well beyond the immediate local area captured by the Greeks. This fourth symmetry breaking made possible the protrayal of spatial depth, of far-reaching background, on a flat surface. Together with: (1) previous symmetry breakings in painting; (2) the restoration, reinvigoration and magnification of classical traditions; and (3) a new dignified humanism that came to synthesize the classical and nonclassical strains in Western civilization, this perturbation led to the brilliant objective, harmonious and spiritual heights of the *High Renaissance* of the late 15th and first half of the 16th centuries.

After the days of Michelangelo, Leonardo, Raphael and Titian, and other titans, Western painting entered a period of relative stasis. Little change transpired in basic style, form and other features despite the signal achievements by particular individuals such as El Greco, Velasquez, and Rembrandt, among others. After at least 150 years of relative quiet there came breaks in tradition that did serve as loci for amplification. As a consequence a fifth symmetry breaking occurred in the late 18th and 19th centuries. It started perhaps with the neoclassicism of David followed by Ingres, Goya, Delacroix, Gericault, Millet, Courbet and other ripple producing artists. Then there followed a rapid ascendancy of individualism in all its aspects: iconoclastic, revolutionary and non-revolutionary, wild and conservative, romantic and ugly, all concomitant with deep concern for and intense concentration on man *per se*, a highly subjective type of development. The first signs of the major amplification associated with this fifth symmetry breaking appeared in the Impressionist movement (Manet, Monet, Renoir, Degas and others) which introduced the non-objective, fleeting momentary or quick-lived visual perception of a person, object or situation. This movement's innovators were soon succeeded by the Post-Impressionists (Gaugin, van Gogh and Cezanne) more attuned to nature, simplicity, structure and form who in turn were followed by: the Expressionists (e.g., Matisse and Kandinsky) exuding a bolder and deeper psychologically-oriented individualism; and Abstractionists and Cubists (Picasso, Braque, Mondrian, Dali, to name a few) who brought the soaring individualism to even greater heights.

It is too soon to speculate when the era of Unbounded Individualism has ended, if at all. There are signs, however, that a period of stasis with respect to major advance, or of modern-day mannerism, has set in sometime after World War II. Whether or not another major amplification of one or more perturbations is on the horizon or is continuing, it is important to recognize that each of the symmetry breakings sketched above was concomitant with an advance, a new basic dimension or element in painting. À la Bronowski's stratified stability, each advance served as a barb in the arrow of painting's evolution. However, in any particular work, the painter may choose not to exploit one or more of these advances—reflecting the use of a principle analogous to Pattee's optimum loss of detail—thereby to achieve effects that otherwise would not be realizable.

In connection with the set of Prigogine-type jumps which led to the sequence of far-from-equilibrium situations discussed above, it is instructive to examine the changes in the nature of the messages which came to be transmitted. As already remarked, these messages, constructed for transmission to those in the local, regional and/or other environments, may be simple or complex. They may be at one extreme objective (scientific at times) or at the other extreme highly subjective and expressive. They may reflect experiences the artist has had, and/or mental models he has come to accumulate over time, in part affected by the hierarchical structure of cultural artifacts and social institutions which have impacted upon him as well as his own accumulated technology, know-how and tools (brushes, paints and the like).

To reiterate, a first level in the capability or potential for message or communication development took place with the use of a line as in caveman art. A second stage had been reached with the imposition of rules and fixed principles within the two-dimensional framework of Egyptian art. A third stage was reached in the Greek Classical Age when the third dimension (in the small and locally) was introduced in depicting bodies and other objects, and doing so in ideal form to reflect beauty and rationality.

These three stages may be depicted in Figure 8.1 by sketches from Dorner (1958, p. 43). The first sketch in the top row indicates the first stage of message development—the use of a line by primitive man to depict a bison or other image. Later in that stage we see, along the first row of the figure, aggregates of such images that represent elements such as planting, sunshine, rain and harvesting.

The sketch in the second row (from Dorner, 1958, p.49) shows a two level, two dimensional hierarchy governing the message delivered by a painting—a stage where rules and fixed principles are set down by a hierarchical social system based on agriculture designed to reflect the existing complex interconnections of people and objects in an early (Egyptian) urban civilization.

The sketches in the third row (from Dorner, 1958, pp. 56, 65 with minor modifications) represent the message of painting reached during the Greek Classical age. The first one shows the capture of a certain number of external forces through reasoning and intellectual activity that leads to the synthesis of these forces—a synthesis associated with rationality. This corresponds to the organized thought of an individual concerning the interplay of these forces—an interplay that takes place in a local three-dimensional world as indicated by the second sketch of that row. The irregular waves coursing in and out of the spheres of organized thought in the third sketch represent the uncertainty of individuals in a society with regard to their relations to each other, their many Gods and Goddesses, and other uncontrollable external forces which underlay the Greek tragic vision of life.

On the facing page, the first sketch in the fourth row (adaptation of Dorner, 1958, p. 69) reflects the overwhelming dominance over remnants of the Greek Classical period of the Christian religion in the Middle Ages—with the result that in general all painting came to convey the basic convergent monotheistic message for humanity—a message, as already remarked, that provided security for the ordinary man harrassed by the many daily uncertainties of life.

The second sketch of the fourth row (from Dorner, 1958, p. 73) depicts the message resulting from the amplification of the spatial perspective perturbation when coupled with the rediscovery and resurgence of Greek Classical art, the application of scientific principles to and careful observation in the portrayal of beings and objects, and the use of color. When combined with the integrating force associated with the continued importance of the church and its religious message and the new authority and power of city leaders, this leads to the third sketch in that row—a depiction of the zenith in message construction attained by the High Renaissance. This last sketch reveals a hierarchical structure of communication—one whose levels are greater in number and more complex than the one attained at the end of the Greek Classical age. That sketch captures: (1) the strong convergent religious force in a communication that substitutes one God for the many Greek gods and goddesses; (2) spatial perspective that captures the full objective portrayal of the third dimension; and (3) the many refinements of the Classical art that were realized—that raised the portrayal of beauty of form and accuracy of composition to even greater heights.

The fifth row brings into the message of art, and painting in particular, the individualism (the subjectivity of the individual artist) which was to turn out to be a truly revolutionary force—which led to the last amplification reached in Western painting. Recall that in the period of relative stasis subsequent to the High Renaissance there had been many perturbations reflecting the struggle for individualism [works of Corregio, Parmigianino and El Greco in the 16th century, of Velasquez and Rubens and the Flemish/Dutch school (Reimbrandt in particular) in the 17th, of Hogarth, Blake and Turner in the 18th]; such did not develop into a major amplification.

Starting with the end of the 18th century and greatly influenced by the French and Industrial Revolutions were the breaks in tradition of message construction associated with David and those who followed which constituted a set of perturbations that did come to be amplified. The first sketch in the fifth row (adaptation of Dorner, 1958, p. 95) showing a set of circles (some distorted) moving away from the center in both directions to different extents does suggest an emerging individualism. The shading in each circle may be taken to represent Impressionism and post-Impressionism, the first major movements in the late 19th and 20th century amplification. The second sketch (from Dorner, 1958, p. 95), taken to represent the further development within each of the circles in the preceding sketch) may be viewed as signifying the emergence of Expressionism—the French Fauve, Die Brüche, Der Blaue Reiter, and other movements in the early 20th century. The bending arrows within the circle signify a much deeper and more emotional subjectiveness than was

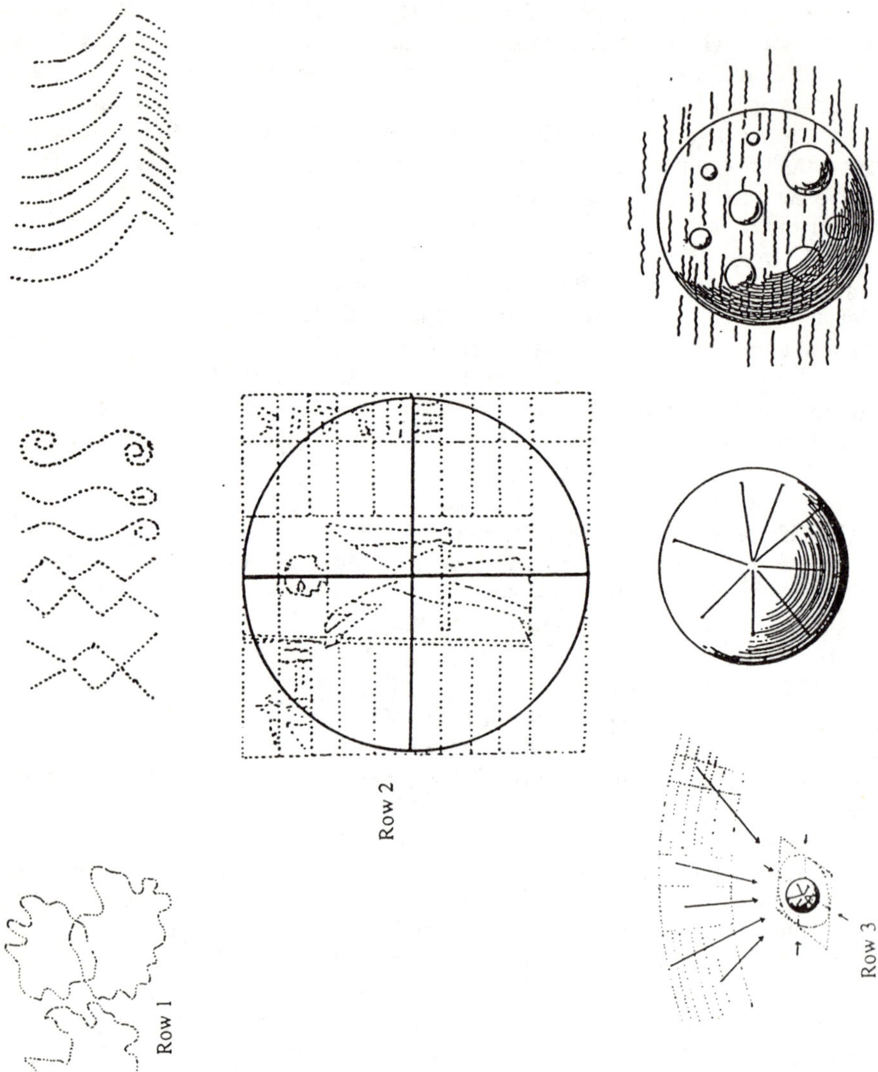
Row 1
Row 2
Row 3

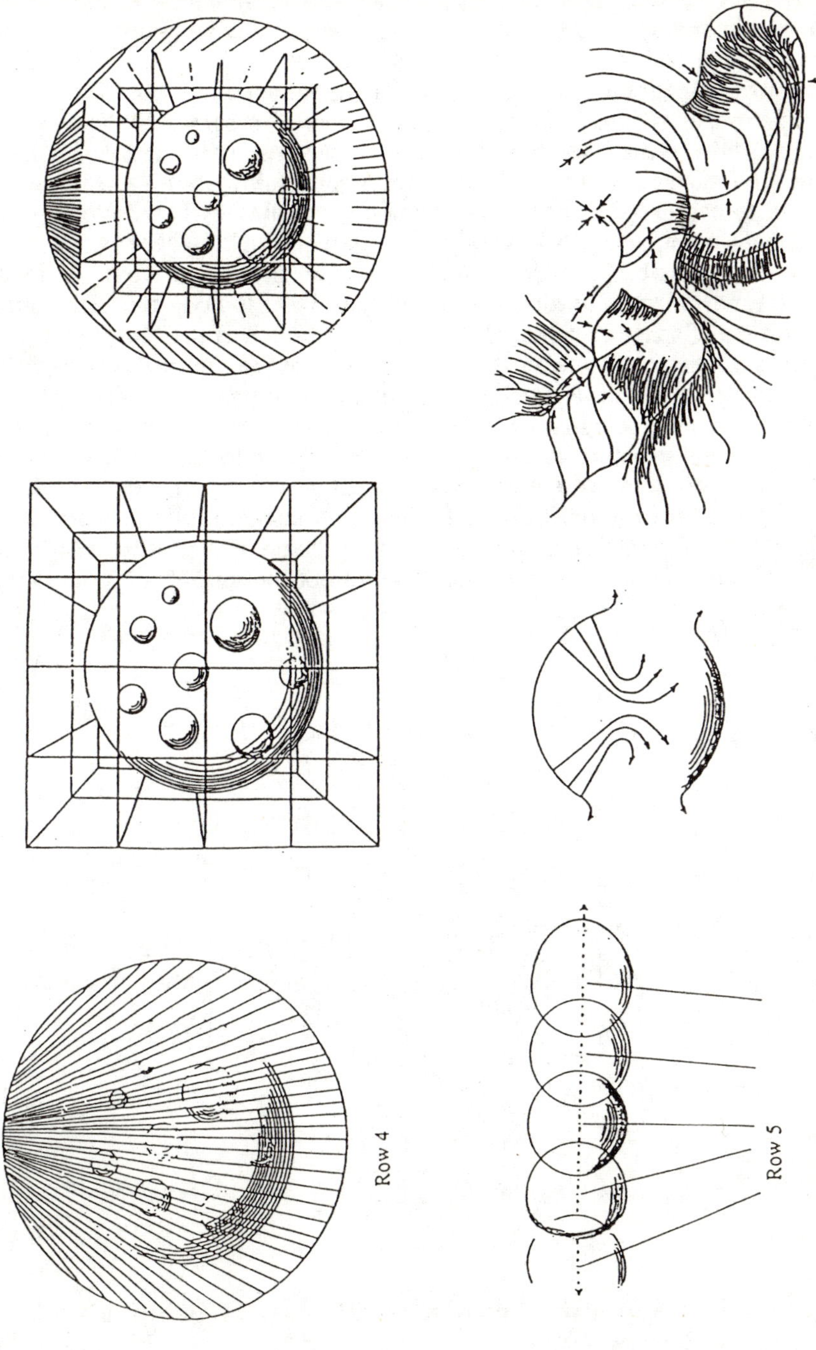

Figure 8.1 Pictorial representation of stages in painting's evolution

present among the Impressionists and post-Impressionists—a much more penetrating and deeper inwardness of one's personality and outlook.

Finally, the last sketch (from Dorner, 1958, p. 117) represents the 20th century culmination of individualism—the unbounded individualism of the Abstractionists (Cubists, Surrealists and others). To some, this last sketch of the fifth row may be viewed as chaos—but it is not. Clear space-time paths are indicated as well as boundaries among subregions—and in such a way as to capture the constantly changing interplay of basic forces—the continually interpenetrating energies that Dorner emphasizes.

It is not possible to depict with a single figure the four or more dimensional hierarchy that exists in painting today—the full three dimensions of space as captured in the middle sketch of the fourth row plus the time dimension (dynamics) as clearly depicted in the last sketch of the fifth row, plus religious spiritual and other emotional elements that one might insist upon designating as a fifth dimension. However, some notion of the figure's structure can be obtained if we were to take the last sketch of the fifth row to represent an irregular section cut out of the structure of the last sketch of the fourth row. To help in visualizing the new structure we construct Figure 8.2. Figure 8.2 may be taken as representing a first, rough approximation of a synthetic depiction of the set of forces operating

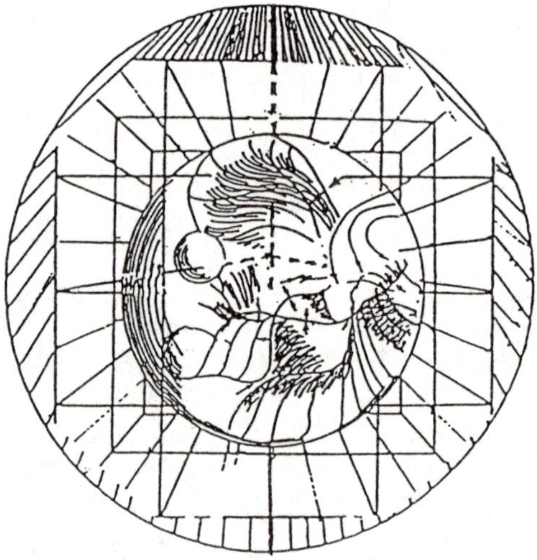

Figure 8.2 **An overall depiction (synthesis?) of painting's perspectives**

upon a painter today who seeks a combination of balance and creativity. A portrayal of the synthesis of forces, where the painter attaches much less importance to religious, spiritual or universal values that are proclaimed by a society's leaders or that govern his outlook, would diminish the number and length of the rays emanating from the top of the figure. Where he is motivated to paint from a pure scientific and objective standpoint, the sketch in the middle of the fourth row of Figure 8.1 can be taken as pertinent. When at the other extreme he is motivated 100% by subjective inwardness, with no concern for relations to and among other members of society, the last sketch of the fifth row of Figure 8.1 may be considered relevant.

The inclination of many on-lookers to view the paintings of some unbounded individualists as chaos needs to be addressed specifically in terms of the findings of chaos and related theory discussed in Chapter 6. These paintings contrast sharply with the order in: Egyptian art; the idealistic forms of Greek paintings on vases; the meticulous detail of Flemish/Dutch portrayals; the scientific and well organized works of the High Renaissance; and the neo-classical school of David, Ingles and others. Yet, as chaos and related theory has established, there can be much order within seemingly disorderly depictions—in the works of Miro for example and those in the many 20th century extremists. Such order may be characterized as deterministic chaos, of the same sort that we find in the strange attractors of Lorenz and Hénon, and in turbulence. Moreover, one can maintain that in painting there can exist sensitivity to initial conditions—see the works of Pollock and others—the particular deterministic chaos realized being highly dependent on the specific initial dab of paint. Stated otherwise, much of abstract expressionistic painting may be viewed as asymptotic behavior of a chaotic system, having a message that appears unfathomable to the uninitiated observer, but in fact one that has considerable depth and complexity. How clear the message of Picasso's *Guernica*. Further, one can claim that a highly abstract painting, just like a very rugged coastline, has a fractal dimension—that is, is not a multitude of random elements but rather a set of elements that can be embedded in a finite dimensional Euclidean space.

Alternatively, one may maintain that current abstract painting is in a noisy, but actively fibrillating (resonating) state approaching an edge of chaos (a concept of noise theorists), from which there will emerge another basic dimension of painting. Such emergence might constitute another symmetry breaking—a transition to another far-from-equilibrium state.

We now wish to relate the materials of this chapter to those contained in Chapter 2. There we summarized the concepts, notions and theories of Teilhard. Clearly, the evolution of painting bears out much of his thinking. We find in painting the increase in what he calls psychism (corresponding to subjectivist and reflective thought) associated with human development—via both jumps and gradual transition from the rudimentary psychism (interior perception) of cave man through that of the Greek idealistic-rationalistic being, up to the most advanced level of a highly individualistic 20th century abstract expressionist (e.g. Kandinsky in his *Small Pleasures,* 1913). One can associate such with the development over time of the increasing cephalization (dominance of the brain over sensimotor activity) of the painter, his increasing hominization, interiorization and personalization (his increasing awareness of nature, himself, and other behaving units) and his growing consciousness of the need for integration with the outer world (cultural melding), wherein differentiation and diversity is preserved. The development of painting (abstract and non-abstract) as a cultural activity manifested in the works we have inherited has been and continues to be consistent with Teilhard's perception of human development. Moreover, it can be easily viewed as one of the basic elements of Teilhard's projected progression from psychogenesis (the current state of man's spiritual development) to his imagined state of noogenesis and ultimately Omega—the ideal state of almost ultrahominization, hyperpersonal integration and infallible spiritual perfection.

As a basic means for expression and message creation, painting can be viewed as a set of memes, as defined in Chapter 3. In large part it constitutes know-how (by and large inherited) about how to use tools (cultural artifacts) to obtain certain effects. Nonetheless, creative activity by the painter as a behaving unit, particularly when combined with purely random elements (e.g. a chance discovery of a new effect), can result in new know-how and tools, which if found of value may become new surviving memes to be added to the current stock for use in subsequent works and by future generations. Michelangelo in his Sistine Chapel depictions was greatly influenced by the religious teachings as well as the painting practices of his teachers and forbears, and in turn devised methods that were able to deepen the emotional character and intensity of religious messages and which served to increase the stock of tools and know-how of those who followed. Discovery of how to capture spatial perspective, innovations in the use of light and coloring, inventions and distortions of geometric figures are all memes that have affected the evolution of painting just as mathematics and language have advanced communication. In general, all analysis relating to the role of memes in the evolution of human society pertains also to the same in the evolution of painting. Even the debate over whether the species as well as the individual behaving organism is a unit on which selection operates carries over to painting. For example, one can argue like Mayr that the survival of a particular practice, tool or element of a style of painting is dependent solely on the qualities (attributes) of that particular element. On the other hand, if that particular element is a constituent of a particular style (the counterpart to a species) and only of that style, the survival of that element is dependent on the survivability of that style. The Egyptian practice of painting a pharaoh twice as large as one of his subjects to reflect an existing power relation, along with many others based on what was known, has in essence died out when the Egyptian style itself was abandoned by painters.

In brief, there is much in the analysis of Chapter 3 on evolution theory that carries over for understanding the development of painting. At the same time analysis of the evolution of painting corroborates many of the findings (thinking) of those who write on biological evolution; for example, it lends support to the co-evolutionary views of Durham and others. Whether one examines the works of the classical Greeks, the High Renaissance titans, the Flemish Dutch masters and those of the current era, one cannot deny the role of genes in making possible survival machines that can turn out to be geniuses who then can have tremendous impact on society, its behavior, its aims, its emotional status and other attributes, which in turn greatly conditions the future surviving gene-constructed machines.

At this point it is useful to view painting, as Bohm does, namely as representing a flow or continuous movement in the stream of consciousness—a continuous unfoldment of what is possible, of the implicate order, as discussed in the Appendix to Chapter 5. Like Kandinsky, Bohm (1981, pp.198-9) compares painting with music. In listening to music, at a given moment a certain note is played

> but a number of the previous notes are still reverberating in consciousness. Close attention will show that it is the simultaneous presence and activity of all these reverberations that is responsible for the direct and immediately felt sense of movement, flow and continuity. To hear a set of notes so far apart in time that there is no such reverberation will destroy altogether the sense of a whole unbroken, living movement that gives meaning and force to what is heard.

Likewise, in closely viewing a classic painting, at a given millisecond or more of time, there is a sensing of an element of that painting and as the eye in successive instants of time moves over the painting, the successive visual images also "undergo active transformation as they enfold into the brain and nervous system (e.g. they give rise to emotional, physical and other more subtle responses of which one may be only dimly conscious as well as to 'after images' that are in certain ways similar to the reverberations in musical notes" . . . it is clear "that a sense of movement is experienced through the intermingling and interpenetration of the co-present transformations to which these images must give rise as they penetrate the brain and nervous system" (Bohm, 1981, p. 201).

For example, take Leonardo's *Last Supper*. As Gombrich (1989, p.224) writes, perhaps the monks of the monastery of Santa Maria delle Grazie

> were first struck by the truth to nature with which all details were portrayed, the dishes on the tablecloth, and the folds of the draperies. Then, as now, works of art were often judged by laymen according to their degree of lifelikeness. But that can only have been the first reaction. Once they had sufficiently admired this extraordinary illusion of reality, the monks would turn to the way in which Leonardo had presented the biblical story.

As Gombrich (1989, p.226) continues, they would call to mind Christ's statement, 'Verily I say unto you, that one of you shall betray me' and in the next milliseconds they would see every one of the apostles exceedingly emotionally affected and saying unto Christ 'Lord, is it I' (Matthew xxvi, 21-2). Then they might also recall from the gospel of St. John the writing: 'Now there was leaning on Jesus' bosom one of his disciples, whom Jesus loved. Simon Peter therefore beckoned to him, that he should ask who it should be of whom he spake' (John, xiii, 23-4); and the monks would observe that most of disciples were asking this question and that those on Christ's sides were shrinking back in terror when they hear this revelation. On closer study they would come to see that St. Peter had rushed to St. John sitting directly to the right of Jesus. As he whispers something into St. John's ear, he inadvertently pushes Judas forward. At still closer study, they would then observe that 'Judas is not segregated from the rest, and yet he seems isolated. He alone does not gesticulate and question. He bends forward and looks up in suspicion or anger, a dramatic contrast to the figure of Christ sitting calm and resigned amidst this surging turmoil'

Now, we cannot know what all this meant to the monks. Probably the reactions of each was at least somewhat different from that of others. Clearly, however, for each it was a process of digestion over time—a continuous absorption of sensations, a continuous movement in emotional response, which at any particular millisecond was affected by reverberations from sensations at previous milliseconds. It may have taken them different amounts of time to recognize and react to the deep spiritual message emanating from this portrayal of the sacred episode, and for some to appreciate Leonardo's deep insight into the behavior and reactions of men and his imaginative genius in putting the scene before their eyes. Notwithstanding, for a number if not all of them it must have been an experience, continuously building up, which was no different than that of many in an audience hearing a grand performance of the Beethoven Fifth.

If we consider other painting, a continuity in the sequence of visual sensation very often obtains. Even in an Impressionist work that yields a fleeting image of a situation, the message takes place over a sequence of milliseconds—and like the sounding of a magnificent chord can generate for even a short period of time the deepening of a most enjoyable experience.

Viewed from the standpoint of historical time, one can state that painting has undergone a rather steady development, with ups and downs of course, that can be considered to be unfolding of the implicate order—unfolding that has yielded the explicate state that exists today. This unfolding has also made possible greater depth and breath of messages that can be communicated by a painting—greater depth and breadth in the sense that the continuous movement in sensations from observing and digesting its meaning can involve ever increasing mental activity and time in effecting a synthesis (partial or complete) of the elements of the painting and their interconnections.

Closer examination of painting as a message or communication mechanism reveals that it is part of Pattee's broadly defined linguistic mode (as discussed in Chapter 4). It is not a rate-dependent process as the pull of gravity upon a falling stone, nor is it an outcome of a computer simulation, whether or not the simulating analyst aspires to replicate intelligence or life (i.e., to achieve artificial intelligence or artificial life). Painting involves subjectivity, inspiration and creativity beyond the (physical) laws of motion, however broadly interpreted. It involves measurement, often highly subjective, as in coloring and shading; and in this measurement and other ways of passing on information, it plays a critical role in formulating initial or terminal conditions of a rate-determined process. For example, the paintings of Hogarth (e.g. *The Rake's Progress*, 1735) and Die Brücke school spoke of social conditions (immorality, impersonality and human suffering) with strong emotional, spiritual and ethical dimensions—in a real sense providing a much deeper type of Census than a set of numbers. Other paintings (e.g. Hicks, *The Peaceable Kingdom,* 1834) relates to what is possible, objectives, goals and even a vision of Omega. At times, a painting (as Picasso's *Guernica*) can be a discrete string of rate-independent messages, which cohere and form a strong statement, or as in the case of Egyptian art a statement about many pieces of relevant knowledge in a situation.

A Teilhard enthusiast might perceive, with further probing, another relationship of painting to the findings of physicists and other natural scientists. Consider the standard concept of entropy (a measure of disorder) where maximum entropy corresponds to complete disorder (e.g. the molecular chaos to which Prigogine refers in his discussion of the Bénard cell experiment and in Figure 6.9). When inputs of sufficient energy from the external world are introduced into molecular chaos, order appears. And, generally speaking, physicists would hold that organisms of increasing order (structurally and otherwise) require for their maintenance increasing energy inputs from the external environment. If one views increasing mental activity as requiring increasing energy inputs and considers such mental activity as psychic energy inputs á la Teilhard, then one can also suggest that the increasing orderly hierarchical structure achieved in painting, at least up to era of Unbounded Individualism, was supported by an increasing influx of energy from the external environment and constituted an increasing utilization of psychic energy by artists. Even further, where abstract expressionistic painting is viewed as just another level in the hierarchical structure of painting know-how (technology), and as representing increasing 'openness' in painting, one may argue that the psychic energy involved in the work of an abstract painter, who has the full technology and know-how of painting at his command, is by and large greater than ever before. If Teilhard were living today he might characterize abstract painting as a phenomenon embodying increased personification (hominization)—another step in the progression to a pure noetics system.

It is instructive to examine the role of painting in another way. We have already characterized it as a means of communicating, of passing on messages. At the simplest level of interaction, the message may be received by an organism (even a nonhuman one), and stop there, with or without a response. If there is a response, from the painter's patron and the public, and one that is positive, such may form the basis for the development of a meme, for example, the acceptance of a new element of practice in painting.

The response to the message, however, may turn out to be more significant; it may not only involve the acceptance of a new element, but also affect the behavior and subsequent work of the painter. This latter outcome may lead to the generation of other memes, some of which as elements of a style may form a new hierarchical structure and, as levels or not of such a structure, may survive for different lengths of time.

Moreover much more complex interaction may be present. Anticipatory behavior both by the painter and his audience may be involved—so that responses and sequences of responses are projected. An interaction might involve a bottom-up communication (as when an artist's painting sends a message to his patron), top-down, lateral or other. At one extreme the interaction may be very simple, and at another, exceedingly involved. The cave man's bison was a simple message—perhaps a message to a God or other powerful figure. The message of the religious Medieval painter was more complex. It was for the lay person, and was one to relieve anxiety and generate faith and spiritual comfort and enhancement. The message of the High Renaissance painter was still more complex. The Gombrich interpretations of the messages of Michelangelo's Sistine Chapel and Leonardo's *Last Supper* (see pp.222, 241-2) make this point clear. And of course the messages of the abstract expressionists, for example Kandinsky's *Small Pleasures* (1913) and Picasso's *Guernica* (1937) are even more complex—and as such of different significance to different groups—the élite, the middle income class and the poor.

While the gravity model of Chapter 7, intended to grasp interaction through the use of symbols representing quantitative magnitudes is not directly applicable to painting, conceptually it is. In the model the different masses are those of the different elements of a painter's public—e.g. religious groups, the establishment, and the voting population. The model may possibly be helpful in indicating how the painter influences those elements, and in turn can be affected by them—all after appropriate specification of the relative weights to be applied to the several publics and the distances separating the painter from these publics.

In fact, a painting may embody a set of messages—e.g. a set of causal models. Take Kandinsky's *Small Pleasures* (1913) against the background of his earlier attempts at this work. One can discern several causal models: the pleasure derived from rowing a boat, the pleasure from galloping up a hill, the pleasure from taking a walk—all these in association with the deep religious spirituality emanating from the portrayal of a hilltop city presided over by a onion-domed structure, surrounded by seemingly chaos generated by the glowing magenta sun, threatening black clouds, agitated waves, and diverse discordant shapes, *integrated, however, by brilliant colors and expressive brush strokes.* In the sense that the painting interconnects several causal submodels to yield an overall warm emotional message—differently sensed and interpreted by his several publics—it serves as a heterarchical model in Pattee's sense. (To Kandinsky, the pleasure was to deluge the canvas with lots of small pleasures.)

In the same manner a more recent painting by Roberta Jean Isard, *Man's Extinction* (1974), conveys an integrated set of messages, a protest of the then young generation of the utter folly and barbaric killing and suffering of the Vietnam war, a reflection of the gloom and despair of that generation's perception of the imbecility and incompetence of the times. In this painting, there are a number of causal submodels: power-driven political leaders taking moves in a game of chess that forces armies of tanks to attack each other; the clashes of tanks that cause deaths and human suffering; the play of time (a fourth dimension captured by the curvature in the portrayed space) that records successive wars leading to the extinction of man (revealed by the gloomy hue of the purple, the sinister red of the receding sun and the crushing spatial convergence). These causal submodels all contribute to the overriding cry of despair and message of the utter irrationality and inhumanity of man causing his ultimate obliteration.

To close, we see how: Teilhard's concept of psychic energy and increasing hominization; Dawkins' concept of memes; the notion of species as a unit in evolution; Durham's notion of coevolution; Bohm's doctrine of unfoldment of the implicate order and consciousness as movement; Pattee's notions of rate-independent processes, linguistic mode, optimum loss of detail, and heterarchical model; the physicist's notion of a sequence of symmetry breakings; Prigogine's jumps to a sequence of successive far-from-equilibrium positions; Bronowski's stratified stability and barbs in time's arrow; deterministic chaos, fractal dimension and noise; hierarchical structure in interaction and diverse causal models; and the gravity model and other notions, relationships and analyses provide deeper understanding of the history of art and Western painting, and in turn how these notions, relationships and analysis find reinforcement and new nuances and insights from this history.

References

Artigani, R. (1991), 'Social Evolution: A Nonequilibrium Systems Model' in Ervin Laszlo (ed.), *The New Evolutionary Paradigm*, New York: Gordon and Breach.

Bohm, D. (1981), *Wholeness and the Implicate Order*, London: Routledge and Kegan Paul.

Clark, K. C. (1969), *Civilization*, New York: Harper and Row.

Cole, B. and Gealt, A. (1989), *Art of the Western World*, New York: Summit Books.

de la Croix, H., Tansey, R.G. and Kirkpatrick, D. (1991), *Art Through the Ages*, New York: Harcourt, Brace, and Jovanovich.

Dorner, A. (1958), *The Way Beyond Art*, New York: New York University Press.

Fleming, W. (1986), *Arts and Ideas*, New York: Holt, Rinehart and Winston.

Gombrich, E.H. (1989), *The Story of Art*, Englewood Cliffs, N.J.: Prentice-Hall.

Hartt, F. (1989), *A History of Painting, Sculpture, and Architecture*, Englewood Cliffs, N.J.: Prentice-Hall.

Honour, H. and Fleming, J. (1992), *The Visual Arts: A History*, Englewood Cliffs, N.J.: Prentice-Hall.

Sewall, J. I. (1961), *A History of Western Art*, New York: Holt, Rinehart and Winston.

Upjohn, E.M., Wingert, P.S. and Mahler, J.G. (1958), *History of World Art*, New York: Oxford University Press.

9 The evolution of Mediterranean and Western religion

9.1 Introduction

As with the previous chapter, limitations of space and time restrict the scope of this chapter. By and large, we shall confine ourselves to the development of the Semitic-type religions (largely in Europe and the Mediterranean area), and ignore important developments in India, China and elsewhere. In sections 9.2-9.4 we cover some of the key concepts embraced in analyses of religion's development, particularly primal religion, and then focus on the polytheistic beliefs of: primitive man; the early urban civilizations in Mesopatamia and Egypt; Persian Zoroastrianism; and Classical Greece. Section 9.5 examines the outburst of major developments associated with the Axial Age, and specifically early Judaism, the first major monotheistic religion, and then moves on to a discussion of early Christianity and Islam as later manifestations of monotheism. In sections 9.6 to 9.8 we examine the growth and breakdown of Roman Catholicism during the Middle Ages and the advent of the Reformation, which was followed by Enlightenment, all leading to the major transition to Individualism at first constrained and later unbounded, but not omnipresent in all faiths in the West.

The closing section contains concluding remarks and fairly extensive examination of the presence in religion's development of commonalities, parallels and relationships embodying concepts and processes discussed in preceding chapters.

9.2 Some preliminary remarks on religion

There exist many definitions of religion and usages of the word religious. For the purposes of this chapter, we may start with a definition of religion which draws upon two of the statements in the Oxford English Dictionary (1989, p.569). Religion is 'Action or conduct indicating a belief in, reverence for, and desire to please, a divine ruling power; the exercise or practice of rites or observances implying this.' Also, religion is 'Recognition on the part of man of some higher unseen power as having control of his destiny, and as being entitled to obedience, reverence, and worship; the general mental and moral attitude resulting from this belief, with reference to its effect upon the individual or the community; personal or general acceptance of this feeling as a standard of spiritual and practical life.'

More generally, and in terms of the concepts and phenomena of interaction and interdependence discussed in previous chapters, we conceive a religious phenomenon to be a combined cultural-social-psychological-economic occurrence. It takes place in historical time and is conditioned by the historical past, immediate or not (see Eliade, 1969). Nonetheless, there are specific elements or dimensions key to religion. First, there is the ritual dimension, one that can be extremely simple and related to an individual's belief in some mystical or divine power whose support, favor or appeasement is sought. It may involve simple prayer, sacrifice or other elementary act, not necessarily sacred. On the other hand, a ritual can be very formal and elaborate as a High Mass in the Roman Catholic church. However, it can become conventional and purely habitual with little inner personal involvement.

Ritual practices are often associated with myths—stories that are handed down by one's associates, group or culture—stories that can for example relate to godly figures or historical events. Myths when combined with sacred rituals are often integrated in an orderly manner or otherwise interconnected in a simple or complex fashion to yield a doctrine or theology that has coherency, clarity and meaning to a follower—sometimes at a low level that can be directly perceived by a simple intellect, and at other times at a complex level that can furnish a total picture of reality to an inquisitive, deep-probing intellect.

Very often where rites are common to a group and associated with myths, a code of ethics comes to exist to provide moral standards to guide the behavior of group members—standards, however, that frequently are violated or given lip service by the members. Nonetheless, while in formal, highly organized religions these standards are forcibly reiterated again and again, again and again they may be widespreadly ignored, particularly in intense emotional situations as when war exists and non-observance of a moral code is permitted in the name of God.

Associated with the individual mystical, psychological and other inner experience are the mass psychological or social experiences that are embedded in religion. Organized religion, even when the bonds tying members together are highly informal, does necessarily require contact by members and typically normal social activities that accompany the ongoing life of an institution. In turn, the economic, political and social dimensions of a community can influence the organization, functioning and even theology of a religion—for example, when political dictatorship obtains.

Finally, most religions rely upon intense personal experiences of prophets and mystics to provoke and sustain a take-off. The experiences of Moses, Zarathustra, Christ, and Muhammad are cases in point.

9.3 Religion of prehistoric (primal) man

While there is little known about the rites and religions of primal man, there is considerable evidence that he did have spiritual feelings and some notion of the sacred. Some 100,000 years ago, Neanderthals practiced the ritual interment of the dead. There is also evidence of the cult of skulls. However, the most unambiguous evidence of the practice of religious rites comes down to us in the cave paintings of the late Paleolithic era. As already mentioned in the previous chapter, there exist in these caves amazing works of high artistic value. Many depict animals (bisons, deer, horses, etc.) that suggest magic or appeal for help from a mystical power. Others, however, which portray human beings disguised as animals, clearly establish ritual practice. Perhaps the most spectacular of these is the *Sorcerer of Trois Frères* (at Ariège, France), the depiction of a human being with large (owlish) dark eyes, wolf ears, lion claws, stag antlers and a horse tail—a portrayal of a shaman, or even priest, or still more unlikely, God himself.

A number of useful concepts in the study of prehistoric religions have been developed by scholars. There is the concept of *mana*. It is an unseen or secret force able to be very powerful that permeates both the spiritual and material aspects of life. This force is present in things (rocks, arrows, and plants) and beings (animals and humans). It can work for good or bad, can protect or destroy, can be manipulated and transferred between persons and objects. It is both magical and part of prehistoric man's spiritual (religious) life, all being inextricably woven into and inseparable from his culture.

Some scholars speak of the *numinous* experience—'the feeling of awe, dread, mystery and fascination men experience when confronted with what is holy, uncanny or supernatural' (Smart, 1984, p. 30). Others emphasize *animism*, the belief that spirits inhere in natural objects, that Nature has a soul.

Then there are the concepts such as: *tabu* (where sacred forces are embodied in persons or acts such that a profane person must keep at a proper distance from them); *totemism* [where a people feel deep kinship with non-human creatures or things (as a serpent, plant, hilltop) and where such are revered and form the basis of ceremonial rites]; and practice of ancestor veneration (where interment of the dead may involve the burial of personal effects and offerings).

Shamanism is another type of phenomena that scholars emphasize. As

Bancroft (1987, pp.20-1) notes:

> The shaman is the oldest human manifestation of spirituality known to us, and one which has apparently continued without a break from the Ice Age or earlier up to the present day. He is to be found in many primitive societies, particularly in Northern Siberia among the Gilyaks, in North American Indian tribes, in Eskimos, and among the Aborigines of Australia. It is known that about 20,000 BC, when shamanism was at its height, there was a wide dispersement of peoples all over the world and it is now believed that where societies and tribes have remained virtually untouched by civilization the ancient religion has remained much as it was. . .
>
> What is a shaman? . . . Shamans are 'specialists of the sacred,' men able to 'see' the spirts, to go up into the sky and meet the gods, to descend into the netherworld and fight the demons, sickness and death.
>
> Shamans have discovered the ways of using ecstasy for the good of the community. Through his ecstatic trance the shaman has an 'experience' of the sacred. He is believed to be able to communicate with the gods and with the dead, with animals and birds. By means of his special techniques—which include fasting, meditation and certain drugs—the shaman attempts to transcend the normal state of man. He wants to rise above what is thought of as the unsatisfactory and fallen state and to find once more the primordial consciousness of purity and innocence. . .

The shaman

> can retrieve the wandering soul of the person who is ill and bring him back to his body; go with the souls of the dead to their next existence; appeal to the gods for their mercy on the soul of a sacrificed animal.

The importance that is attached by scholars to shamanism in the religious life of primal man undoubtedly differs, and should differ, in detail from tribe to tribe—owing to differences in tribal physical environments and historical experiences. The differences in importance also reflect the different disciplinary approaches among scholars.

When we move on to examine the role of myths and their development, we find a more widespread scholarly recognition of their central position. Campbell (1972, pp.19-20) in fact sets forth strong claims. First, he maintains 'the fundamental themes of mythological thought have remained constant and universal, not only throughout history, but also over the whole extent of mankind's occupation of the earth.' Furthermore, in 'every one of the mythological systems that in the long course of history and prehistory have been propagated in the various zones and quarters of this earth, these two fundamental realizations—of the inevitability of individual death and the endurance of the social order—have been combined symbolically and constitute the nuclear structuring force of the rites and, thereby, the society.'

Whether or not, one defines myths, rites and religions in the manner that Campbell does, and agrees with his view that mythical aims and laws were of primary importance, while economic were secondary, the role of myths—legends of how the earth came into existence, stories about ancestors, etc.—must have been central to primal societies, and in many cases, if not all, bound inextricably with religious rites and practices.

There, of course, have been many theories about religion and spiritual practices of primal man. At one extreme is that of Schmidt (1931) who argued that first there was monotheism, a belief in a single High God, which later became overlaid by animistic, polytheistic and other notions. At the other extreme are the logical, evolutionary theories that suggest a kind of natural theology. Such theories, according to Smart (1984, pp.49-50),

> first infer the existence of a soul, then infer that natural objects also have souls, and then connect up different phenomena, such as rain, or fire, in order to ascribe them to a single deity working behind and through them. The theory also presupposes that a sharp distinction between spirit and matter can be drawn in primal thinking. But we have seen in contemporary preliterate cultures the personal and impersonal shade off into one another and that the visible and the invisible are inextricably interwoven. Moreover, the use of red ochre in primitive burials to symbolize blood and life suggests that prehistoric men did not think simply in terms of an invisible and immaterial soul-force.
>
> However, it is perfectly true that the earliest human remains indicate a cult of the dead, while the cave paintings which hint at mythological beliefs about divine beings must be dated later; and this is consistent with the view that ancestor worship precedes polytheism. But the evidence is not at all conclusive, simply because it is more likely that modern archeologists will find much more evidence of interment than of other features of religious belief—skulls and bones survive a long, long time, while the language and thoughts of the earliest men are lost forever.

Yet despite our extremely limited knowledge of prehistoric societies, we may hold that at the very start of reflective capability in homo sapiens (and primal societies), rites, tabu, totemism, shamanism, myths and other spiritual and religious aspects of his group life, however simple, were all unified and inseparable from what we might designate his culture. Whether one takes the position consistent with the thesis of Campbell (1972) that asserts the dominance of myth and the universality of its basic features, or accepts the contention of Eliade (1974) that regards shamanism as the dominant religion of present as well as archaic hunters, one can hypothesize that there existed a religious- or spiritual-type symmetry among the primal societies oriented to hunting activities—a symmetry to be broken with the emergence of agricultural civilizations.

9.4 Ancient and Mediterranean polytheistic religions

a. *Mesopotamian*

While little can be firmly stated about the practice and development of religion in primal societies (many critically important questions remain unanswered), much can be said beginning with the era of agricultural life and urban settlement. Such took place between 4,000 and 3,000 BC. From the available written records the fertile land of the river valleys (the Nile, Tigris and Euphrates) enabled population to abandon the hunting and nomadic existence. Early on, surplus food production permitted the formation of population clusters, and later villages and cities. Settlement of a much more fixed and permanent character resulted. The symmetry of a mobile hunting population was broken. Especially in the most fertile valleys where food surpluses of large magnitude were possible, large numbers of nonagricultural population—fighters, administrative officials, craftspersons and religious leaders—could be supported; centralization and formation of administrative structures of several levels in civil and religious activities became possible. In time, empires were built up, accompanied by the growth of elaborate temple cults and organized priesthood. All this development was accelerated by the increasing replacement of oral transmission of knowledge with written statements.

In the early empires of Menes (Egyptian), Hammurabi (Babylonian), Sargon (Assyrian), Darius (Persian) and Alexander the Great we find hierarchical structures that imposed restrictions on relatively immobile agricultural and urban populations—restrictions that were far cries from those few that primal society imposed on the wanderings of small hunting populations. Also, in the evolving structure of mythologies incorporated in polytheistic religions, we see an increasing hierarchical organization. Witness the top God Marduk of the Mesopotamians, the all-powerful Zeus of the Greeks, and the hosts of gods of lesser rank—all this being a prelude to monotheism.

The earliest written documents which go back to the third millennium BC are Sumerian. From them and other evidence it can be inferred that a population speaking Sumerian, a non-Semitic language, came down from Central Asia or the Siberian steppes and settled in Lower Mesopotamia. Later, Akkadians, from the neighboring area to the North and speaking a Semitic language, infiltrated the Sumerian cities in successive waves; and under Sargon, they eventually conquered these cities. In time the religions of the two groups were fused. There resulted a set of diverse polytheistic beliefs, in part stemming from the fact that in the earlier times, before strong unification from conquests, each city had its own favored gods. Yet from the time of Sumer, there were three dominant gods who ruled the earth: ***An***, the sky god; ***Enki***, the water god; and ***Enlil***, the wind god. They were later replaced by ***Marduk*** associated with the Sun and with vegetation.

The religion that resulted from the fusion of the Akkadian and Sumerian myths, legends and cultures contained several aspects significant for later developments. One was the doctrine that the divine and human realms are separate. Only in the divine realm is immortality to be found. Recognition of the distance between heaven and earth led to the belief that the good man finds his sustenance from the holy things or beings that he worships.

A second basic aspect was the belief that Marduk, the dominant God, controlled the destinies of humans. As Smart (1984, p.249) states:

> This sense of fate or destiny gave the Mesopotamians a strong interest in prediction of destiny. They believed that they could predict not merely by earthly methods of divination, but also by a study of the stars and of planets and the moon. Men gained an understanding of the will of the celestial deities through astrology. Accurate records kept by the Babylonian astrologers paved the way for later developments in scientific astronomy. Their capacity for divination was supplemented by various other magical skills, which helped to promote the power and prestige of the priesthood, characterized by an elaborate hierarchical structure.

Related to social consciousness and concern with morality was control that was attributed to the Gods. Such became explicit in the Code of Hammurabi, Hammurabi being the all-powerful sovereign of Babylon who succeeded in unifying the several Mesopotamian states.

Other basic aspects of later significance were (1) the passing on, with adaptations, of basic myths and legends—for instance the Babylonian flood story which, with some changes in detail, is embodied in the myth of Noah in the Old Testament—and (2) the slight movement in the direction of monotheism taken by the cult of Marduk who was addressed as *Bel*, the supreme Lord.

But the powerful Hammurabi was not a god. He represented the gods; he mediated between the world of men and the world of gods, and thus served as a ritual union. Even *Gilgamesh*, the fabled king of Urek, failed to achieve immortality. Thus, there were limits to human, even kingly possibilities. Nonetheless, it was presumed that man has spirit, can obtain the blessings of gods through rites and prayers, and in ziggurats, which for the Babylonians represented centers of the world where the gods came down to earth, man could insure communication with heaven and gods.

The synthesis of the Sumerian and Akkadian cultures in the fertile valleys of Mesopotamia resulted over time in major social, political, economic and religious organization. Such occasioned significant contributions to later religions—the notion of everlasting life for gods but not men, the exaltation of certain gods to a supreme rank, the development of the personal element in religious experience, and the recognition and acceptance of the limited capabilities of each individual, which in turn permitted a tragic view of human existence so strongly present in later Greek religion and culture.

b. Egyptian

The Egyptian civilization took root during the fourth millennium BC. This civilization was completely oriented to the valley of the Nile river whose fertility rested on the mud-strips left behind from the river's annual inundations. As in Mesopotamia, surplus food production permitted the formation of small population clusters, villages, towns and ultimately a capital city.

Eliade (1978) states that it was the contacts with the civilization of Sumeria that brought about a major mutation in the previous Neolithic cultures situated along the Nile. The cylinder seal, brick construction, boat-building know-how and above all writing were imported from Sumeria at the beginning of the First Dynasty, about 3000 BC. Unlike the Mesopotamian society, the Egyptian was isolated by geography and so not subject to invasions by outsiders. Coupled with the navigability of the Nile, this enabled an increasingly centralized administration to develop over the years and to be ruled by a pharaoh. The pharaoh was assumed to have divine status, thereby, according to belief, being the meeting point between the invisible and visible worlds, the point of contact between heaven and earth, and consequently the pivot of Egyptian economic, social and cultural development.

Menes, in year 3200 BC, founded the united state, thus becoming the first pharaoh. According to Eliade (1978, p.86),

> the pharaoh, incarnate god, established a new world, a civilization infinitely more complex, and far higher, than that of the Neolithic villages. It was supremely important to insure the permanence of this work, accomplished in accordance with a divine model—in other words, to avoid crises that could shake the foundations of the new world. The divinity of the pharaoh constituted the most potent guarantee. Since he was immortal, his death meant no more than his translation to heaven. The continuity from one incarnate god to another incarnate god, and hence the continuity of the cosmic order, was insured.
>
> It is remarkable that the most important socio-political and cultural creations took place during the earliest dynasties. It is these creations that established the models for the following fifteen centuries. After

> the Fifth Dynasty (2500-2300 BC), almost nothing of importance was added to the cultural patrimony. This 'immobilism,' which is characteristic of Egyptian civilization but which is also found in the myths and nostalgia of other traditional societies, is religious in origin. The stability of hieratic forms, the repletion of gestures and exploits performed at the dawn of time, are the logical consequence of a theology that considered the cosmic order to be supremely the divine work and saw in all change the danger of a regression to chaos and hence the triumph of demonic forces.

As with a number of other mythological-religious systems, the specific myths and rites that were based upon events that took place in the 'First Time,' the fabled time of the beginnings, represented the perfection of all things. Although disorder and evil subsequently intruded, such could be and were overcome; and the successful practices in the ensuing resurrection efforts came to have the status of sacred rites to supplement the rituals and formalities of life (the calendar, royal emblems, etc.) established at the First Time. The rigidity and prescribed forms of Egyptian art, already mentioned in the previous chapter, was an instance of the inflexibility within the culture. This is not to deny that some changes did occur. Advancing knowledge and technology, and cumulating experience with administration inevitably led to change—but in the Egyptian case, slowly and at times even imperceptibly. For instance Menes, the first sovereign, built Memphis as the new capital for the united Egypt, and there he was crowned as were all later pharaohs for more than 3,000 years.

Yet, there were perturbations during these thousands of years. We have already mentioned in the previous chapter the short-lived effort of Akheniton at monotheism and a more humanistic way of life. There also were a number of adaptations regarding the assembly of Gods and their world. Initially there were: *Horus*, associated with the divinity of kingship; the great sun god *Re* of the 'First Time' myth, who was later reconceived at times as *Re-Atum-Khepri*; the vegetation fertility god *Osiris*, who like vegetation dies but is resurrected as vegetation is in the springtime, and who served as a paradigmatic model for the pharaohs (their death and passage to eternal life); and others, some of whom were adaptations of the numerous deities associated in the early days with the different rural settlements, each having its own gods.

Central to the Egyptian religion was not only the belief in the immortality of their divine beings, the pharaohs, but also the personal faith inspired by the myth of *Osiris*—a myth which allowed the Egyptian to associate himself with the dying and rising god. This was different from the Babylonian religion wherein a divine being could not achieve immortality, no matter what; the distance between heaven and earth was insurmountable. This difference led in the Egyptian civilization to the cult of the dead—concentration on magical ways to preserve and enable the deceased human on earth to attain life in heaven, in contrast to the Babylonian concentration on understanding the will and predicting the actions of deities who controlled matters.

c. *Persian Zoroastrianism*

As already remarked, the monotheism of the Egyptian pharaoh, Akheniton, represented a short-lived perturbation, a set of memes which constituted a short-lived survival machine. Another perturbation which early on promised to be a long-enduring and major religious force—a long-enduring survival machine of major import—was Persian Zoroastrianism. This religion emanated from the teachings and life of Zoroaster, more properly designated Zarathustra. While there is uncertainty about the time and place in which he lived, they are usually taken as the early half of the first millennium BC and the Eastern part of Iran, respectively. Zarathustra's beliefs were in strong contrast to those of the polytheistic, animal-sacrificing tribal society ruled by a militaristic aristocracy into which he was born.

In a mystical experience, Zarathustra receives the revelation of a new religion directly from the Lord, *Ahura Mazdã*—an inward-seeking ethical and more abstract religion compared to the archaic ritualistic one he derided. In one sense, it is a monotheistic religion. This comes about because *Ahura Mazdã,* is the father, among other Entities, of twin Spirits. One, *Spentya Mainyu* (the Beneficent Spirit), chooses goodness and life. The other, *Angra Mainyu* (the Destroying Spirit), chooses evil and death. These two options are available to each individual who has full freedom to choose either one. Zarathustra proclaims that all should choose to follow the paths of *Ahura Mazdã* and *Spentya Mainyu* and lead the good life. For in the Indo-Iranian tradition of the journey of the dead, there will be a sorting out. Those, who in their life on the earth choose the good and just ways, will be admitted into the House of Song (paradise); those who choose sin and wildness will remain forever guests of the House of Evil.

While the above proclamations of Zarathustra can be considered an advocacy for monotheism, careful analysis and later revisions and revaluations when widely accepted preclude a monotheistic interpretation. As Eliade (1978, pp.313-4) writes:

> In short, the point of departure for Zarathustra's preaching is the revelation of the omnipotence, holiness, and goodness of *Ahura Mazdã*. The Prophet receives the revelation directly from the Lord, but it does not found a monotheism. What Zarathustra proclaims, and gives as a model to his disciples, is the choice of God and the other divine Entities. By choosing *Ahura Mazdã*, the Mazdean chooses good over evil, the *true* religion over that of the *daevas*. Hence every Mazdean must fight against evil. No toleration is allowed in respect to

> the demonic forces incarnated in the *daevas*. This tension will very soon harden into dualism. The world will be divided into the good and the evil and will end by resembling a projection, on all cosmic and anthropological levels, of the opposition between the virtues and their contraries. Another opposition is barely indicated but will have a great future in Iranian speculation: that between the spiritual and the material . . .

Zoroastrianism reached a high point when it was adopted by the Persian monarch Darius, if not earlier by his father Cyrus, and became the dominant religion of the Persian empire. However, association with political affairs and authorities contributed to the degrading of its monotheistic aspect. Later after the breakup of the Persian empire by Alexander the Great, there was decline as a consequence of Hellenistic influences but also a spreading of the Zoroastrian doctrine far beyond the Iranian world. Still later in the Arsacid and Sassanian dynasties (226-642 AD), Zoroastrianism became a hierarchically organized religion in the service of the ruling classes, warring aristocracy and crown. In the more modern period, the fervor of Islam has pushed Zoroastrianism out of the Persian world and most of western India which Zoroastrianism had penetrated. Today, the practice of Zoroastrianism exists, but only on a relatively small scale in India with its main center in the Bombay area; it is practiced by prosperous Parsees in a non-proselytizing manner partly to preclude influx of lower caste Indians.

While Zoroastrianism, as a set of memes and thus a survival machine, has lost out in the long run to other religions in competing for world constituency, it has left behind a rich heritage. Early Zoroastrianism made vivid the association between religion and virtue—the rewarding of the good and the punishment of the wicked—and made explicit the options that man faced throughout his life—in Christian religion, for example, between Christ's salvation and Satan's hell. Its conceptions had a forceful impact on the pre-Christian Judaism, which in turn fed into the succeeding Christianity.

d. Greek and Other

In the Greek Archaic and Classical periods, the polytheistic beliefs that characterized Greek religion was significantly different from Zoroastrianism. There was no prophet, no form of revelation; there was no sacred book, no specialized clergy, no church. The religion was not strictly defined, nor did it make available a creed or coherent set of beliefs to which a faithful follower could subscribe. It comprised a set of myths handed down: (1) orally in each household in stories related to children; and (2) through the voices of the poets, especially Homer and Hesiod. While each city-state of Greece had its own set of gods, there were still by consensus common ones: Zeus, the all powerful father of gods and men; Apollo, the god of light who represented poise and harmony; Dionysius, a vegetarian deity, symbolizing passion and earthliness. Perhaps more than other polytheistic religion that were its contemporaries, the Greek religion was inextricably bound to all the elements of a containing intellectual culture and civilization. It was subject to a free and critical examination by the Greek citizen, but one that did not lead to disbelief. However, it made clear that man could not achieve immortality, and that death and likelihood of tragedy awaited him in the end.

Overall, the Greek religion was not a deep, penetrating spiritual movement dominating the existing cultural system. Yet, among the Greeks, there were some cults—Dionysus, Orphism and Eleusinian mysteries—which did provide a more personal relation to a faith. This was also true of the Isis, Mithra and other cults that sprang up in the Hellenistic age. These together with others developed in the times of the Roman Empire—the state cult, Epicureanism, Stoicism, the Plotinus cult—were all part of the smorgasbord of diverse religions, cults and deities and jumble of religious rites through and out of which strong monotheistic religions emerged.

Classical Greece has been categorized as having, like Ancient Rome, a primordial religion. There were theological-philosophical theories constructed by intellectuals but their ideas and notions remained outside the realm of religious practice and influence. Classical Greece without a well-developed religious doctrine was not able to generate a world religion (see Shils, 1987, pp. 259-63).

9.5 The Axial Age and the 'Great' religions

The period 800-200 BC has been designated the Axial Age. During this period, there emerged 'great' religions as opposed to the existing 'pagan' ones. The religious belief systems that developed have shaped the contours of the respective civilizations in which they were embraced and came to exercise major influence. According to Jasper (1949), the Axial Age civilizations included 'ancient Israel, ancient Greece, the early Christian world, Zoroastrian Iran, early Imperial China, the Hindu and Buddhist civilizations, and, though post-dating the Axial Age proper, the Islamic World' (Eisenstadt, 1987, p. 313). What distinguished the 'great' from the pagan religions was the nature of the perception and definition of the relation of the given 'mundane' order and the otherworldly 'transmundane' order.

As Eisenstadt notes:

> In all human societies, the transmundane order has been perceived as somewhat different, usually higher and stronger, than the mundane one. In pre-Axial Age pagan civilizations, this higher world was symbolically structured according to principles very similar to those

> of the mundane or lower one; in other words, there existed a high degree of homology between them. Relatively similar symbolic terms or connotations were used for the definitions of both gods and man and for both the mundane and transmundane orders, even if there always was a stress on the differences between them. In such societies the transmundane world was usually equated with a concrete setting, 'the otherworld' which was the abode of the dead, the world of spirits, not entirely unlike the mundane world in detail.
>
> By contrast, in the Axial Age civilizations, the perception of a sharp disjunction between the mundane and transmundane worlds developed. There was a concomitant stress on the existence of a higher, transcendental moral or metaphysical order that is beyond any given this-worldly or otherworldly reality (Eisenstadt, 1987, p. 313).

And, generally speaking, if one sought salvation and immortality, as was and is the case in many if not all societies, it was no longer possible just to fulfill one's current obligations to one's group. One needed to reconstruct his behavior and personality on a higher moral, spiritual and more uplifting level. It would be necessary for one to be compassionate, to have deep concern for the suffering and poverty of others, to behave consistently with a sincere consideration of the welfare of each and every member of one's world community—thereby to pursue principles of a higher ethical and sacred order that for the person could bridge the chasm between the transcendental and mundane worlds.

How come there was during the relatively short period designated the Axial Age such a concentration of religious developments which had major impact on succeeding civilizations? There is no answer to this question. As Armstrong (1993, p.27) carefully speculates, the new ideologies and religious systems

> reflected the changed economic and social conditions. For reasons that we do not entirely understand, all the chief civilizations developed along parallel lines, even when there was no commercial contact (as between China and the European area). There was a new prosperity that led to the rise of a merchant class. Power was shifting from king and priest, temple and palace, to the marketplace. The new wealth led to intellectual and cultural florescence and also to the development of the individual conscience. Inequality and exploitation became more apparent as the pace of change accelerated in the cities and people began to realize that their own behavior could affect the fate of future generations. Each region developed a distinctive ideology to address these problems and concerns.

In China there were: (1) Taoism (with emphasis on the goals of spontaneity, naturalness, and techniques of self-transformation, and on the play of the opposite [yin and yang] leading to harmony for man); and (2) Confucianism (an all-encompassing humanism, concerned with self-culture and filial piety to achieve a desirable ethical ordering for society). In India, there was Hinduism, a complex religious tradition which has no prophets (or particular founder) and no set creed. Within it: the sacredness of the ancient texts, the Vedas, is generally accepted; belief in reincarnation and transmigration exists; and the right way of living is given critical importance with the consequence that diverse religious beliefs and practices came to coexist. Also in India there developed Buddhism; it advocated the extinction of desire, craving and clinging which is the source of suffering and pain, and urged a search for complete detachment and equanimity except for the promotion of good, thereby to achieve release from all opposition and the realization of the state of *nirvana*.

In the Middle East, there was the occurrence of Zoroastrianism with monotheistic elements, which we have already discussed and of monotheistic Judaism which we shall examine in the next section.

Whatever the causes or conditions that did give rise to the recorded concentration of religious developments in the Axial Age, one can maintain that there was a breaking of symmetry in the preexisting pagan world—the 'great' religions being offshoots which subsequently came to dominate civilizations. Put otherwise, one can maintain that these 'great' religions were perturbations that came to be amplified into survival machines that are with us today in a major way, except for Zoroastrianism.

The 'great' religions can also be categorized as: (1) those which are monotheistic in which there is a concept of God external to the universe but who is able to guide it; and (2) those like Hinduism and Buddhism with metaphysical nondeistic bases. Since we are concerned with the Middle East and Western civilizations, where religions have embodied, almost without exception, the presence of Gods, we shall not consider the nondeistic religions of the Far East.

We have already noted that Greek religion, by and large, did not have a deep religious spirituality, was not inspired by a prophet, had no form of revelation, no sacred book, no specialized clergy, no church. Yet the Greek society during Greece's Classical age belongs to the Axial Age civilizations. In the preceding chapter we noted how this society was oriented to the search for rationality and the explicit capture of beauty of form, especially of gods and goddesses, and how its emergence represented a major break of symmetry—a Prigogine jump—in the arts and also in the field of philosophy. But this break and major advance did not see any corresponding development in spirituality, although many centuries later Greek philosophy would significantly impact the introduction of reason and rationality into theology.

a. *Early Judaism and the outburst of surviving monotheism*

One cannot assign a definite date to the appearance of monotheism. To do so would require a precise definition of this phenomenon, and even then there might be considerable disagreement among scholars. We have already mentioned that Schmidt (1931) maintained that among primitives there existed first a belief in one god, the High God (the Sky God), which later became overlaid by animistic, polytheistic and other elements. In much later societies, there was an all-powerful top god who ruled over other gods and/or spirits, as in the case of Marduk of Mesopotamia, and in still others there was the cult of the Mother Earth Goddess, another strongly dominating figure. All this suggests an early existence of elements that could lead to a simple monotheism.

We also observed that around 1375 BC the Egyptian pharaoh Akheniton embraced monotheism and a real sense of humanity characteristic of the early stages of the 'great' religions—but this early perturbation in the development of religion which was present, too, in Egyptian painting was short-lived.

Lastly, we have noted that Zoroastrianism, which was widely adopted as the religion of the Persian empire in the middle of the first millenium BC and which gave primary emphasis to a dualistic theology, could be claimed by some to have introduced a well-constructed monotheism. But even though it was a survival machine of great influence for many centuries, its significance ebbed, being only a small dribble today.

The first full expression of monotheism must be attributed to Judaism—a perturbation that was amplified and currently continues to be strong. There stemmed from Judaism and its Semitic origins, two offshoots, Christianity and Islam, which in time also came to be major religious survival machines.

Perhaps it can be said that the primacy of monotheism was first clearly established when the Israelites who had been exiles in Babylon were able to return to Judah in the middle of the sixth century BC. The Babylonian exile had convinced the Israelites that there was one and only one mighty God, Yahweh (YHVH, an alternative designation). The exile was viewed as having resulted from their sins and transgressions—their failure to abide by the covenants handed down to Moses.

The embracing of monotheism by the Israelites was not a sudden phenomenon. The process involved a gradual transition with ups and downs. According to the Bible, Abraham is the founder of the belief in One God. He is told by Yahweh that he will become the father of a far-reaching mighty nation. The evidence that we have is that Abraham led his tribe into Canaan, the Promised Land. There, he, his son Isaac and grandson Jacob lived supposedly as monotheists, but more likely they would have believed in some of the deities of their neighbors in Canaan, and in particular in *El*, the High God of the Canaanites (see Armstrong,

1993, pp. 11-13.) This would havc been so even though the Bible tells us of the trust that Abraham and Jacob placed in the High God and thus could be considered to be men of faith. Yet, the beliefs of the Israelite leaders and tribes continued in a polytheistic manner through the period up to the Exodus from Egypt. At that time, the early years of the 13th century BC, Moses leading his people out of slavery enters into a covenant with God at Mount Sinai. This and other events of the Exodus establishes Yahweh as the definitive God of Israel; and Moses was able to convince his followers that Yahweh is the same High God in which Abraham, Isaac and Jacob believed.

Although Yahweh was the primary god of the Israelites, they continued after Moses' time to worship other gods, likely with fervor steadily diminishing over time and with ups and downs through the period of Solomon's reign (tenth century BC), and through the times of prophecies (by Ezekiel, Isaiah and others), of King Joshua's reforms, and of Babylonian exile (587/6-538 BC).

In the years prior to resettlement in Judah, the prophets played a most significant role in the attainment of the primacy of monotheism among the Israelites and its subsequent ethical development. As Smart (1984, pp.293-4) states:

> the utterances of the Hebrew prophets radically altered the history of Israel. One reason for this was the fact that their attention was directed to the great God Yahweh, the God of their fathers, who was more than a localized deity or a 'first' god among many others. He had their sole allegiance, not just because they were professionally associated with his cult, but more importantly because of the profound and overwhelming character of their religious experience . . .
>
> A further factor in the prophets' revolutionary reform was that each felt himself to be the intermediary between Yahweh and the children of Israel. Just as Moses was both the potential leader of the people and the leader chosen by God to bring them to himself, the prophets were the mouthpiece of God, possessed by him; but their gaze too was turned in the direction of the people and of the society about them. God's words were addressed to the peoples' situation. This concern for society was not just political, though it was often that; it was also strongly ethical. The prophets saw in the troubles besetting Israel a warning of the misuse of the people's talents and opportunities. Injustice, aggrandizement, and false religion combined often to set Israel along the wrong path—a path, the prophets felt, that would lead to disaster. Thus they repeatedly emphasized the call to moral holiness. God was not only powerful; his nature embodied justice and goodness; hence his worshipers must themselves embody justice and goodness. Here then was a powerful development of ethical monotheism, foreshadowed and founded by Moses, and carried forward by the prophets.

Moreover, the prophets were highly intuitive about the future, could sense emerging social, economic and political problems, and thus had visions of the direction in which a successful religious movement must go. They were capable of sound judgement that leaves well behind resort to magic, ritual practices, sacrifices, and the giving of oracles.

For example, immediately before the fall of Jerusalem in 586 BC and the exile of Israelites to Babylon and elsewhere, Jeremiah was intensely at work adding a new, ethical dimension to the Hebrew religion. He preached that it was incumbent not only for each community of worshippers to abide by the covenant, to be moral in a collective fashion—as was the then current view—but also for each individual worshiper to be free of sin and lead a highly moral life—a life involving concern for the welfare of each and every person surrounding him. The direct relation to God involved not only the community but also the individual; and it was the duty of each prophet, as the intermediary between Yahweh and the children of Israel, to inform each Israelite of this relation, and to teach and lead him into moral and ethical behavior. Note that this meant that the Israelite must eschew a wholly inward, monastic type of existence. Rather he must go out and face the world, to provide help to others. His life is to involve a non-static, dynamic interaction with others.

Further, in some of the teachings of the prophets, the need for an aggressive dynamism is implied. In the period of Babylonian exile, an unknown prophet (sometimes called the Second Isaiah) preached that the Israelites, as the Chosen People, have further responsibilities. Their current suffering as exiles is a result of not only their own sins and transgressions, but also their failure to live up to their worldly obligations. God, said the second Isaiah, has ordained that the Israelites provide 'light' to all nations. Once the required cleansing and purification of the sinful Israelites was completed, God would see to it that they return to Jerusalem and there establish a society so fruitful and joyous that nations in all regions of the world would look upon the society with admiration. The people of these nations would be not only sympathetic and favorably disposed to this development, but also enlightened, able to perceive the righteousness of God, and see the way to salvation. All the world's wickedness and evil would come to an end. God would reign in full glory.

Other aspects of early Judaism are significant. There were the reforms instituted by King Joshua. One was to concentrate the priests of the land in Jerusalem to rid the worship of gods other than Yahweh prevalent throughout the land. Local areas, deprived of their religious leadership, developed a more informal way of synagogue worship under local control—the forerunner of the common form of service prevalent today in Judaism and Christianity.

Another aspect was the partial or complete replacement of military-political leadership with a government led and organized by priests—a development fostered in the middle of the sixth century BC by Cyrus the

Persian ruler who judged such to be beneficial for the preservation of his empire. As a consequence, there was established a tradition of law promulgated by the priesthood with interpretations and reformulations by the rabbi of the local synagogue consistent with the local environment—a development which proved to be so essential for the survival of Judaism in later years when Israelites were dispersed throughout the Roman empire.

A third aspect concerned the conceptions of a Messiah and Apocalypse. There had always been thought, undoubtedly partly influenced by Persian and Zoroastrian beliefs, that a Great Day would come when all enemies of the people of Israel would be overcome, when a grand new kingdom under Israelite leadership would emerge. Toward the end of the pre-Christian era, this belief was transformed into the projection of a Messiah descending from heaven who would restore the nation of Israel and lead it and the world which it would come to influence and/or govern to a glorious and prosperous eternity.

Classical Judaism came to an end and the hopes and visions of the Israelites were shattered when in 70 AD the Romans squashed the attempted rebellion organized by a group of political zealots. The temple in Jerusalem was burned to the ground and again the Jews had to flee; and when another, the last revolt, in 132 AD was put down, the country of Israel was depopulated.

The exile and scattering of the Jews throughout the Roman empire initiated a new development in Judaism—a development which was the outgrowth of the need for the effective fusion of basic beliefs and the practices already established and which consequently strengthened even further Judaism's capabilities as a culturally-live survival machine. One element was the monotheistic belief in a strong personal God, accessible to each and every individual in ways that could be *diverse* to reflect the different experiences of individuals. With reference to the Pharisees, the most respected and important sect of Palestinian Judaism at the end of its classical period, Armstrong (1993, p.72) writes that they believed that:

> God could be present in the humblest home as well as in the Temple. Consequently, they lived like the official priestly caste, observing the special laws of purity that applied only to the Temple in their own homes. They insisted on eating their meals in a state of ritual purity because they believed that the table of every single Jew was like God's altar in the Temple. They cultivated a sense of God's presence in the smallest detail of daily life. Jews could now approach him directly without the mediation of a priestly caste and an elaborate ritual. They could atone for their sins by acts of loving-kindness to their neighbor; charity was the most important *mitzvah* in the Torah; when two or three Jews studied the Torah together, God was in their midst.

A second basic element was the Torah. As the embodiment of the covenant handed down by God at Mount Sinai and the expression of God's will explicitly set forth in all the interpretations of the sages over the centuries, the Torah was able to and did provide guidance on how to live for all the Jews no matter where dispersed and no matter what their adversity. As a book of law and principles regulating everyday living and nurturing a person's spiritual needs it preserved the cultural identity and common interests of a geographically spread-out population.

A third basic element was the traditional practice of Torah reinterpretation. Despite the extensive cumulation over the centuries of interpretations by wise and renowned scholars, at any time the existing interpretation of Torah doctrine was subject to change—change reflecting the changing times and the evolving environment—change that could be and was instituted at the local synagogue level by the rabbi and his congregation—change that could be and did come from continuing Talmudic study by pious members of that congregation.

These three elements, God, Torah, and practice of Torah reinterpretation constituted for Judaism a heritage that was dynamic—a heritage with remarkable capacity for adaptation to the times and environment over the centuries to follow—all without the need for a central Temple and an extensive multilevel hierarchical organization.

b. Early Christianity: An outgrowth of Semitic-Judaiac culture

Less than a century before the destruction of the Temple in Jerusalem and the forced exile of the Jews, there was born in 4 BC in Nazareth a charismatic preacher and faith healer, Jesus, later given the name Christ (the Greek equivalent of the 'Anointed'). He and more importantly his devoted followers, the Apostles and their associates, were to institute in important ways a more simplified form of monotheism—a form that was to find great appeal to and capture the imagination of the masses, pagans as well as Jews and others. It was in later times capable of successful competition against another simplified monotheism which was also to have appeal to and capture the imagination of the masses—those belonging to the Arab culture and cultures subjugated by the devout followers of Islam.

Many hypotheses have been and can be advanced to explain this outcome. Involved was an intricately interconnected set of forces that clearly cannot be completely unravelled. In the background was advancing technology in transportation, communication and production that greatly facilitated the diffusion of ideas and the travel of charismatic and evangelical preachers carrying the message. At the same time, as we shall see, Christianity embodied a more simplified message—one that was easy to respond to impulsively, to be embraced immediately and subsequently to incorporate into one's behavior, however uneducated and simple one might be. In other respects, too, the time was ripe. There was high expectation among the Jews that a Messiah would soon appear to rule as a king over a perfect world that was promised. The turbulence created in the established Jewish communities by the burning of the Temple at Jerusalem, the center of Judaism, and the exile of Jews to disparate parts of the Roman Empire created a partial vacuum that cried out loudly for new leadership. Additionally, the accumulated social and political know-how with respect to the exploitation of hierarchical organization for achieving efficient and far-reaching effective control over vast areas—gained from the experience of organization within the Roman empire over the centuries—was later to become available for disseminating efficiently and effectively an appealing religious message over populations in many distant areas. Lastly, the official adoption by Constantine of Christianity as the religion of the Empire, and later by Charlemagne of the Holy Roman Empire were, partly reflecting the influence of the above factors, clinching events for the widespread adoption and strength of Christianity.

These points need further elaboration. First, one may advance the hypothesis that a primitive's failure to understand thunder, or the eclipse of the sun, or the sudden death of a loved one, or any of a number of other events may each be associated with a spirit, power, force, or a particular God controlling that event. Yet this many-God polytheistic perspective can and most probably did come to constitute a perplexing, overcomplicated, and/or confusing situation for many. For many it may also have involved inconsistencies. The notion of a single God, in control of all these events, forcefully and persuasively set forth by a charismatic, mystical figure, can easily simplify the situation for a primitive or uneducated human and have considerable appeal for him. That such a charismatic mystical figure comes to emerge is not at all unlikely. History records the not infrequent appearance of such individuals. And while these individuals may have had more than one intense mystical experience, and associate each with a different God, it would seem equally likely that one intense experience would be adequate to set them off on their godly evangelistic journey. Even when some individuals had more than one such experience, it is likely that one or more of these experiences could be associated with the same unique God.

Jesus was one such figure—as were Abraham and Moses and later Muhammad. According to the Apostles, Jesus' baptism by the ascetic, John the Baptist, was a deep mystical experience, revealing to him "the image of the Spirit of God descending in the form of a dove and saying: 'this is my Son, the Beloved' " (Eliade, 1982, p. 331). After being driven by the Spirit into the wilderness, there to be tempted by the Satan, Jesus emerges untainted, so it is said, and begins his practice of baptism.

It is known that Jesus was brought up as a Galilean Jew, who probably was well acquainted with the Torah and the Jewish law. While he was respectful of the substance of the law, he was certainly critical of the detailed and fine interpretations that the Pharisees gave to it. Also, he did not go along with the views of the other Jewish factions—with the hatred of the Zealots for the Romans and their call for violent rebellion, with the compromising and appeasement-like attitudes and approaches of the Sadducces, and with the withdrawal and monastic propensities of the Essenes. His life views and approaches were unique and independent of the ways of these groups. In the uncertain and troublesome times when the Jewish nation was crumbling under the might of the Roman imperialistic forces, he recognized, with the penetrating vision of a prophet, that there was a great need by 'all' for a comforting message *and* that the Jewish doctrine inextricably bound to the existing Jewish law could appeal to only a limited population. To him, 'all' embraced the law-abiding Jews, respectful public servants and citizens of all beliefs, terrorists (rebels), uneducated pagans, aliens, social outcasts, and even the hated tax-gatherers. To achieve salvation, to attain the spiritual comfort of being among those under the protecting hood of a warm personal God, it was not necessary for an individual to have led a lawful, righteous life, to have abided by the requirements of popular and official religions. This comfort was obtainable immediately. All that was needed was faith—faith that harks back to the faith of Abraham—faith to be expressed right here and now—faith embracing an inner attitude of surrender and openness to God.

This was the essence of his message, which he proclaimed to all with unswerving confidence and tremendous zeal. There were, of course, other aspects of this message and, later, extensions and restatements of it in the writings of the Apostles, especially Paul. Early on these followers developed the belief in the resurrection, which made of the Apostles a group of resolute men, certain that Jesus was the long-awaited Messiah, and thus, through devotion and his mediation, certain of their own God-given powers and the invincibility of faith. These beliefs persisted whether the kingdom of God was proclaimed to be actually present already accessible to the devout through Jesus, or imminent in the near future.

Of course, numerous rites developed. Among others were: (1) *Baptism* (representing death and resurrection by immersion in water, associated with the death and resurrection of Christ and providing identification with Him—an extension of the practice of immersion of the Judaic Esseneses sect in order to cleanse one of one's sins); (2) the *Eucharist* (the sacrament of Holy Communion in which bread and wine are consecrated and received as the body and blood of Jesus or as symbols of them—an extension of the Jewish domestic liturgy of the blessing of the bread and wine); and (3) the *Pentecost* (a festival which commemorates the descent of the Holy Spirit among Jesus' disciples). Moreover, there emerged from Paul's teaching the concept of *grace* (the freely given favor and love of God available to all with faith who exude the love of God and the love of neighbor).

The Pentecost, as an historical event—a phenomenon of glossolalia, a speaking with many tongues—gave birth to the Church. Having received and internally embraced the Holy Spirit, Jesus' disciples were inspired to begin preaching the Gospel and performing miracles. Their successes abroad (outside Jerusalem) generated disputes about the Law, as kept by the

Judaea-Christians of Jerusalem, and led to a meeting of a Council in Jerusalem in 49 AD. A compromise solution was reached which required converted pagans only to abstain from unchastity and from sacrificing to idols food from blood and from what is strangled. Circumcision and other requirements of the law were abandoned.

The above makes clear the greatly reduced requirements of the new church for redemption and salvation—those of faith, love, and the sacraments. When compared to the requirements of Judaism, it is clear why the new church would have so much greater appeal to the masses. But a potential for mass appeal by itself does not insure the realization of the mass appeal. A feasible organizational structure must also be realized. Moreover, political, economic and other social events and technology must provide a favorable environment for such.

Christianity is a historical religion. It locates within man's historical record. The first revelation for Christianity is also the one for Judaism—the declaration to Moses of God, as the God of Abraham, Isaac and Jacob. There follow the revelations of the prophets of Judaism, the birth of Jesus in 4 BC, and subsequently the Christian beliefs regarding redemption and salvation. Gradually, the influence of Judaism declined, and more and more Christianity with its new dimensions came to the fore of monotheism. By the end of the 1st century AD Christianity had transformed itself from a Jewish sect to a gentile movement—a movement that challenged the tradition and conservative paganism of the Roman empire. While persecution and martyrdom followed during the 2nd and 3rd centuries, and still later in the West of the empire, they were relatively limited in magnitude. In the East, according to Armstrong (1993, p.105), by 235 AD Christianity

> had become one of the most important religions of the Roman empire. Christians now spoke of a Great Church with a single rule of faith that shunned extremity and eccentricity. These orthodox theologians had outlawed the pessimistic visions of the Gnostics, Marcionites and Montanists and had settled for the middle road. Christianity was becoming an urbane creed that eschewed the complexities of the mystery cults and an inflexible asceticism. It was beginning to appeal to highly intelligent men who were able to develop the faith along lines that the Greco-Roman world could understand. The new religion also appealed to women: its scriptures taught that in Christ there was neither male nor female and insisted that men cherish their wives as Christ cherished his church. Christianity had all the advantages that had once made Judaism such an attractive faith without the disadvantages of circumcision and alien Law. Pagans were particularly impressed by the welfare system that the churches had established and by the compassionate behavior of Christians toward one another. During the long struggle to survive persecution from without and

> dissension from within, the Church had also evolved an efficient organization that made it almost a microcosm of the empire itself: it was multiracial, catholic, international, ecumenical and administered by efficient bureaucrats.

How had the efficient organization of the Church evolved? One of the factors provoking organization was the refusal in 66 AD and subsequently of groups of Judaea-Christians to support the messianic war of the Jewish nation against the Romans. Consequently, these groups had to seek refuge in diverse cities and regions outside Jerusalem—an action that can be said to mark the beginning of the breaking away of the Church from Judaism. Initially, the Eucharist, the sacred meal, a memorial to the death of Christ, being a function fostering solidarity, continuing faith and community love, was presided over by a bishop. With growth in the number of Christian communities, the bishop (or an equivalent leader) was unable to perform this function at each locality. Presbyters (priests) had to take over the bishop's functions at the many meeting places at which the bishop, in time designated the leading religious figure, could not be present. Still later his mounting pastoral and administrative functions (disciplinary, financial and care of the needy) had to be increasingly delegated to others who came to be called deacons. With still greater growth, it became necessary for control to be exercised by metropolitan areas (e.g. by the bishop of Rome over Italy, of Alexandria over Libya and Egypt, and of Antioch over Asia Minor). The requirement for still greater centralization coupled with excellent leadership at Rome first by Peter, Jesus' chief disciple, and later by such personages as Clement, Stephen, Pope Leo I and Pope Gregory I, caused Rome, with its advantage of being the location of a top political administrative center of the times, to become a first order node of the developing church hierarchy. Also, presence at Rome of the early bishops obviously gave them an opportunity to observe how Octavius, later Augustus, the first Roman emperor, was able to institute a most orderly and efficient administration that persisted until at least the waning years of the 2nd century AD. (There were of course some years in which a certain amount of disorder occurred—e.g. during part of Nero's reign when we have the first record of Christian persecution.)

The emergence of the effective, powerful and far-reaching dominance of Christianity was not, however, independent of particular events—perhaps characterizable by some as random or happenchance, but undoubtedly to some extent influenced by the force of Christianity itself. One was the legalization of Christianity by Constantine in 323 AD, perhaps the most portentous event in Christianity's history. (As a result of this and other actions of Constantine, Constantinople became a second first order node.) Later Constantine intervened to dampen disputes within the church, believing that a strong empire required a peaceful and united church. With legalization of and lessened internal dissension within the church, the influence and power of the Christian bishops came to equal the heights that the Roman senators once attained. Still later, in 496 AD the conversion of the Franks to Christianity and more important the anointment in 800 AD of Charles the Great (Charlemagne) by Pope Leo III of Rome as Emperor of the Romans, thereby creating the Holy Roman Empire, established the unquestioned primacy of the papacy in Western Europe and also its independence of Constantinople's Christian emperor.

We need not deal here with the further growth of the power of the church. The Eastern church under the Byzantine emperors reached a peak in development in the tenth century. The Western Church entered its Age of Glory and Power in the late eleventh.

c. *Islam: The third and final (?) thrust of the Abrahamic monotheistic cycle*

There has been one other successful thrust leading to a major development of monotheism as a Prigogine-jump—a transition to a new far-from-equilibrium situation in the development of religions originating in the West and the Middle East. This has been Islam, a religion whose force and relevance has been inadequately appreciated in the West and much distorted in studies. One relevant perspective on Islam is given by the Muslim scholar Sayyid Hossein Nasr (1993, pp.427-8):

> Islam considers itself as the final plenary revelation in the history of present humanity and believes that there will be no other revelation after it until the end of human history and the coming of the eschatological events described so eloquently in the final chapters of the Koran, which is the verbatim Word of God in Islam. That is why the Prophet of Islam is called "the Seal of Prophets". Islam sees itself as the final link in a long chain of prophecy that goes back to Adam, who was not only the father of humankind. . .but also the first prophet. There is in fact but a single religion, that of Divine Unity . . . which has constituted the heart of all messages from heaven and which Islam has come to assert in its final form.
>
> The Islamic message is, therefore, none other than the acceptance of God as the One . . . and submission to Him . . . which results in peace . . . To become a Muslim, it is sufficient to bear testimony before two Muslim witnesses that 'There is no god but God . . . and that Muhammad is the Messenger of God.'

As well as being considered by the Muslims as the final religion of humanity, Islam is also viewed as representing a return to primordial religion, the diverse polytheistic and other pagan doctrines being directions which were mistakenly taken in the past. Islam states the truth that alway was. Moreover Islam is an inclusive religion; it does not relate to a single people, as Judaism does, but is for all peoples. As a final revelation it is considered to be a universal one that can encompass all religions. It holds that all rightly guided religions derive from the one and only one God, Allah. It stresses the continuity of religious experience and thus recognizes as prophets all those preceding Muhammad—the prophets of Judaism back to Abraham, and later, Christ. However, it does not view a prophet as any one other than a messenger of God, and thus denies the incarnation of Christ and the Trinity.

Like Christianity, Islam is a simple religion that can appeal to the masses. It does not have the extensive requirements for godly behavior set down by Judaic law. Also, it does not have the relatively elaborate rituals that had come to develop within Christianity. A good Muslin: prays five times a day, facing Mecca; is compassionate and gives alms to help the poor, thereby contributing to the solidarity of the community; fasts during the month of Ramadan; and if his circumstances permits, undertakes a pilgrimage to Mecca at least once in a lifetime. These practices, together with belief in one all-mighty God, constitute the five pillars of Islam.

The greater inclusiveness of Islam and its closer approach to universality than Judaism's and Christianity's is seen in its conception of religion as a total way of life, embracing not a small part but all of one's daily activities—one's job, political work, social engagements, and cultural involvements. It does not recognize a secular/sacred dichotomy; the sacred engulfs what the non-Islamic Western world considers secular and includes, guides and stimulates the development of literature, science, and all other intellectual activity. Moreover, from this view, the commitment to work for a just and equitable society pervades more extensively the life of a 'good' Muslim than that of a 'good' Jew or Christian.

Islam is the outgrowth of the work of the prophet Muhammad. He was born 570 AD in Mecca and grew up in its environs. He lived in a time when Mecca, being a strategic location along trading routes, was growing rapidly. As a consequence, certain clans of his tribe, the Quraysh, were becoming wealthy, egotistic and greedy and less concerned with the needs and survival of the much poorer clans and their members. The tribe was disintegrating morally and politically—a real concern of Muhammad. Further, the outlook for the other tribes of Arabia was bleak. Though there was a strong communal spirit within each, there was constant warfare among them.

The religious environment surrounding Muhammad involved a combination of Semitic polytheism and animistic elements. He also had some exposure to Christian teaching and came to know the ideas and practices of the Jews. Before his revelations he would regularly take his family to Mount Hira to make a spiritual retreat during the month of Ramadan. Undoubtedly he must have given much thought to, and was disturbed by, the problems within his tribe and among the Arabian tribes, and by the inconsistencies among the religious beliefs of his world. Undoubtedly, these problems and inconsistencies were present in his mind when he had his mystical experience and communicated his first divine messages: the existence of one and only one God; the imminence of the Judgment; and the resurrection of the dead that furnish to the faithful the blessings of Paradise, and to the miscreants the discomforts (but not the Christian tortures) of Hell. Other divine messages followed and in time these oral messages were written down by his companions to constitute the Koran. Its verses and chapters are the substance of the daily prayers of the Muslim; they have provided the basic principles (ethical and moral) which have guided his behavior from birth to death, have served as the roots of Islamic law, and have greatly influenced the development of the Islamic intellectual tradition. While there is no chronological order to the revelations recorded in the chapters of the Koran, the beauty of its language when recited aloud and its numinous, transcendent effect upon the average Arab has been a potent force in the spread of Islam.

Eliade asserts that Muhammad was a religious genius. It was evident in his 'will to assimilate and integrate traditional mythic-ritual practices, ideas and scenarios into a new religious synthesis' (Eliade, 1985, p. 71). For example: (1) he proclaimed his celestial ascension (nocturnal voyage) to the throne of Allah (by tradition a visit to Heaven and return with a Holy Book being an absolutely essential experience to validate a claim to be a prophet); and (2), via a reinterpretation of the pagan and immemorial Abrahamic tradition, he incorporated as a basic practice a pilgrimage to and rituals to be pursued at the rock of ka'ba where Abraham received the word of God, a practice essentially contradictory to his doctrine of absolute monotheism.

Muhammad was also a political genius. His establishment of *ummah*, the all-embracive community of the faithful, was a means of countering and eliminating tribal loyalties and the feuding they generated; additionally, it greatly facilitated the fusing of Muslims and non-Muslims into one society with a strong sense of unifying sisterhood and brotherhood. He also was extremely skillful in negotiations and via verbal accords won more in campaigns against nomad tribes than by military action. Yet, he skillfully associated his military missions with godly service, as so many previous and subsequent prophets and religious leaders have; war motivated to establish a monotheistic godly kingdom was to him to be preferred to apostasy and anarchy.

During Muhammad's charismatic leadership, Islam was able to spread throughout Arabia and be organized around Mecca as the religious center. Under the caliphs who followed, Islam spread rapidly throughout the Mediterranean area; in less than a quarter of a century it had become a world power to be reckoned with. We need not go into any detailed discussion of this occurrence. Reasons included not only the religious fervor of the tribal forces in pursuing Holy Wars, but also skillful exploitation of sudden, swift-moving, unexpected attack strategies implementable from desert bases. Moreover, the populations of the conquered territories were not in general treated harshly by the Arabs, and being disaffected by previous rule did not need to be forcefully compelled to become Muslims; persuasion, direct and indirect, was adequate. As already noted, Islam, as a simple religion, and the purest form of absolute monotheism—one disembodied of Torah law and elaborate Christian ritual and imposed hierarchical structure—was more understandable than pagan and other existing doctrines and hence easy to accept in their place.

It must of course be noted that Islam, like Judaism and Christianity, was not devoid of internal dissension. Major divisions developed, the most significant one being between the Sunnis and the Shi'ites. The Sunni branch has and still constitutes the orthodox one, following closely the revelations in the Koran, and having a strong egalitarian ethic in that decision making (policy) is based on the concensus of the community. In contrast, the Shia branch displays a messianic belief. It holds that leadership should be hereditary, to be associated with the descendants of Ali (Muhammad's cousin and son-in-law). The descendant in each generation is to be chosen by God, not by the community, who thus would be divinely inspired and like Muhammad perfect, sinless and infallible. And as with other major religions, fierce religious wars have not been uncommon in Islam, resulting nonetheless in what some would designate as progress within the ummah, which survived internal violence with vigor and strength.

9.6 The Reformation and the emergence of constrained individualism in the West

As already noted, the Jewish nation was thoroughly defeated by the Romans in years 70 AD and 132 AD. The Jews were dispersed throughout the Roman empire, and until the 20th century existed without nationality. Christianity, a simpler monotheistic religion, replaced Judaism as the dominant religion, reaching its peak of authority and power around the end of the 12th and beginning of the 13th centuries. And as we have just noted, Islam emerged in the 6th century AD as the purest of the monotheistic religions, and by the end of that century was challenging Christianity as the dominant world religion in the Mediterranean and the West.

We are not as interested in the details of history once Judaism, Christianity and Islam became embedded in society as in the elements, internal and external, of each religion which contributed to the transition to the next major regime—to a Prigogine jump to the succeeding far-from-equilibrium situation that has been characterized *as individualism, at first restricted and later unbounded in the West.*

Clearly by the 15th century AD monotheism was dominant and polarized in Christianity and Islam. Judaism comprised a relatively small, scattered constituency of little influence—and on a number of occasions it served as a scapegoat for Christianity. Other sects were plentiful, but of negligible power. Christianity and Islam were the potential seedbeds for major change.

As a religion, Islam emphasizes the community, seeking institutions concerned with the whole life of a community. It does not require an institution like the Christian church devoted to the ends of the faith alone. Moreover, to reiterate, each member of the ummah does not have a need for a priest or other religious figure to mediate his works with Allah, or to administer sacraments since there are few if any in his religion. Each member confronts God directly. Hence, there was little if any hierarchical structure embodied in Islam up to and well beyond the 15th century.

In contrast stood Christianity of the West. In its Golden Age, whether regarded as beginning with the reign of Pope Gregory VII at the end of the 11th century, or reaching its peak of power in the reign of Pope Innocent III during the 13th or later, there emerged a top-heavy, elaborate and complex hierarchical structure. As is well known, any such hierarchical structure becomes in time inflexible and subject to arteriosclerosis. This certainly was true of the Christian church of the later Middle Ages whose leadership was so concerned with its survival and power. And as can be expected when the right set of external and internal forces appeared and interplayed in an autocatalytic and crosscatalytic fashion as they did in the Reformation period during the 16th century and later, the Christian hierarchical structure toppled.

What was this set of external and internal forces? There were many, and scholars differ greatly in assigning relative importance. We can start with medieval scholasticism (the development of theological systems designed to bring into accord revelation and reason, faith and intellectual analysis) which reached its peak with the writings of Saint Thomas Aquinas (1182-1226) and which were strongly attacked by Scotus (1266-1308) and William of Ockham (1285-1349) in the 14th century. The works of the latter provided the basis for later views by Eckhart (1260-1327), diverse mystics, and those associated with the popular piety movement. These views suggested or, more strongly, insisted that spiritual uplifting and perfection no longer required a monastic life, but was also possible for the man on the street through contemplation and acts of charity and love—a democratization force. Associated with this development was the despair brought on by the 'Black Death' plague (from 1347 on), the hundred years war between France and England (1337-1453), and other pestilences and calamities which led to a need (an implicit demand) by the masses for a closer and more direct connection to God or a divine figure. Such a need came to be felt even more with translation of the bible (1385), which with the advent of printing (1450) transformed the bible into a revolutionary document for the ordinary layperson; he/she could now question the authority of the Church leaders. Wycliffe (1330-1384), the promoter of the English translation of the bible, and his followers, who preached far and wide, were highly critical of the papacy and the Church's biblical interpretations, stirring up still more popular unrest already generated by economic hardships and social injustices of the times. For example, in

Moravia, a fierce but unsuccessful religious rebellion sparked by Hus occurred in 1416.

While all these forces were at play, there was in process a major development of humanism within the European West. Already, the pre-Renaissance advances in painting were trickling down into the church-controlled society as early as the 14th century, assigning greater importance to man *per se*. Later, the mounting Renaissance interest in classical literature, philosophy and art coupled with the resulting neoplatonism and infusion into Western thinking of ancient scholarship by refugees from Constantinople (when conquered by the Turks in 1453) induced further skepticism about Christian dogma and criticism of its supernatural orientation. Increased attention was given to the original biblical texts with a new creative interest in human values. Valla (1407-1457) and later Erasmus (1469-1536) were leading figures in this movement.

Moreover, forces for nationalism were stirring and gaining strength, while a number of cities were becoming centers of wealth and economic power, all tending to oppose the univeralism of the Roman Catholic church and fostering the taking of anticlerical stances. These developments were reinforced by the discoveries of new lands and cultures by Columbus and Portuguese navigators during the 15th century, and by the subsequent influx of wealth from America.

The forces just noted and others were in the background when Martin Luther (1483-1546) came on the scene as a major perturbation—one that set in motion snowballing and mutually reinforcing drives in the social system. The outcome over the decades that followed was what must be regarded as a revolution, a drastic modification of Western Christian structure and doctrine—although Luther himself had not anticipated such radical change. Briefly put, Luther vehemently preached a doctrine of justification by faith, that God could forgive sins and unjust acts and offer other gifts provided they were accepted with faith in Him; such forgiveness and gift-giving had nothing to do with works. Luther also inveighed against elaborate sacraments (considering only baptism and the Eucharist as proper), and against indulgences, financial abuses and other corruption of the papacy, in essence taking exception to the infallibility of the pope. As is well known, his forcible preaching instituted and reinvigorated evangelical reform movements throughout the European West, although he disagreed, often heatedly, with other strong figures in the reform movement: for example, with the Dutch Scholar Erasmus who was critical of scholasticism and sought reform through tolerance, dialogue to achieve mutual understanding, and education and not through the vigorous attacks of Luther; and with the more radical Swiss theologian Zwingli (1485-1531), who insisted that not the real (as proclaimed by Luther) but only the spiritual essence of Christ was present in the Eucharist. Another reformer, Calvin (1509-1546), whose impact came later, and who like Zwingli had to battle with the Anabaptists (a reform movement insisting upon the rebaptism of adults), preached the Bible as the sole authority for all religious beliefs and practices; he maintained that man is always a sinner being involved in original sin, and sternly proclaimed predestination, yet good works are required of man.

While the above and other inspired reformers were proselytizing and feuding among themselves, the Roman Catholic church took little action. By the middle of the 16th century it had failed to subject itself to a thorough reexamination of its functions and behavior. Then, at the Council of Trent (1543-63) only mild reforms were effected: limitations were imposed on papal power; bishops were required to reside in their dioceses and their power was reestablished; theological instruction of the clergy was made incumbent on the church everywhere; and a number of abuses were eliminated. Much more important for the limited Catholic renewal that transpired were the inspirational and highly admired missionary, educational and social welfare works of the Jesuit order founded by Ignatius Loyola (1491-1556).

The end of the 16th century saw the Protestant movement firmly rooted and as a whole flowering within the Christian West. There were of course the different nationalistic brands of Protestantism—the Anglican church of England, the Huguenots in France, and the Reformed Church of Scotland. Then there came forth diverse sects: Mennonites, Hutterites, Amish, Unitarians, Puritans (separatists in England from whom emerged Baptists, Congregationalists, and Quakers), and others that existed in the 16th century and also later. No longer could one claim, as one could before the 16th century, that there was *diversity within unity* in Christianity of the West. The toppling of the unified structure with all the persecutions, inquisitions, battles and wars that ensued led to drastic societal changes. As already remarked, Luther's vitriolic criticisms of the vulnerable Roman Catholic hierarchy can perhaps be said to be the first of a number of devastating attacks. By the end of the 16th century that structure had toppled. Sacramentalism had been reduced for many and the ethical monotheism of the Old Testament came to be practiced widely. Political authorities had in many instances replaced papal authority in the control of religious practice. The notion of priesthood of all believers had been widely acclaimed, and the sanctity of common, ordinary life had been affirmed. All kinds of jobs and vocations could be claimed to be consistent with a worthy, spiritual life. Moreover, the way was open for new literate, creative and energetic persons to emerge and assume a niche in society. In short, the Reformation set in motion the emergence of the individual as a truly relevant behavioral unit in society, which some may claim to constitute the first step, the start, of the Revolution that was to lead to the current state (21st century) of the full autonomy (liberty) of the individual or 'individualism unbounded.' However, it was only a first step. There still remained religious structures in what was the Roman Catholic world; now, however, there was a variety of such structures and there still were repressive political bodies constraining individual liberties. In considering the possibility of atheism in this period, Armstrong (1993, pp.286-7) [with reference to Febvre (1982)] has observed:

> . . . a full-blown atheism in the sense that we use the word today was impossible. As Lucien Febvre has shown in his classic book, *The Problem of Unbelief in the Sixteenth Century*, the conceptual difficulties in the way of a complete denial of God's existence at this time were so great as to be insurmountable. From birth and baptism to death and burial in the churchyard, religion dominated the life of every single man and woman. Every activity of the day, which was punctuated with church bells summoning the faithful to prayer, was saturated with religious beliefs and institutions: they dominated professional and public life—even the guilds and the universities were religious organizations.

9.7 Enlightenment to individualism unbounded in the West

During the heyday of Reformation and even earlier the seeds were planted that would eventually flower into Enlightment—seeds that would lead to: (1) an outburst and steady advance of technological development and scientific inquiry; (2) the French Revolution of 1789; and later (3) massive industrialization (the Industrial Revolution). The advent of printing, the heliocentric views of Copernicus and Gallileo expounded in the 16th century and on, other advances in the experimental and theoretical sciences and mathematics by Francis Bacon (1561-1626), Pascal (1623-62), Descartes (1596-1650), Newton (1642-1717) and others, and the humanism of the Renaissance (which fostered scholarly interest in classical texts and values), all contributed in a mutually reinforcing manner to the Enlightenment movement. This movement in Europe and America may be said to have taken hold during the last two decades of the 17th century and early years of the 18th and to have dominated the 18th. It was largely initiated and driven by French scholars and writers.

The Enlightenment movement represented another forward step in the emergence of individualism within the orbit of religion. The philosophical outlook that came forth no longer took for granted the Catholic tradition which still governed the Protestant as well as the Roman Catholic church. Intellectuals, political leaders and in time even some ordinary laypersons came to place more emphasis on reason, rational thought, and scientific explanation of religious thinking. They seriously questioned, if not discarded, the sacraments and purely mystical experiences of the church. Descartes, the distinguished mathematician, maintained that all phenomena, including every *valid* religious belief and principle and even God's existence, could and should be explained in terms of cause-effect relationships. Later Newton deepened the scientific perspective on religion and, for example, maintained that the amazingly complex system of celestial and terrestrial bodies that defied collapse from gravitational forces could not have existed except as a creation of a divine mechanic, the sovereign God.

These and other incursions of mathematics and science into the religious sphere was a precursor of Deism, an attempt to establish a rational religion based on self-evident propositions and not on inherited sacraments and Old and New Testament doctrines. Such appealed to many who as a result of creative advances in science, art and literature came to be skeptical of religion and its professed miracles.Voltaire (1694-1778) was one who passionately warred against superstition, institutional Christianity and the belief in the omnipotence of God, especially in the day-to-day flow of events. Kant (1724-1804) emphasized experience (ordered by categories that man constructs) as necessary for the attainment of knowledge and therefore questioned man's ability to prove the existence of God (as the First Cause of the world) since man in no way has relevant experience. However, since cause-effect scientific determinism cannot provide a basis for morality and freedom of will, he allowed that there remains a basic need here for religious guidance.

The outbreak of the French Revolution in 1789 cemented for many the view that God is not omniscient, omnipotent, and omnipresent; he does not intervene in events. The revolution gave more credence to: (1) psychological and ethical analyses, such as Locke's, holding that individual character grows out of experience from birth onward and that notions of original sin or goodness (or predeterminism) are meaningless; and (2) the materialistic belief that man is strictly concerned with survival and maximizing pleasure (or minimizing pain). The latter led to: (1) the radical political theory that the role of the state was not to create and perpetuate a city of man that resembles that of God but to provide a harmonious organization among men for their protection and welfare; (2) the unorthodox notion of society as a social contract; and even (3) atheism.

As could be expected, reaction to Enlightenment occurred—reaction that took diverse forms, in particular to Deism and the related Enlightenment rationalism of an increasingly sophisticated and rarified character, understandable by intellectuals only. Such theology and erudite thinking came to have little appeal to the mass of ordinary Christian people, and catered too little to their need for ritual and spiritual guidance to address problems of day-to-day living. Thus, one type of reaction in the 18th century and early 19th was a renewal of evangelical fervor. The emotional Moravian Brethren movement gained strength, and through its influence upon the fervent preacher Wesley (1703-91) led to the widespread growth of Methodism, characterized as a religion of the heart. To the working class and peasantry, it espoused the simple message that through one's own experience one could gain true faith and be assured of salvation. Within the Anglican Church, too, and elsewhere evangelicals were effecting revivals;

and greatly influenced by this movement was the enactment of social legislation to improve the lot of the common man subject to the vicissitudes of industrialization—reflecting another aspect of the growing individualism of society.

Associated with this development in the 19th century was an attempt of a number of romantics—Schleiermacher, Ritschl, and others—to combine reason and feeling, for example the feeling associated with the awe stemming from the universal human experience of the numinous (the mystery of life). In his dialectics Hegel embraced both reason and non-reason (imagination) giving more weight to the former. Still others, the poets Keats, Wordsworth, and Coleridge, linked the numinous to the wonders of nature, while in contrast the Oxford movement and its luminary Newman were effective in reaffirming basic features of the Roman Catholic high-church—a restoration of the sacrament, liturgical life and elaborate rituals and ceremonies—in certain respects a return to the Middle Ages.

At the other extreme, the thesis 'God is dead' and related hypotheses had many strong proponents. Schopenhauer had proclaimed that there was no God, no Spirit at work in the world, just the brute instinct to live; each individual must himself create his own life and motivation. His view was strengthened by the findings from the fresh scientific approach to history which seriously questioned the veracity of many Biblical statements. Fuerbach, Comte and other positivists downplayed God's role in man's life, since so many negative aspects (sin and struggle) were associated with Him.

To Marx, God was irrelevant. To Nietzsche, not only was God dead, but He had imposed unnecessary restrictions on and committed crime against life. To Freud, belief in God was an illusion, and while it served and still does to combat helplessness on the part of the individual, people must outgrow this untruth.

Scientific advance and especially Darwin's theory of evolution in his *Origin of Species* (1859) continued to nurture a diversity of views concerning the role of religion and to shake Christendom's foundations in the 19th century and throughout the 20th. To many, religion and science were necessarily in sharp conflict, being diametrically opposed to each other. The two World Wars introduced still more turmoil into society and still more deviance in religious thought.

While it is too early to acquire an adequate historical perspective on developments in the 20th century, it can be said that in Europe the movement toward individualism and conflict in thinking about religion was on the increase, certainly up to World War I. Churches were emptying; atheism and agnosticism were becoming pervasive. All this was taking place while a tremendous and successful missionary effort was spreading Christianity throughout the world—resulting in an increasing variety of Christian doctrine since each culture impacted came to blend its own unique features into the Orthodox, Roman or Protestant faith presented.

In North America, during the 19th and early 20th centuries Christianity may be said to have been faring better than in Europe. Against the background of the Catholic and Protestant mainstreams, and in particular the Puritan faiths that arrived in the 17th century and prevailed in the 18th, there was in the 19th and early 20th centuries an outburst of revivalistic and, more broadly, evangelistic movements, all variations on a much more explicit theme of inward experience. From this outburst came such sects as the Seventh-Day Adventists, Jehovah's Witnesses, the Latter Day Saints (Mormons), and Christian Science.

As the 20th century comes to an end, there are many possible views as to what has transpired since World War I, although a trend toward unbounded individualism can be easily claimed. Witness, from the mid-20th century on, the Roman Catholic church's tolerance of other religious ways and, more significant, the admission of the importance of the worshiper's experience. Yet one development which runs counter to this hypothesis is the Ecumenical movement. Shortly after the turn of the century, various American churches came together in a National Council of Churches, later to become a World Council of Churches once churches from other lands joined the group. The objective of this organization is to share ideas and cooperate on various projects—social welfare and missionary—in ways that their differently held truths permit—a movement that may be regarded as a definite step toward Christian unity. Whether this movement, if joined by the conservative Catholic and Orthodox Christian and other non-Christian faiths, can link effectively with political, economic and social forces that call for world government in which the element of nationalism has been greatly subjugated and even eradicated, and in which excessive individualism is constrained and individual energy harnessed for the welfare of a more heavily populated world is an open question. This outcome, the dream of many idealists, which might give birth to a new religion, would involve a more macro-oriented society. It would constitute a return swing of the pendulum to a new regime—to a new Prigogine-type jump.

All in all, the individualism of the Christian West achieved in the Reformation and which was highly constrained, became unfettered, blossomed and by the end of the 19th century and at least well into the 20th may be said to have reached, as in the arts, unbounded individualism.

9.8 Individualism in other faiths

As already remarked, Christianity and Islam have been, since the 7th century, the two dominant religions in the Mediterranean and the West. And since the Middle Ages, Christianity in the West has undergone a dramatic revolution—from a unitary Roman Catholic hierarchical structure with tremendous power and on top of its society to a position of relatively weak or modest power, its authority being seriously questioned by the forces of modernization and the outgrowth of strong rampant individualism. This phenomenon has been in effect a clear transition to a new regime (a new far-from-equilibrium situation) in the society affected. The question that then arises is has such a transition also occurred in the Eastern Orthodox and other major faiths of the Mediterranean and West (inclusive of the branches of these faiths that have successfully penetrated the Americas).

a. *In the Eastern Orthodox faiths*

Up to about year 1000 AD the Eastern Orthodox Catholic church was the leading force in Christianity, reflecting the origin of Christianity in the eastern Mediterranean basin and the more creative intellectualism emerging there, in large part traceable to Greek philosophical and related inquiry. From the 7th century onward and throughout the 10th, Orthodox Christianity successfully brought the Balkan nations, the Slavs and finally Russia under its umbrella, and engaged in fruitful intellectual exchange with the Arabs. Thereafter the dominant role in Christianity shifted to the West, Eastern Orthodox becoming increasingly subject to encroachments by Islam and mounting socio-economic problems. And the fall of Constantinople to the Turks in 1453 and later Ottoman domination precluded significant progress in the Eastern churches, except in unconquered Russia. Until the dismemberment of the Ottoman empire after World War I, a period of conservatism characterized the theology and functioning of these churches, the only significant movement being Hesychasm, a monastic discipline emphasizing still more the prevalent notion of deification. This notion conceived each person as possessing within himself an image or icon of God, and thus capable of continuing transformation via this divine presence—which sharply contrasted with the Western Christian dogma claiming that human perfection is attainable only through discipline, humility and sacrifice.

The organizational structure of the Eastern churches was significantly different from that of the pre-Reformation Roman Catholic church characteristic of the Middle Ages. From the very beginning and up to the present time Eastern churches never have had the centralized organization of the Medieval Roman Catholic Church. There was always the commitment among Eastern missionaries to preserve the indigenous culture of each converted people; accordingly, local church autonomy was prevalent. As an instance, the bible was translated into the Slavic (the people's) language; its use contrasted with that of the Latin bible by Western missionaries that as a consequence required the teaching of Latin to each converted community. Moreover, while for many years Constantinople was the center of the Eastern churches, Byzantine Christianity developed the principle of administrative independence or 'autocephaly' of each culture's and later each nation's church. Thus there emerged in time the autocephalous churches: Greece (1833), Romania (1864), Bulgaria (1871), Serbia (1879) and Albania (1937).

Once the Turks conquered the Byzantine empire which resulted in a state of depression in the Greek-speaking churches, independent Orthodoxy was able to survive only in Russia (where a separate Patriarchate was established in 1589) and some sections of the Slav area. But, partly as a result of internal dissensions and the constraints imposed on the Russian patriarchate by the secular authorities (in particular, Peter the Great), no basic change in theology and church organization emanated from this patriarchate. Thus, Eastern Orthodoxy continued in much the same way as it was in the days of the Byzantine empire, despite the many splits and sects that formed in Russia and the works of distinguished figures like the Patriarch Nilson, and the influence on theological thinking of great literary figures such as Dostoievsky and Tolstoy. Also, there did not appear in Eastern churches the indulgences and other extremely corrupt practices of the Western churches that made Western reform urgent. But perhaps a more important factor underlying the above continuity was that the deification, a direct connection to God, which was attainable in the Eastern theology by the peasant and ordinary lay person met in large part his spiritual needs—a direct connection which the elaborate and rigid hierarchical structure of the Roman Catholic papacy precluded.

Hence, though the Eastern churches had contact with the Reformation, albeit limited, there was no urgent need for basic change. This was still so at the time of the Russian Revolution of 1917 and the dissemblement of the Ottoman Empire. Since the end of World War I, communism placed a lid on basic change in the Eastern Orthodox church of Russia and related churches in nations subject to control by the Soviet Union. While thus far there has been no major transition in these churches, one can yet expect the element of individualism to grow within them as they throw off the yoke of the former Soviet communism and its direct and indirect constraints, and perhaps as they participate more extensively in the ecumenical movement.

b. In Judaism

As already remarked, Judaism survived in the dispersion (Diaspora) that followed the crushing defeats by the Romans in years 70 AD and 132 AD. Following the fall of Jerusalem a long line of scholars worked on the compilation of the voluminous Talmud completed in the 6th century AD—a compilation comprising detailed and sharper definitions and explications of Hebrew beliefs, laws, ethical principles, ritual requirements and folk lore. Among the outstanding leaders of Judaism in the pre-Reformation era were the philosophers Saadia (892-942 AD) and Maimonides (1135-1204 AD). Saadia, the so-called 'father of Jewish philosophy,' attempted to bring together reason and revealed truth, maintaining that both emanated from God. Maimonides, emphasizing that knowledge of God is available to all morally righteous Jews striving for truth, sought rational principles underlying God's ethics and the commandments. Then there were mystics, active ever since the Roman expulsions, of whom the leading lights were Rabbis Akiva, Ishmael, and Yohannan, who placed emphasis on experience and contemplation and whose views by the end of the 13th century were embodied in a Kabbalist doctrine that came to prevail in every country of the Diaspora. During the Reformation era, Luria (1514-72) developed this doctrine further, in a way that embodied an element of reincarnation. Each person through prayer and good deeds, and with a spirit of joy and kindness, can help rebuild the God of Judaism; and by pursuing a pure life the divine spark that each individual possesses can lead to his salvation. Later in the 17th and 18th centuries there appeared other mystic movements. A grandiose one which was enthusiastically and with great expectations embraced was centered around the figure of Sabbatai (1626-76) who convincingly claimed that he was the Messiah of Israel. Later, when Sabbatai under threat of death converted to Islam, there resulted a most devastating and embarrassing collapse of the movement. A final mystic development, Hassidim, which is with us today, was founded by Baal Shem Tov (ca.1700-60). Designed to bring the spiritual messages of the Kabbalah to the common people, this movement provided a social function for mysticism and helped to foster inner renewal and regenerate religious values among the Jews.

It was not until the 19th century that reason and rationalism entered extensively into Jewish thought. This phenomenon was in part an outgrowth of the freedom from ghetto life (its restrictions and isolation) which the social and ideological revolutions (the French in 1789, the American in 1776) made possible. Jews were granted full citizenship in France and America, and Napoleon opened up a number of ghettos during the days of his successful European campaign. Jews could now interact fully with leading figures shaping European culture and all its scientific and intellectual advances.

Mendelssohn (1729-86) helped plough the way. Within the broader German Enlightenment movement, he philosophized that God was a personal and not a metaphysical reality, knowledge of whom was attainable by reason. He conceived a universal religion of reason for all mankind. The ideas of Kant (1724-1804)—for example that the concept of God was for some people a convenient and for others a necessary strategy to enable them to function efficiently, even though we cannot prove His existence—also gained ground among the Jews. As a consequence of the thinking of these and other intellectuals, new divisions appeared within Judaism. Early in the 19th century, a Reform Temple was established in Westphalia and later Reform Judaism found very fertile ground in the United States where the Union of American Hebrew Congregations was established in 1873. In the light of scientific history the Reformist finds that the old Talmudic Orthodoxy is no longer believable and in fact repressive. He comes to believe in a Personal God in an individualistic manner, and seeks cooperative relationships respectful of others with different religious and cultural themes.

Opposed to the non-mystical, non-traditional, individualistically-oriented doctrine of Reform Judaism stands that of the Orthodox group which, perhaps in part as a reaction, proclaims more vehemently than ever traditional scriptural interpretations and ritualism. Other believers, however, seek a reconciliation of the new critical history and questioning of the Bible and the Talmud. One can claim that these books must have incontestable elements of truth and eternal values since their scriptures have been able to nurture Jewish life over so many centuries and so many periods of extreme hardship; the Bible and Talmud have stood the test of time. These believers form the Conservative group which established the Jewish Theological Seminary of America in 1886.

Finally, the aggressive nationalism of the Zealots of old has once again emerged and reasserted itself as a Zionist movement; its adherents together with members of the other groups comprise the constituency of the current state of Israel. The strength of the Zionist movement, and for many its justification, were significantly enhanced by the ungodly genoicide of World War II.

In sum, Judaism has come to be significantly affected, albeit late, by the revolution of individualism, the effect most clearly seen in the emergence of Reform Judaism, and to a lesser extent in Conservatism—and of course to varying extents in the diverse subgroups of the Reform, Conservative, Orthodox, and Zionist divisions that have emerged. Because of the rich traditions of Judaism, which have for so long served its people through extreme hardship (a most telling instance being the Holocaust, an unimaginable hardship magnified by the almost complete absence of concern by the outside world), one could not expect the extensive discarding of the old as has occurred in Western Christianity. Moreover, there has never emerged a corrupt, inflexible multilevel hierarchical structure which could fan forces diametrically opposed to individualism, as was the case in pre-Reformation Roman Catholicism. Thus the spread of individualism within Judaism has been on a more modest and constrained basis, and in a much less violent fashion. However, this more modest incursion of individualism into Judaism does not count much in any test of the hypothesis of a late 20th century unbounded individualism stage in religion; the Jewish constituency comprises such a small part of the Western world's population and is so small compared to the Christian and Islam followings.

c. *In Islam*

As already noted, politics, economics and cultural activity were inextricably interwoven with and subservient to religion in Islam. Accordingly early on, in the struggle for power (the Caliphate) between the Umayyad family and that of Ali (a member of the family of Muhammad), the three divisions of Islam emerged which are with us today: (1) the Sunnis, the orthodox, founded on the traditional practices and ways that evolved during the first century of rule by the Umayyad family and which has constituted the mainstream of Islam over the years and still does; (2) the Shi'ites who early on insisted that Ali (the cousin and son-in-law of Muhammad and father of his only still-living grandson) was the first true caliph and that only descendants of the family of Muhammad can rightfully serve as caliphs; and (3) the Kharijites, today a small sect, who held that only the community has the right to elect its leaders and who professed piety and egalitarianism, taking the position that there should be no laxity in the observance of the faith as it was laid down by Muhammad.

We have already remarked that within Islam there has always been emphasis on the community, ummah, and a role for the individual within that community. Also, the individual alone has to stand before God for his or her own fate; this fate was not predestined, nor determined by the actions of others. Moreover, the individual was expected to give alms to the poor whenever he could (an outgrowth of the communal sharing of wealth in the earlier Arabian tribes). In these and other related ways, the individual existed as a more recognizable entity and was accorded much more respect than in the medieval days of Christianity and in the industrial communities spawned by the Industrial Revolution. Hence, there was no need for political revolution, such as the French and American, and religious revolution—no need to gain freedom from extreme subjugation so often present in the West. The non-existence of this need is reflected in Islamic writings. What we see in Islam are different theological hypotheses and spiritual movements introducing rationalism to greatly differing extents. The oldest theologians, the Mutazilis within the Sunni community, were the first to inject elements of rationalism in their thinking about God and his actions. Later, Ash'ari (878-941), taking a middle course between reason and tradition, between the Mutazili rationalists and the more traditional literalists, maintained, without achieving an effective compromise: (1) the indisputable existence of an indispensable, mysterious, invisible religious element of life but yet; (2) that it was the duty of each Muslin to use *interpretive tools* to achieve a full comprehensive concept of God. Based on his thinking the Asharite school emerged which has provided the dominant theological framework for Sunni Islam over the centuries.

Specific to Shi'ism was the esoteric dimension—that of inner comprehension and interpretation of the Divine Law enunciated in the inspired texts of the prophets and Muhammad. From this dimension together with the ascetism stemming from the Kharijites (who decried the luxuries to which the Arab victors were succumbing) emerged Sufism, the most well known mysticism of Islam. While in its early years (8th and 9th centuries) it primarily served the intellectual élite and was strongly resisted by the orthodox theologians and authorities, through restatement of its doctrine in a much less abstract manner it came to have wide appeal to and acceptance by the laity. It met the religious needs of the people who could not absorb the sophisticated and impersonal thinking of the orthodox theologians. The religious chants and sacred music and dance of the Sufis were able to establish rapport with the ordinary individual and provide him with spiritual substance. In effect, it came to satisfy the esoteric claims of Shi'ism better than Shi'ism itself and led to Shi'ism's decline.

Two of the leading lights of Sufism and related doctrine were Al-Hallaj (857-922), who expounded the thesis of deification through direct mystical experience; and al-Ghazzali (1058-1111) whose writings provided one way of justifying, in a complementary side-by-side manner, both the beliefs of the Sufis (often tempering their extravagant claims) and the orthodox doctrine of the Asharite school—in effect making Sufism much more acceptable to the orthodox.

Parallel with the writings of the Sufi theologians were those of the philosophers. Drawing upon Greek works of philosophy and science were: (1) al-Kindi (796-873) who was perhaps the first Muslim to examine the Koran with a thorough rationalistic approach, taking the position that human thought and reasoning could discover revealed truths; (2) al-Farabi (872-950) who considered philosophy as a way of understanding the truths expounded by *all* the prophets of *all* religions; (3) Ibn Sina, also known as Avecinna (980-1037), who sought a proof of God's existence—a proof that could refine the conception of God and for which each intellectual could aspire to achieve—but who in later life also allowed that prophets could obtain an intuitive, imaginative knowledge of God without resort to reasoning; and (4) Ibn Rushd, also known as Averroës (1126-98), who took issue with al-Ghazzali's polemic, maintaining that between religion and rationalism there was no contradiction, that the same truths emerged from each—a conclusion that all philosophers (the intellectual élite) should be able to reach.

Then there were: (1) Ibn al-Arabi (1165-1240) who argued that human reason cannot provide understanding of mystical phenomena, such phenomena being the result of experience only; (2) Suhrawardi (1155-91) who further elaborated the mysticism of *Light* and emphasized the role of creative imagination and intuition in the approach to truth, without, however, ever abandoning reason; and (3) Rumi (1207-73) whose mystic poetry (and philosophy therein contained) and emphasis on sacred music and dance (he founded the *Whirling Dervishes* brotherhood) played a major role in the renewal of Muslim religious experience.

With the Mongol conquest in 1258, which ended the Abbasid dynasty, the flowering of classical Islamic civilization was soon to be brought to a halt. (In Spain, there had already been a gradual retreat of Islam for many decades.) The centuries-long age of outstanding theologians and philosophers was to fade. While major Islamic empires flourished up until the 19th century, cultural advance and progress in philosophical and religious thought was decidedly slowed and perhaps at times brought to a standstill. In brief, beginning with the 14th century Islam entered a conservative era with restatement of themes of days gone by and with literalist interpretation of scripture. Nonetheless, there were some significant developments: the Safavid dynasty in Iran (1501-1736) nurtured a cultural rebirth within its domain; and Mulla Sadra (1571-1640) and others emphasized the role of nature and orderly processes and transformation therein, adding still another approach to mysticism as a way to come to know God. Akbar, the Mogul emperor, (1560-1605) fostered religious tolerance and cooperation, contending that the one God was revealable by any rightly guided religion and urging unity among believers of different doctrines. Yet, during these days there was little absorption by Islam of the major advances of the Italian Renaissance of the 15th century and later years.

The conservative nature of Islam continued during the 18th and 19th centuries, with most Muslim thought concerned with the immanent, subjective approach to God of the mystics. There was the conservative Walli-Ullah of Delhi (1703-62) who wanted Muslims to return to the firm sources and theology of old. There was also al-Wahhab (1703-92) who opposed the prevalent mysticism and urged going back to the pure and unadulterated type of monotheism of the Koran—a renewal of what had been abandoned, in particular the prophetic messages of Muhammad.

During the later 19th and 20th centuries there was a mounting concern of Muslim intellectuals for catching up with the West in technological and economic development, and for adopting the ways that seem to have led to what was perceived as progress in the West. Al-Afghani (1838-87), while holding to the mysticism of Suhrawardi, considered social and political reform necessary. Abduh (1849-1905) urged the use of mathematics and science as an aid to contemplation and renewal of faith in God. Iqbal (1877-1938) revitalized some of the notions of the Mutazilis, reemphasizing the need for reasoning by the individual; he reasserted the ability of man to distinguish between good and evil and to assume responsibility privately and publicly for good works, while at the same time affirming the validity of the original, uncomplicated beliefs and practices of Islam.

Simultaneous were strong political efforts at modernization. Ataturk (1881-1938) tried to make Turkey a modern nation, making religion a private affair, and forcing Sufi orders to go underground. Shahs Kahn and Pahlavi were motivated to move in the same direction during the period 1925-79. Their repression of religion led to their overthrow by the revolutionary forces inspired by Ayatollah Khomeini. Khomeini proclaimed that political forces must be ordained by and subjugated to the faith. His inspiration has had a widespread effect and has served to revitalize Islamic Fundamentalism.

In sum, during the first period of its cultural flowering and world dominance and the later period when the West came to economic and political ascendancy and saw great cultural advances, Islam never developed the intensity of individualism that the West has witnessed. Even in the most recent decades, the focus on religion inclusive of its communal aspects is present. It is there among the resurgent Fundamentalists. It is there among the modernists who from 20th century experience associate, like the Fundamentalists, decadence with Western practices and traditions. However, the modernists still consider the use of science, mathematics and reason as necessary complements to the theology and philosophy of early times in order to achieve vital social, economic and political advance for the Islamic people. The focus on religion remains, of course, for the masses who are still infused with the mysticism of the long past.

9.9 Concluding remarks and commonalities, parallels and relationships with concepts, processes and analysis in other areas of knowledge

Ever since nature endowed man with a brain and the power of reflection (however limited)—and even well before—man confronted various mysterious phenomena: thunder, sun blackout, volcanic eruption, forest fire, etc. These happenings must have generated fear at times, aroused curiosity at other times, and so forth. Man's wondering about their causes undoubtedly led to the belief that some powerful spirit, God, or other unknown element was in existence. And at times because of fear, anxiety and helplessness, the primitive may have reached the conclusion that the spirit, god or other element must be placated if angered with him, or must be appealed to for help to avoid an approaching calamity (a wild fire) or threat of one, or otherwise looked up to in one way or another—in effect required some form of reverence, prayer or the equivalent.

On the basis of what little we are able to infer from archeological findings and scientific studies, it would seem most reasonable that man associated a different spirit, animal, being or unknown element with each of the unexplainable phenomena—or at least with some—despite the single High God theory of some scholars.

Hence, there must have arisen in man a polytheistic belief—belief in many spirits, animals, or beings—often different from one clan or subgroup to another. Such belief also existed under shamanism and became inextricably bound up with mythology that accumulated over time. Even after the socio-economic transition from a hunting society to one based on agriculture—a transition that can be rightly regarded with respect to the river valley civilizations as a Prigogine jump to a new far-from-equilibrium situation—polytheism not only endured but also prospered. It became much more sophisticated and elaborately ritualized, and culminated in a hierarchically organized priesthood, the ziggurat and pyramid as massive and magnificent structures epitomizing this development.

The Prigogine break (major transition) with respect to religion came later, involving a much more reflective realm. This break can be associated with the Axial Age—the period within which Zarathustra and the second Isaiah came forth with their inspirational and forceful messages. Zarathustra put out front the struggle between Good and Evil, with monotheistic implications in his discussion of the benefits of admission into the House of Song. The Second Isaiah proclaims, during the Jews' Babylonian exile, the revolutionary concept of one Almighty God, high and exalted, supremely personal and merciful, who permits persecution in order to rid a chosen people of its sins and unrighteousness, but who also promises redemption and glory within a flourishing nation for rightful behavior.

In this Age, there were major advances elsewhere. In Greece, during its Classical age, there were tremendous advances in philosophy, the arts, reasoning and rationality; however, polytheism persisted, concern for spiritual development being greatly overshadowed by the philosophical-rationalistic thrust. In China there were Taoism and Confucianism, and in India, Hinduism and Buddhism, which while having major impact, were not of the monotheistic character of Judiasm and the later dominating religions of the West.

Once the major perturbation of Axial Age monotheism took hold among the Jews in explicit and vigorous fashion during and after the Babylonian exile, there subsequently emerged first Christianity and then Islam, both of which can be regarded as further manifestations of this One God belief. One aspect of these manifestations was the reformulation of the Judaic monotheistic doctrine to be more accessible to the ordinary layperson and more responsive to his spiritual needs. One reformulation was achieved in early Christianity which held that spiritual comfort and the attainment of salvation was possible by faith alone when associated with an attitude of surrender and openness to God, thus dispensing with the more restrictive Jewish law. Another reformulation (some 500 years later) achieved in Islam was also simple; not only did it dispense with most of the Jewish law and its principles, but also required much less attention and devotion to ritual and sacraments than the then current Christianity. Within a short period of time, Islam and Christianity came to be the two dominant religions in the Mediterranean and West, the growth of Judaism having been stymied by its dispersion throughout the lands consequent to its devastating defeat at the hands of Rome.

The particular doctrine and set of institutions that evolved in Islam during the ensuing centuries did not lead to an elaborate hierarchical structure as came to characterize the Roman Catholic hurch in the West. Always in Islam the notion of ummah (world community) played a central rule, and within the Sunni branch, the mainstream of Islam, policy was viewed as based on the consensus of the community, implying some say by the individual. Moreover, the individual himself had to face God directly for his or her own fate; it was not predetermined, nor mediated by a priest or other religious official. In short, the individual was afforded much more respect, and was a much more recognizable entity than in the Age of Glory of the Roman Catholic church in the West. And this role of the individual in Islam, admittedly not strong, exists today despite attempts at modernism.

Not all Christianity underwent the major revolution initiated by the Reformation. The Eastern branch, centered at Constantinople, never developed the highly organized elaborate hierarchical structure that the Western branch attained on the eve of the Reformation. While the Eastern branch dominated the West in the early centuries AD and reached its peak in the 9th to 10th centuries, there was always an element of autocephalism within it, though at times weak. And this element persisted when states nationalized their religious institutions, each nation's organization being basically independent of Constantinople.

The Almighty Roman Catholic church of the West started to topple during the Reformation. Its power declined and continued to do so during the Enlightenment period and subsequent years. This in effect constituted a relatively sudden transition to a new far-from-equilibrium situation, a Prigogine jump from a stage of highly repressed individualism to one of strong, if not unbounded, individualism. This transition can be said to have reached a late stage, if not an end, in the 20th century. Whether or not: (1) this vigorous individualism which is widespread in Christianity in Western Europe and North America will in a major way permeate Judaism and Islam and other religious groups and organizations of the world and become even more dominant, or (2) a return swing of the pendulum will see the emergence of much more constrained individualism subject to a new ecumenical theology, and a globally-oriented socio-politico-economic world culture are each a phenomenon which remains to be seen.

Before closing this chapter, we wish to relate its findings to those of other chapters. Teilhard's views, as expressed in Chapter 2, in general find much support in the writings covered in this chapter. His view that science and religion can effectively complement each other, or be fused, is present in the thinking of many religious leaders and scientists since the Enlightenment era—Descartes, Newton, Mendelssohn, Abduh and Iqbal, among others. Whether or not, they have actually complemented each other in an effective manner is an open question. Certainly for the majority of the world's population, whether primitive-indigenous, or sophisticated and highly intellectual, the answer is that they have not. Science, whether theory-oriented and based on reason, or experience-generated in combination with religion, has had, according to one hypothesis, a net negative impact on spiritual enhancement.

On a more specific level, one can clearly discern, in this chapter's presentation of religious development, albeit skewed, a history of increase in psychic energy of intellectual thought and order—even, one might claim, a history of decreasing entropy broadly speaking. One need only compare Teilhard's thinking or that of more modern theologians such as Tillich and Niebuhr with that of the early figures—the second Isaiah, Muhammad or the Apostle Paul. Within both modern Islam and modern Judaism one finds the mystical experiences of the major religious figures much deeper than those of the leaders in the early river valley civilizations or early Judaism and Islam. With the increasing concern in the late 20th century of many national populations with human rights, global pollution, and nuclear warfare and their destructive potential, which is increasingly reflected in the non-orthodox branches of Western religions, we see evidence of increasing hominisation, hyperpersonal thought and organization and perhaps shades of Teilhard's noogenesis. Certainly, one can claim that the current ecumenical movements are steps in the direction of convergent integration.

Chapter 3's materials on the selfish genes and memes, as they operate and interplay within the world and human society, is central to the analysis of this chapter. After all, what are myths, rites, sacraments and theologies but memes, simple and complex, or mutually compatible combinations of memes. These memes, particularly in combinations, compete for constituencies (just as firms do for markets), each member of which represents an organic entity formed by selfish genes behaving within the constraints of a system of coadapted memes. On a more concrete level, Christianity and Islam are each a mutually compatible complex of memes, each competing (in a Social Darwinism sense) for converts in their missionary and related efforts. Moreover, the notion of coevolution, wherein the individual phenotype and the species interact and determine their joint time-path, applies also to socio-religious development. This was very evident in the emergence of unique leaders and their religious sects (orders) in the Reformation era and subsequent years—emergence that often occurred (as in the case of species) at peripheral locations of an existing framework (territory). Both the genes that so rigidly govern the development of the characterstics and personality of a sect's (order's) leader and the attributes of the very sect (order) intertwinably affect the time-paths of their mutual development—of course within a coadapting (competitive-cooperative) framework involving other phenotypes, sects (orders and religions) and their socio-politico-economic and physical environments.

Religion, and the various activities associated with its development, reflects the operation of a nondeterministic, linguistic mode discussed in Chapter 4. This means that the time, place or other aspect of religious and associated phenomena are not predictable in any precise sense. While one may conclude at a particular time and location that a basic change in a religious structure is highly probable, or even inevitable, as one may have in 1500 AD with regard to Western European religion, or today with regard to the role of women in orthodox Judaism and other religious orders, Fundamentalist or not, one cannot predict the when and where of the change. Moreover, many changes with regard to religious-oriented phenomena, whether minor or revolutionary, are irreversible. In this respect, the clock of religion cannot be turned backward. Further, religion and associated activity operate in ordinal time, and certain of their aspects, as with some cultural traits, may remain frozen for extensive periods of time. Often in a culture, religion sets goals and objectives, which can be viewed as initial, intermediate or terminal conditions for basic rate-determined processes such as economic production. Also, religious activity can be involved in measurement and evaluation, for example in judging whether an economic system is performing adequately in meeting the needs of a society. In this regard and others, religious activity functions as part of the cognitive system. Such invokes reflective thinking, involving cognitive processes of individuals, but despite recent developments in brain theory and concurrent distributed network models that try to replicate the performance of the brain, a satisfactory level of understanding such performance is yet to be reached. We do not understand, whether in religious activity or not, how the brain chooses to predict and measure, how it learns, discovers previous non-observables, and develops new doctrines and carries on other creative activities.

The work of Bohm (in the Appendix to Chapter 5) relates to the understanding of mystical, spiritual and in general esoteric experience. Typically, such is not a static phenomenon, but is as thought (consciousness) a process of continuous movement and unfoldment, at times unending and never complete. While we may frequently apprehend things (an icon, a cathedral, a painting) and other elements of reality as a series of static images or occurrences, such apprehension is upon deeper analysis an unbroken, undivided process of flows—a continuous absorption of sensations. Any development is in fact a coherent body of such unbroken, undivided processes; and religious development of a person and community is no exception, though our limited capabilities at perception and reflection may lead us to believe otherwise. Further, religious development such as the current and future growth of the ecumenical movement may be viewed, for enhancement of our understanding, as explicate order emerging (unfolding) from an existent higher-level implicate order that is being or is yet to be revealed.

While some scholars are optimistic regarding the use of artificial intelligence methods, computer simulations, cellular automata, concurrent distributed network models, and adaptive fuzzy systems for understanding the processes of the brain and thus its creative thinking, the promise of these approaches lies mostly with regard to decision-making analysis (policy formulation) where quantitative data or equivalent symbolic representation is possible—and not for understanding the highly subjective and qualitative aspects of mystical and spiritual experience. However, for help in studying the spread and diffusion over geograpic space of religious ideas and theology, and the competition among ideas and theologies for converts, these methods and models may be of considerable use.

The processes and phenomena in physical sciences and cosmology discussed in Chapter 5 are paralleled and thus contribute to the understanding of the stages and history of religions. Some relevant basic concepts are: symmetry breaking and the sequence of symmetry breakings (for example in the original Ur matter); (2) the build-up of the hierarchical structures (for example of chemical elements and of diverse bodies of matter in the physical world); and (3) the formation of stable building blocks (à la Simon), or preferred configurations reflecting stratified stability à la Bronowski. Certainly symmetry breaking and hierarchy-organizing processes have been present in the development of religion. Because of lack of archaelogical evidence, we do not know exactly how mysticism and spiritual beliefs developed among early hunting life and societies when brain development enabled humans to begin to reflect and reason. We may speculate that contemplation of catastropic events (thunderstorms, earthquakes, and the like) broke the symmetry of a primitive's initial stage of noncausal, nonreflective and pure instinctive behavior to yield a many-spirit, many-god conception of the forces at play. As these primitive societies advanced, cumulating know-how on the construction and use of tools, and later agricultural technology, their first simple polytheistic or equivalent mystical frameworks evolved into organized hierarchies of gods (as with Mesopatamian and Egyptian civilizations).

Whatever the resulting pattern of many gods generated, we have speculated that in order to avoid confusion over the authority or function of each god and inconsistencies, civilized man of the West came to adopt, fervently in many cases, the conception of one God (or nature itself), all-powerful, omnipresent, the creator of all things—one to whom to make sacrifices, pray, and/or appeal for forgiveness, assistance, salvation, and so forth. As a consequence, monotheism came to be dominant, breaking the symmetry of polytheism—with Judaism, Christianity and Islam (all of Semitic origin) coming to rule the Mediterranean and West.

The belief in one God, almightly and omniscient, did require the notion of faith. All the monotheistic religions, Judaism, Christianity and Islam, pointed to Abraham as the first religious leader to manifest faith in one God. In time, the conception of monotheism came to couple morality with faith—e.g: in the *moral precepts* of the Commandments handed down to Moses, later espoused in a more extended form by the Second Isaiah; in the teachings of Jesus; and in the concept of ummah and messages of Muhammad as embodied in the Koran. Faith *and* morality became basic to monotheism, even though extensive moral degradation was present in the late Middle Ages in the Roman Catholic church.

Already, by the 6th century BC, the factor that was to initiate still another symmetry breaking in religious development had come to the fore in the Greek Classical age, namely reason (rationality) which was later to be backed up by science, especially starting with the Enlightenment era of the 17th century. There had been, of course, attempts at injecting reasoning into religion ever since Philo (13 BC - 50AD) tried to fuse the revelations of the Bible with philosophy, and the Mutazilis tried to bring rationality into Islam theology; but these did not yield effective and convincing results. Once science had firmly established itself and the scientific approach in historical studies came to question the truth of many of the statements in the Holy Books, the symmetry breaking based on reasoning and individualism in religious belief and interpretation which characterizes the current situation came to be realized in a widespread fashion—today with extensive atheism and agnosticism at one extreme and orthodox fundamentalism at the other.

To gain some insight and better understanding on how order is reestablished once a symmetry is being or has been broken and a system moves on to a higher level or adds a new higher level to an existing set, we can refer to the work of the Prigogine School and chaos theory briefly discussed in Chapter 6. Perturbations, the onslaught of random elements (noise), are always present in religious development, and these perturbations are of different magnitudes. Those associated with Akheniton and Sabbatai, we may recall, were two that were significant but did not survive. That associated with Zarathustra survived in a major way for a considerable time, but suffered severe dampening effects and exists at a relatively small scale today. And there are many other perturbations that can be cited that have experienced some amplification but which have not survived. In contrast, those associated with the appearance of Moses, Jesus and Muhammad were greatly enhanced and have had enduring lives.

Recall (pp.146-47) how Prigogine illustrates the process whereby a perturbation under appropriate environmental conditions amplifies and takes a biological system to a new far-from-equilibrium situation. He considers a population of unicellular amoeba (*Dictyostelium discoideum*) that grows in number in a landscape of limited resources and in time confronts starvation conditions. Sooner or later a cell begins to secrete a chemical substance, (cAMP), which generates a set of forces of attraction and autocatalysis that results in density gradients and a snowballing process forming a multicellular aggregate effective in confronting the starvation situation. In similar fashion, under appropriate conditions characterizing a stressful religious situation, a perturbation, within it or its environment, can generate forces of attraction and social cohesiveness. Such can result in density gradients and a combined autocatalytic-crosscatalytic process, thereby defining for a given population a new cultural-social-political-economic system with a new religious configuration—one that is effective in relieving the initial stress. There is sufficient historical materials to suggest this was the case when Muhammad appeared on the scene in Arabia and inspired the Islam development—or more recently when Ayatollah Khomeini set in motion with his exhortations and charisma the process that resulted in the religious revolution within the Islamic community and the spread of a revived Fundamentalist theology.

In physics, chemistry and biology the transition from a given state (under stress) to a new far-from-equilibrium state in general occurs much more rapidly than in human society. This can easily be accounted for by the much greater speed with which a message pertaining to a density gradient can be transmitted over distance via electromagnetic waves than a message on a social density gradient via man-made communication facilities—although advance in communication technology is rapidly reducing this differential.

As with the natural sciences, when a point of bifurcation is reached in religious and associated development, often chance alone determines which path is taken. Moreover, there can be seemingly disorderly processes which do exhibit order and may involve phase transitions—as the strange attractor of Lorenz (with regard to his weather model) or that associated with the Red Spot of Jupiter. So also, it may be contended, there can be seemingly disorderly (turbulent) behavior within a religious structure, whether minor or major, which reflects a strange attractor—which in effect is bounded despite what appears as random gyrations. Perhaps, the experience of Judaism over the last centuries, or centuries prior to the Zionist movement, may be speculatively characterized as governed by a strange attractor. Still more, just as a rugged coast line has a fractal dimension—a dimension that can be embedded in a finite dimensional Euclidean space—so may a religion have a fractal dimension, for example in world-wide Islam, in its rugged up and down, and back and forth behavior. Even more, and again as with a rugged coast line, or with the Hénon phenomena, when the ups and downs, and back and forth experiences of a religion are viewed at ever finer scales, one may discover self-similarity.

Other aspects of chaos and related theory may be found of significance. Sensitivity to initial conditions (the Butterfly Effect) can also be pertinent to explanation of the evolution of a religion. One wonders what the history of Judaism would have been had not Sabbatai converted to Islam. Irreversibility, self-organization and loss of detail especially in hierarchical structures are other factors associated with chaos theory, the loss of detail being particularly relevant when we turn to discuss connections of this chapter with the contents of Chapter 7.

Chapter 7 presents some general theory and findings about hierarchies. Hierarchical structure characterizes most if not all religions—but there is considerable variation among the hierarchies that exist. There are extensive, elaborate and complex hierarchical structures such as the Roman Catholic, with many levels, with on the average many nodes at each level, and with major global extension serving huge population numbers. At the other extreme are simple hierarchical structures, such as the Society of Friends consisting primarily of local monthly meetings (the principal units) that combine to form quarterly meetings, which in turn form annual meetings—all together having limited global extent and constituency. The hierarchical structures may be fairly regular or highly irregular, having different sets of functions and activities (functions being scripture interpretation, theological reconstruction, church social policy and decision making, adjudication, information storage, communication, education, counseling, spiritual guidance, ritual performance and so on). The allocation (assignment and sharing of these functions among levels and nodes may also differ. Exercise of authority may be top-down (as in the Roman Catholic church) or bottom-up (as in the Society of Friends), or both ways and so may the constraints that each node and its constituents impose on other nodes and constituents. In particular the influence of the individual at the grass roots level may vary greatly, from one extreme where it may be little, if any, to the other where it along with the influence of others at the same level may have almost complete, if not complete, control of decision making and policy. More generally, this aspect relates to the potential for participation in the functions and activities of an hierarchy. To illustrate, in 1952, the Catholic liturgy primarily in Latin militated against participation of the members of a local church in individually attending to their own spiritual needs. In contrast, in one prevalent branch of the Quaker faith, the liturgy is 100 percent decentralized. Each member may testify with regard to his religious experience and provide spiritual advice to other members present at a meeting for worship. Moreover, each member can be fully involved in biblical and other theological interpretation, and in the meeting's business and decision making, having in effect through verbal opposition a significant veto power on any proposal. Such participation may constitute a very positive force in building up morale and increasing the likelihood of local meeting survival, but then on the other hand may interfere with the attainment of other desirable and even exceedingly important survival objectives.

Generally speaking, the patterns of interaction within and among religious hierarchies may be quite diverse. Interaction among individuals, nodes, levels and with the environment may take place in a multitude of ways, as it does among existing religions; and selection may take place on individuals, nodes, levels and entire hierarchies themselves, again as is evident in religious history.

As with all hierarchies, and as described in Chapter 7, the same set of forces are at play—those associated with specialization, scale economies, agglomeration, participation potential re: morale building and so forth. (The specific play of deglomeration, disintegration and related forces will be spelled out in the ensuing chapter with regard to the decline of the papal empire.) It is unnecessary to detail them here. However, one element that is significant for the *major* religions which we should illustrate here is the principle of the optimum loss of detail, also related to Simon's near-decomposability, in association with the limited capability of the human mind. To point this up, consider that part of the Roman Catholic hierarchy in the United States, which is subservient to the world node at Rome. In the mid-20th century (1952), the spiritual sustenance of the 27 million or so of Catholic faith could have been handled by neither the three cardinals (in New York, Chicago and Detroit), nor the apostolic delegate (in Washington, D.C.), nor the 27 archbishops, nor the 158 bishops, nor all of these combined. Spiritural sustenance required the functioning of the 44,439 priests and other officials at the 15,653 parishes. Almost all the information on the needs, conditions, and problems at these 15,653 parishes had to be extensively filtered in order not to overwhelm the 158 bishops and their assistants and cause gridlock or debilitating (even catastrophic) congestion. In turn the information possessed by these 158 bishops had to be filtered (judiciously or injusticiously) to be passed on to the 27 archbishops, and later filtered again for consideration by the 3 cardinals and the apostolic delegate, and even more for consideration by Rome. Clearly, this filtering process has been basic to the hierarchical organization of the Roman Catholic religion and continues to be a *sine qua non.*

One dynamic factor significantly affecting most organizations is the advance of technology. We need not belabor the point about how advance in production technology has resulted in finer and finer occupational specialization, greater and greater economies of scale and agglomeration, and transitions to more and more complex hierarchies and states farther and farther removed from Prigogine equilibrium. At the same time, advance in communications technology together with increasing education has steadily increased the potential of a constituent at the grassroots or other level to exercise influence on policy and decision making at levels above the one he/she occupies, as explictly defined in the participation potential model of Chapter 7. Also, with this model we gain additional insight for understanding the extreme individualism in the religious sphere of the agnostic and atheist. The former through flow of information (including diverse inspirational, religious messages and scientific findings) is able to obtain spiritual sustenance (confront the existing mysteries and/or unknowables of life) with any of a number of religious doctrines. The latter can find in the mysteries, and as yet unknowns of the existing physical and social orders, a direct source of sustenance for whatever his spiritual needs may be. Thus, communications advance, which increases enormously the potential flow of information to individuals, can lead to not only increased participation by individuals and groups in the development of a new theological belief and reinterpretation and reformulation of an existing one, but also increased capability of an individual independently to determine his/her own pertinent religious beliefs, rites, and sacraments.

A question for the future is whether or not the current soaring growth of communications technology will lead to much more extensive individualism inclusive of agnosticism and atheism in the religious domain or to still more complex and elaborate hierarchical structure. The latter, some may claim, may well be the outcome of a rising tide of ecumentalism that requires the appearance of another Messiah of almost infinite inspirational ability, an outcome which could well turn out to be another symmetry breaking, one that could possibly take society much closer to Teilhard's Omega or its equivalent.

While religions, as sets of memes, have had a major impact on social development, so have painting and other arts, as sets of memes. Both these sets developed at roughly the same time as part of the total system of culture. How have these two sets been interconnected and how have they influenced each other?

An extremely close connection that may be claimed to be a common development occurred in the cave man era. Recall that one hypothesis concerning cave man art holds that it was to work magic, appeal for support in hunting from some unknown power, serve as a religious rite or spiritual action, or some other related element. Certainly, the spectacular *Sorcerer of Trois Frères,* represented both a step forward in cave-man depiction and a deepening of religious ritual practice. The very closely connected or common development of painting and religion is also apparent in the Egyptian civilization, at least to the extent that the Egyptian tomb painting is representative of what Egyptian painting was as a whole. The tomb painting was governed by the mystic-oriented purpose it was designed to serve, in particular the transmigration to heaven of the pharaoh at his death. This painting was of a highly rigid character, reflecting the immobility of Egyptian religious development—an immobility dictated by the dominance over the years of 'First Time' events and practices—a dominance which was facilitated by the geographic isolation of the Egyptian civilization. Note also that the bubble-type Akheniton perturbation was in both painting and religion .

This very closely connected or common development of painting and religion was brought to a halt, at least in part, in the Greek Classical age. In that age there was a tremendous development of sculpture, architecture and painting along with major advance in reasoning, philosophy and related intellectual pursuits. But, although the gods and goddesses were placed on pedestals and depicted most harmoniously and gracefully, there was no spiritual deepening in religious life or theology. The pre-Classical polytheistic framework persisted, the only change being in the glory of its depiction. In this Classical age, art and that religion required for spiritual sustenance became two different subcultures. Likewise, during the Hellenistic and Roman eras these two subcultures developed along different paths, and in large part independent of each other until the disintegration of the Roman empire. Also, the monotheistic development resulting in Judaism, during these and preceding times, did not seem to have any impact upon painting, and vice-versa.

With the collapse of the Roman empire and the ensuing uncertainties, fears and anxieties of the common person, and with the legitimization of Christianity by the Emperor Constantine, the Christian church came to provide a spiritual message that offered hope and security for the laity. The resulting emergence of Christianity as the overriding force in Western society had a major impact upon painting, an impact that dictated that painting should concentrate on a visual depiction of the Christian message, all other aspects to be subdued. Thus, there resulted the antimaterialistic and highly ritualistic painting of the time, with the beauty and harmony of the Classical age works set aside and forgotten.

With time, however, various elements reappeared in paintings, for example the reality and importance of man himself; and then new elements were added, especially spatial perspective which served to deepen the religious message. With further increases (albeit small) in the artist's independence of church authority, as well as the appearance of new economic and political forces in society—such as the growth in economic status and power of the Italian city-states—humanism became more prevalent in painting and the advances of the Greek Classical age were rediscovered. There emerged the major development in painting that culminated in the High Renaissance. While religious influences and objectives were still present, they no longer dominated. They persisted in a forged partnership with the creative urges of the artist himself, and in the case of the genre works of the Flemish-Dutch school, depicting insightfully the untarnished reality of feasting, dancing, and grieving common folk, they were no longer present.

The almost complete emancipation of art from religion's influence came later with the appearance of new wealthy commercial and business leaders who could serve as patrons of the artist together with the toppling of the almighty Roman Catholic church sparked by the Reformation forces. And subsequently, the rapid development and creative character of painting and the art subculture together with scientific advance came to be two of the leading factors forcing church reorganization and the restructuring of religion as a subculture, both throughout the Enlightenment era and the current era of Individualism. For several decades, according to one hypothesis, artists and other individualists together with scientists have been in a dominating position in society. If so, does this portend a social system in which art and religion as two sets of memes will be largely divorced from each other? Or is there some basic interconnection between the two that is essential for social development, which in part underlies the current swing of the pendulum toward ecumentalism?

References

Armstrong, K. (1993), *A History of God,* New York: Knopf.

Bancroft, A. 1987, *Origins of the Sacred,* New York: Routledge and Kegan Paul.

Borowitz, E. B. (1987), 'Jewish: An Overview' in Eliade, M. (ed.), *The Encyclopedia of Religion*, New York: Macmillan, Vol. 8, pp. 127-49.
Bowker, J. (1978), *The Religious Imagination and the Sense of God*, Oxford: Clarendon Press.
Brown, L. (1993), (ed.) *The New Shorter Oxford English Dictionary*, Oxford: Clarendon Press.
Campbell, J. (1972), *Myths to Live By*, New York: Viking.
Cragg, K. (1987), 'Christianity in the Middle East' in Eliade, M. (ed.) *The Encyclopedia of Religion*, New York: Macmillan, vol. 3, pp. 362-68.
Eisenstadt, S.N. (1987), 'Religious Diversity' in Eliade, M. (ed.) *The Encyclopedia of Religion*, New York: Macmillan, vol. 12, pp. 312-18.
Eliade, M. (1969), *The Quest: History and Meaning in Religion*, Chicago: University of Chicago Press.
Eliade, M. (1974), *The Myth of the Eternal Return*, Princeton, N.J.: Princeton University Press.
Eliade, M. (1978, 1982, and 1985), *A History of Religious Ideas*, Chicago: University of Chicago Press, Vols. I, II, and III, respectively.
Eliade, M. (1987), (ed.), *The Encyclopedia of Religion*, New York: Macmillan, Vols. 1-16.
Febvre, L.(1982), *The Problem of Unbelief in the Sixteenth Century, the Religion of Rabelais* (translation by Gottlieb, B.), Cambridge, MA: Harvard University Press
Fishbane, M. A. (1993), 'Judaism: Revelation and Traditions' in Earhart, H.B. (ed.), *Religious Traditions of the World*, New York: Harper Collins, pp. 373-48
Frankiel, S.S. (1993), 'Christianity: A Way of Salvation' in Earhart, H.B. (ed.), *Religious Traditions of The World*, New York: Harper Collins, pp. 485-601.
Frederick, M. D. (1993), 'Islam and the Muslim Community' in Earhart, H.B (ed.), *Religious Traditions of the World*, New York: Harper Collins, pp. 603-711.
Gnoli, G. (1987), 'Zoroastrianism' in Eliade, M. (ed.), *The Encyclopedia of Religion*, New York: Macmillan, Vol.15, pp. 579-91.
Gove, P.B. and Merrian and Webster (eds.) (1991), *Webster's Third New International Dictionary*, Springfield, Mass: G.& C. Merriam Co.
Harakas, S.S. (1987), 'Christianity in Eastern Europe' in Eliade, M. (ed.), *The Encyclopedia of Religion*, New York: Macmillan, Vol. 3, pp.372-79.
Hastings, J. (ed.) (1951), *Encyclopedia of Religion and Ethics*, New York: Scribner & Sons, Vols. 3, 4.
Hilderbrand, H. J. (1987), 'Reformation' in Eliade, M. (ed.), *The Encyclopedia of Religion*, New York: Macmillan, Vol. 12, pp. 244-54.
James, G.A. (1987), 'Atheism' in Eliade, M. (ed.), *The Encyclopedia of Religion*, New York: Macmillan, Vol. 1, pp. 479-90.

Kassis, H.E. (1987), 'Islam in Spain' in Eliade, M. (ed.), *The Encyclopedia of Religion,* New York: Macmillan, Vol. 7, pp. 336-44.
MacGregor, G. (1987), 'Doubt and Belief' in Eliade, M. (ed.), *The Encyclopedia of Religion,* New York: Macmillan, Vol. 4, pp. 424-30.
Nasr, S.H. (1993), 'What is Islam' in Sharma, A. (ed.), *Our Religions,* San Francisco: Harper San Francisco.
O'Collins, G. (1987), 'Jesus' in Eliade, M. (ed.), *The Encyclopedia of Religion,* New York: Macmillan, Vol. 8, pp. 15-28.
Pelikan, J. (1987), 'Christianity: An Overview' in Eliade, M. (ed.), *The Encyclopedia of Religion,* New York: Macmillan, Vol. 3, pp. 348-62.
Pelikan, J. (1987), 'Christianity in Western Europe' in Eliade, M. (ed.), *The Encyclopedia of Religion,* New York: Macmillan, Vol. 3, pp. 379-87.
Plott, J. C. (1977), *Global History of Philosophy: The Axial Age,* Delhi: Motilal Bararsidass, Vol. I.
Rahman, F. (1987), 'Islam: An Overview' in Eliade, M. (ed.), *The Encyclopedia of Religion,* New York: Macmillan, Vol. 7, pp. 303-22.
Random House, (1987), *The Random House Dictionary of the English Language,* 2nd ed., unabridged: New York, Random House.
Schmidt, W. (1931), *The Origins and Growth of Religion,* translation by Rose, H.J., London.
Seltzer, R.M. (1987), 'Jewish People' in Eliade, M. (ed.), *The Encyclopedia of Religion,* New York: Macmillan, Vol. 8, pp. 30-41.
Shils, E. (1987), 'Intellectuals' in Eliade, M. (ed.), *The Encyclopedia of Religion,* New York: Macmillan, Vol. 7, pp. 259-63.
Smart, N. (1984), *The Religious Experience of Mankind,* 3rd ed., New York: Charles Scribner's and Sons.
The Oxford English Dictionary, (1989), 2nd ed., Vol XIII, Oxford: Clarendon Press.
Vail, A. and Vail, E. (1970), *Transforming Light: The Living Heritage of World Religions,* New York: Harper and Row.
Vermant, J.-P. (1987), 'Greek Religion' in Eliade, M. (ed.), *The Encyclopedia of Religion,* New York: Macmillan, Vol. 6, pp. 99-118.
Ware, K. (1987), 'Eastern Christianity' in Eliade, M. (ed.)., *The Encyclopedia of Religion,* New York: Macmillan, Vol. 4, pp. 558-76.

10 Commonalities and rise and fall of civilizations and empires

10.1 Introduction

Against the background of the previous chapters we now wish to examine the presence of commonalities—in particular their pervasiveness in, and implications for, the life of past major Western civilizations and empires. We shall do so chronologically in order to enhance our insight for projection into the future. However, we lack sufficient historical perspective to go beyond the 18th century in this chapter. In the subsequent chapter we will engage in some speculation.

Because of limited time available for careful study and research, we have precluded any but minor reference to the great Eastern civilizations and empires. Also, since now we wish to consolidate the set of concepts, processes and analyses from which we draw our commonalities, we necessarily need to repeat in sections 10.2 and 10.3 much material of preceding chapters. The reader may find a quick glance through these materials adequate for the subsequent discussion.

10.2 Forces and processes in the emergence of the physical world

The best point at which to start is the Big Bang. Recall that at terrifically high temperatures both a common (composite) force—the Ur force—and a common matter—the Ur particle—emerged. As the force dispersed matter and temperature dropped, a *symmetry breaking* occurred—the force of gravity broke off of the common force and the graviton (yet to be empirically verified) separated from the common Ur particle. Another symmetry breaking led the remaining force (after gravity split off) to disintegrate into the strong and electroweak forces, and the remainder of the Ur particle (after the graviton had separated) into quarks, leptons and gluons. The quarks and gluons in time came to compose the hadrons—the protons and neutrons. With further reduction in densities and temperatures, the next symmetry breaking saw the decomposition of the electroweak force into the electromagnetic force and the weak force, and the electrons, muons, and tau-leptons each took on an identifiable form while photons proliferated and the heavy particles W^+, W^- and Z^0 appeared. In effect there was realized a *hierarchy of successively broken symmetries.*

Alternatively, the physical system that came to exist and may be said to exist today can be viewed as a system or set of *fields*, the intensity of a given field at any given point being the probability for finding there its associated quantum—the associated fundamental particle observed by experimentalists.

With continuing drop in temperature, it became possible for protons and neutrons to remain together to form nuclei—the beginning of a period of *hierarchic construction* via nucleosynthesis. The first nucleosynthesis (an energy absorbing, neutron capturing process) yielded nuclei of deuterium (heavy hydrogen) and helium. More temperature drop enabled the electromagnetic force between electrons and protons to take effect resulting in the chemical elements (atoms) of hydrogen, deuterium and helium. Later, heavier chemical elements were cooked up, for example carbon from helium. And so on. There thus evolved the set of chemical elements, hierarchically structured in terms of (a) number of shells (spherical layers) of electrons, and (b) number of electrons in the outermost shell. Elements having the same number of shells form a group that corresponds to a row of the periodic table; elements having the same number of electrons in their outermost shell form a family.

The different kinds of forces, particles and chemical elements associated with the above hierarchies and differences in their behavior and functioning already implied differences in the behavior and functioning of differently composed matter, inorganic or organic, that would further evolve.

Note that the existence of chemical elements together with the property of *stratified stability* clearly implied the formation of a sequence of stable configurations, of greater and greater complexity, formed from less complex elements and thus gave direction to evolution, each stable configuration acting as a barb on time's arrow. This property of stratified stability is found in the evolution over time of a number of physical systems—providing thereby a strong force that counters *reversibility*. The formation of an hierarchy of nested self-gravitating systems among the cosmological bodies—the stars, star clusters, galaxies, galaxy clusters and so forth—has also been asserted.

The emergence of order within the total physical system as already depicted has also been subject to noise and other elements leading to disorder. Recall that successive *period doubling* occurring at points of *bifurcation* that ultimately leads in the May population model to a situation where population can take on an infinity of values that cannot be predicted (namely chaos), points up the phenomenon of disorder that is inherent in physical processes. Turbulent fluid flow behavior is an instance. Yet there can be *order within seeming disorder*. The onset of successive period doubling phenomena (the sequence of points of bifurcation in the May model, and also the advent of certain stages in turbulent processes) can be predicted in accord with a universal constant, the *Feigenbaum number*.

From chaos and related theory, other statistical regularities are obtainable. While the behavior of orbiting stars around their galactic center in accord with Hénon's model or of other physical phenomena in accord with Lorenz's system of equations cannot be predicted, yet it can be established that there exist bounds to the wanderings of the variables involved, that is that *strange attractors* exist. Moreover, a system, such as the weather, may be extremely sensitive to initial conditions; its unpredictable time-path may never go through the same point twice (and thus be *aperiodic*). Yet this time-path may be subject to constraints and other statistical regularities. It may be found to meander within a minimal embedding finite dimensional Euclidean space and to have a *fractal structure*. Still more, in reflecting *deterministic chaos*, it may be found to possess *self-similarity at different scales* upon magnification—as do coastlines and Lorenz mappings.

Further understanding of actual structural change and pure dynamics comes from the *far-from-equilibrium analysis* of the Prigogine school. A steady flow of energy (say heat) into a basic equilibrium state of *molecular chaos* (one of maximum entropy) can lead a system to a *nearby equilibrium* state such as *thermal conduction*. A further increase in flow which produces a critical temperature difference between a layer at which the heat enters and that at which heat is transmitted to the outside world leads to a far-from-equilibrium state, as in *thermal convection*—a state where the different parts of the system have organized themselves (a *process of self-organization*) into an orderly and coherent pattern. The process of self-organization reflects the existence of correlations between relatively distant parts of the system via information transmitted by *density gradients*. Far-from-equilibrium states are thus obtained from the imposition of different kinds of *nonequilibrium constraints*, where again both random elements and determinism govern the specific pattern of order and coherence realized, and where *autocatalytic* and *self-reproducing processes* may be involved .

Hence, to symmetry breaking, hierarchical construction, stratified stability and irreversibility, we add from chaos and related theory: (1) bifurcation, period doubling, Feigenbaum number, strange attractor, deterministic chaos, sensitive dependence on initial conditions, aperiodicity, fractal structure and dimension, self-similarity at different scales; and (2) from Prigogine-type theory, molecular chaos, entropy, nearby equilibrium states, far-from-equilibrium states, convection, self-organization, long-range correlations associated with density gradients, nonequilibrium constraints, autocatalysis and self-reproduction, and the joint play of chance and determinism as factors in evolution and the associated set of ongoing processes prior to the advent of man.

10.3 Forces and processes in the emergence of the biological world

Much is now known from particle physics, chemistry and associated fields of science, about how there evolved the various combinations of protons, neutrons, and electrons and subsequently chemical elements and their hierarchical structure. And under given and changing conditions and with increasing knowledge from chaos, noise and related theory and from far-from-equilibrium analysis we have come to know much about the behavior and functioning of many of these chemical elements and their compounds that have come into existence. However, we know much less about how living, replicating matter evolved. At some time, a primordial soup of organic matter (the diverse compounds of carbon) emerged, dominated by the stable forms of matter. Purely by happenstance, an amazing molecule was formed, a *replicator*—a stable molecule that could create copies of itself, and it did. (Of course, if the first replicator had been unstable, the replicator would have disappeared. But eventually a stable one did appear.) In time, the population of stable replicators grew until its increased demand for organic material in the soup when coupled with that of other non-replicating molecules left no unused supply. With further increase in the population of replicators, the phenomenon of *competition* and the *problem of allocation of scarce resources* would have emerged then if not earlier.

The process of replication involved copy-errors, that is the formation of mutated replicators, most of which were eliminated in the competition for resources. But sooner or later some of the mutated replicators came to be endowed with attributes making them more effective in the competition for the scarce resources—the more so, the greater their copy-fidelity and fecundity— thereby causing the elimination of the less effective replicators and other resource demanders.

Dawkins expounds lucidly on how the replicators also came to aggregate and link together to form along with *mutation* still more effective survival machines, with still more complex and elaborate structures. *Specialization* and exploitation of *scale economies* increasingly occurred among the replicating parts; *coordinative conditions* were adhered to, and in general increasingly involved hierarchical structures emerged embodying ever more differentiation. Cells appeared, followed in succession by multicellular organisms, organelles, tissues, organs, systems of organs, and so on, each adding a new level to and thus elaborating an existing hierarchy.

Somewhere along the line the gene emerged. Being a replicator, it was unawaringly selfish—concerned only with its survival. So were the combinations of genes, chromosomes and other elements that emerged as more efficient *survival machines* and ultimately plants and animals—each a complex configuration of highly specialized parts. Also, within species that emerged, *altruism* of different degrees appeared—the altruist being unawaringly machined to increase the welfare and survivability of its fellow reproducing members and the selfish genes of which they were composed—a higher form of *cooperation* involving fully constituted phenotypes of a species. Both competition and cooperation (locally and over geographic space) appeared among the *extended phenotypes* of different species—and the fusion of parasites and hosts into a superior survival machine was a common phenomena.

Mutual cooperation of members of a species frequently led to the development of *memes*—units of cultural inheritance of a species, for example bird songs, which as language or otherwise often come to have a positive effect upon the survivability of the phenotypes of that species. In the minds of some scholars the unit on which natural selection operated could be not only the gene, but also the species with its full array of mutually compatible memes—an hypothesis of *coevolution*. *Coadapted meme complexes* came to supplement and fuse with *coadapted gene complexes.*

Further, on occasion, sudden bursts of speciation, that is formation of new species over short periods of geologic time occurred, generally speaking at peripheral (isolated) locations of existing species—bursts associated with the concept of *punctuated equilibrium.*

In all the above there was symmetry breaking, hierarchical construction, stratified stability and irreversibility, bifurcation, period doubling, presence of the Feigenbaum number and strange attractors, deterministic chaos, sensitive dependence on initial conditions, aperiodicity, fractal structure and dimension, self-similarity at different scales, molecular chaos, entropy, nearby equilibrium states, far-from-equilibrium states, convection, self-organization, long-range correlations associated with density gradients, non-equilibrium constraints, autocatalysis and self-reproduction, and joint play of chance and determinism. To these we now add replication, mutation, competition, cooperation, specialization, coordination, allocation of scarce resources, selfish gene, coadapted gene complex, survival machine, altruism, extended phenotype, meme, coadapted meme complex, species as an evolutionary unit, coevolution, punctuated equilibrium and peripheral speciation as elements and processes present upon the advent of man as a species.

While we have discussed to some extent hierarchies—in particular hierarchical construction—it is useful at this point to examine key aspects of hierarchical processes that affect evolution. One is *interaction*, of which there are many kinds and forms. Within a given hierarchy there may be interaction among the units (nodes) at a given level, but also among the levels and the units at different levels. Interaction may be top-down, bottom-up, lateral, or a combination of these kinds. Further, there can be interaction: (1) of a unit at a given level with all or part of the environment surrounding the hierarchy; or (2) of a level of the hierarchy, or the whole hierarchy as a unit, with the environment; or (3) of one hierarchy with one or more other hierarchies; or (4) among one or more levels in each hierarchy; or (5) among the hierarchies each viewed as an entity. Moreover, such interaction can yield over one or more rounds *feedback* effects which can be autocatalytic or *cross-catalytic*. In the biological world, there are: (1) hierarchies of replicators (those comprising for example the initial replicator of the primordial soup, genes, chromosomes, cells, etc., each of which may or may not be or have been an interactor, that is have an effect on other behaving units); and (2) hierarchies of non-replicating survival machines (those of phenotypes). And selection can of course occur at a number of levels as these hierarchies and their parts simply or complexly interact, not only among themselves but with their environments and parts of their environment. In hierarchies of highly elaborate and complicated survival machines, acting both competitively and cooperatively, internal coordinative conditions requiring a balance between stability and flexibility need to be met, incorporating perhaps low-level *anticipatory behavior*.

Social hierarchical structures form to take advantage, among others, of: (1) economies of specialization, occupational and geographic; (2) *scale, localization, urbanization, agglomeration and spatial juxtaposition economies*; and (3) desirable alliance and coalition features. *Information flows* of all sorts are involved—the pheronomes of ants, songs of whales, fluorescence of fireflies—and in a very broad sense *matter/energy transfers* and fields of them.

Of course associated with specialization and the inability of any one human being to store and employ the full set of knowledge within his/her society are: (1) the principle of *optimum loss of detail* that then permits the behaving unit to act upon only relevant data; and (2) the feature of *near-decomposability* within structures whereby information flows from one level up to the next via one or a few channels (linking pins). This in effect reduces use of resources in digesting less essential information and releases resources for information storage and attainment of different objectives.

We view man's appearance as a biological perturbation (mutation) that was amplified and resulted in a relatively sudden and dramatic increase in the size of the brain, reflective and reasoning capability and consciousness awareness. Biologically speaking, a punctuated equilibrium may be said to have occurred, and a new far-from-equilibrium state viewed as having been reached. More specifically, in man's activity the *linguistic mode* of Pattee (broadly defined as the cognitive-linguistic-spiritual mode) came to assume a major role and to complement in a major way the *deterministic (dynamic) mode*. Within the linguistic mode memes and culture along with subjectivity operate. This mode covers the *measurement-control* and interpretive processes which provide description of what is going on—information necessary for evaluation aimed at insuring satisfactory or optimal performance for survival. It covers rules for measurement and a language or symbol system. Moreover, in order to store information to be used later in making decisions and policy and executing orders and demands, or merely to pass on information based on experience, it embodies syntactical rules for combining symbols into meaningful statements.

What goes on in the linguistic mode is rate-independent. Its processes occur in ordinal time only and are irreversible; they are datable, i.e. historical. Thus the processes in the two modes, deterministic and linguistic, are not reducible to each other—and in particular the processes in the linguistic mode are not reducible to rate-dependent processes as in the deterministic mode. For example, although the brain is composed of billions of neurons, each subject to rate-dependent operation, yet the *reflection* and the creative activity that goes on in the brain cannot be designated a rate-dependent process. In effect, so far as man is concerned, processes in both modes are only complementary, but yet each is essential in the functioning of the unitary system of the reflective human being. Processes in each mode constrain or effect the operation of processes in the other, necessarily in a coordinative way for a surviving human being, subject of course to the unpredictable play of chance (noise).

There have been attempts to replicate the brain's operation by computer simulations. Recall that computers can: (1) perform feats of calculation and logic well beyond the realm of the brain; (2) spell out implications of different decisions and policies hardly reachable by the brain using the existing set of mathematical methods; (3) interact with an investigator and put their capabilities at his/her hands to unearth the magnificent displays of chaos theory and the amazing findings obtained in this and other areas hitherto unattainable; (4) enable an investigator with his brain-oriented theories and corresponding models to form classifications and find hitherto unobservables, therewith to help spark the creative activity and the brain's emergent behavior; and (5) provide some insight (though admittedly highly limited) into how the brain in its reflective activity confronts uncertainty and imprecision and generally functions. However, they cannot *per se* produce *creative thinking* and *emergent behavior*.

Such thinking and behavior coupled with man's sensimotor system and *consciousness* leads, in the words of Bohm, to a continuing *unfolding* over time of the *implicate order*. This then comes to be man's ever-increasing *explicate order*—the reality that man perceives and defines and that embraces his technology, know-how, awareness (appreciation and recognition) of interdependence among things and forms, and in general his culture. There are, however, limits to man's knowledge of and ability to understand interdependencies—not only reflected in the number and extent of unexplainables that he characterizes as belonging to a random order, but also because of the increasing complexities of reality (the implicate order) which in itself appears to evolve. Here the concept of a *heterarchical model* can be helpful, one that comprises an ensemble of coordinated causal submodels that pertain to the several pertinent interactions in a given complex situation. These causal submodels are coordinated in the sense that each imposes constraints on the operation of the others, and in turn the operation of each one is contingent upon the constraints (including measurement-control processes) which the others impose on it. Conceptually, the heterarchical model goes beyond a complex dynamical systems model with its rate-determined causal processes and coupling functions in that the former embodies measurement-control and other processes which are rate-independent.

Within its framework an hierarchical model may specifically contain a *participation potential* submodel—a model based on the gravitational field concept of physics—to provide a quantitative element for the investigation of interactions in a complex social situation. Thereby impacts of actions of one behaving unit upon those of other units can be captured (but only in part), whether via phenomenon such as sound and light waves, or other facilities for communication.

Hence, to the set of elements and processes noted on p. 306 affecting the evolution of man and his society, we should add: mutual interaction within and among hierarchies, and between hierarchies and their elements with the environment; anticipatory behavior; scale, localization, agglomeration and spatial juxtaposition economies; information flows; matter/energy transfers; optimum loss of detail and near-decomposability; rate-determined processes in the deterministic mode; measurement-control and other processes in the linguistic mode; reflection and consciousness, creative activity and emergent behavior; unobservables; unfoldment from the implicate to the explicate order; interconnected causation via an heterarchical model; and interconnected behavior (in part) via the concept of participation potential.

10.4 Forces and processes in the development of man's society and culture

a. *Some general considerations*

We now wish to reflect upon the development of man's society and his culture, and do so in a more general fashion than in Chapters 8 and 9 on painting and religion, respectively. We will refer to materials on his economy, polity and social structure, and on his music, sculpture, architecture and other cultural elements to complement materials already covered in Chapters 8 and 9. We shall speak of change and transition more generally and have more specific reference to periods of time during which little or modest change occurred, and even periods of stasis. In fact, it is clear that since the Late Pleistocene era which began some 100,000 or so years ago, stasis as well as all types of change and transitions may have been realized. At times change has been regular, irregular, slow, rapid, exponential, and has involved outbursts of development (Prigogine jumps and punctuated equilibria) on the one hand, and retrogression on the other. Many patterns (combinations) of changes can be found among periods examined, all subject to random events.

Further, when we examine outbursts of development, these may cover all, many, a few or even only one of the subsystems of a society's activities and culture. Because of the interconnection among hierarchical structures, the outburst in one subsystem may induce an outburst in another, or an outburst in one may await (require) an outburst in another, and so forth. As already remarked, the Greek Classical age was characterized by outbursts in development of many subsystems, but clearly not in the religio-spiritual one.

With these general statements in mind, we turn to all the factors, concepts, relationships and other elements already set down in this chapter and inquire how their presence and operation as commonalities impacted several classes of activities (subsystems) of mankind and his so-called civilization.

b. Man in the Pleistocene Age

Begin with the activities of man in the Late Pleistocene period when Neanderthals roamed the earth. It is believed that recognizable modern humans had not anatomically evolved much before this time. We also are able to date intentional burial of the dead around 100,000 years ago, which as a ritual provides the first evidence of an awareness of the possibility of a future life. At about that time, too, we see increased sophistication of toolkit and technology (compared with the definite preconception of tool form previously); however, the rate of change in tool assemblage was still slow.

Giant strides seem to have been achieved in Europe and specifically southwest France and northern Spain once the Upper Paleolithic stage was reached about 40,000 years ago—the time by which a fully modern Homo Sapiens had evolved. Between 30,000 and 40,000 years ago, major innovations in stone- and bone-tool making (e.g. blades and bone points) occurred leading to a great variety of specialized and sophisticated products. In Chapter 8 we have already remarked upon the extraordinary cave-man paintings of the Upper Paleolithic times. Such was associated with fine sculpting and engraving and production of skillfully ornamented jewelry, these too representing a sudden, abrupt advance in artistic expression. Concomitant with these developments were major increases over time in the efficiency and extent of hunting, foraging and other subsistence-driven activities, and changes in family and social organization. There ensued: (1) more division of labor between men and women (the former involved in making and using tools and weapons as well as hunting and fishing, the latter in gathering nuts, berries and wild grains, weaving baskets and making clothing); and (2) population concentration, emergence of larger settlements, better organized use of dispersed resources, and planned subsistence patterns.

Among scholars, however, there is considerable difference in views regarding the geographic extent of transition, its timing and its abruptness—that is the extent of divergence from a pattern of more gradual change. Trinkhaus (1989, pp. 42-3) comments upon the transition between late archaic humans (the Neanderthals) of the Northwestern Old World and early robust versions of anatomically modern humans, and between the side-scraper-dominated industries of the Middle Paleolithic stage and those that were end-scraper- and worked-bone-dominated of the Upper Paleolithic. He states that this transition

> has usually been seen as either a major watershed in human evolutionary history or an artifact of our typological approaches to the prehistoric record, which obscure gradual continuities in human anatomy and behavior. Recent research involving both new discoveries and reanalyses of previous existing remains, is painting a picture of this transitional period that is both more complex and more realistic.
>
> This Upper Pleistocene transition is increasingly being shown to be real, in the sense of being a period of accelerated change relative to rates of change prior to and after the transitional period. However, there were also both considerable regional diversity in the nature of this transitional period and a lack of direct chronological concordance during the transitional period between various human biological changes, between archaeologically documented shifts, and between aspects of the biological and behavioral spheres. What emerges is an image of a period during which many aspects of human biology and culture changed and established new relationships with each other to produce adaptive systems similar to those known ethnographically for recent human hunter-gatherers. The actual natures of the populational changes involved in this process are becoming clearer but remain controversial. And the forms and extents of the associated behavioral shifts are still disputed.

Yet many scholars do hold that there was major transition at least in the limited area of southwest France and northern Spain during the Upper Paleolithic period. Here, then, was *significant symmetry breaking*, certainly in artistic expression, technology and related phenomena.

The above discussion suggests the hypothesis that abrupt transitions, when they do occur, can be fairly isolated geographically, even though the anatomical and brain development of humans may be similar among geographic areas.

c. *Man in the Neolithic Age and the early urban civilizations*

The initial permutation that was to be amplified and herald the next major transition—namely the Agricultural Neolithic (new Stone Age) revolution—may be dated somewhat later than 10,000 years ago, the end of the Pleistocene era. Again, there is little knowledge concerning the circumstances under which homo sapiens discovered how to plant and care for seeds and how to domesticate animals. One theory is that a woman in gathering food randomly came upon cultivation know-how—perhaps when a family from an overpopulated region wandered into an area which turned out to be an inhospitable environment for hunting, fishing and foraging. But in addition to chance there were undoubtedly basic factors underlying the major transition that was to occur, over whose relative importance there is much disagreement among archaeologists and anthropologists. Some emphasize the environmental changes that took place after the Pleistocene—changes in temperature, sea level and the consequent distribution of plants and animals and their speciation. Others emphasize the attainment of a threshold of knowledge about plants, their treatment and storage such that as the stock of knowledge mounted so did the probability of discovery of cultivation know-how. Others appeal to population pressure upon resources. And so forth.

Whatever the factor or combination of factors that led to the discovery, there is evidence of Neolithic village societies in: the Near East, about 8,000 BC (at Jericho and Jarmo and even earlier in present-day Turkey); China, about 4,000 BC; India, by 3600 BC; and Mesoamerica, about 4,000 BC. These societies gradually grew in size and at different times came to replace a number of the Paleolithic settlements. However, the sudden amplification of the agriculture discovery perturbation did not take place until many centuries later, and when it did it gave birth to urban civilization—far-from-equilibrium situations in the Near East at the confluence of the Tigris and Euphrates rivers in Mesopatamia, and along the Nile in Egypt.

In Chapter 8, we discussed the great advance in tomb painting of the Egyptian civilization when contrasted with cave-man works of the Upper Paleolithic. There is only some evidence to support the thesis that the tomb painting and other art constituted a Prigogine jump as some might claim and that it was not the outcome of a gradual transition in visual expression. However, as indicated in Chapter 9, religious organization did undergo drastic change leading to an hierarchical order of priests in the temples of the urban societies. The hierarchical order was in fact closely linked to the hierarchical structure of social classes: the king at the top, often asserted to be divine; a class of hereditary aristocrats, some being men of military stature; the priesthood; merchants; farmers, mostly peasants; and serfs and slaves. The nonagricultural population—the key officials, their underlings and servants, the army, the craftsmen, the scribes, and other urban population—was supported by agricultural surplus. The agricultural surplus in turn was dependent upon: tight and efficient flood control and irrigation management and allocation of water by central authority which in turn required documents prepared by the scribes; the provision of more efficient tools from skilled stone and later metal workers; production of pottery vessels to store grain and other commodities; and so forth. Accompanying all this development were: the invention of writing (around 3,000 BC); the domestication of bigger and stronger animals (bulls and horses) for farm work; smelting of copper and tin to produce bronze (about 3,200 BC); and the development of the wheel (by 3,000 BC), the plow (about 3,000 BC) and other technology. In turn the growth of population at the urbanized centers necessitated: a much more formal set of laws (the Hammurabi code being a classic set); a governmental structure to administer them; and a commercial system to promote exchange of goods and trade with other regions. This exchange was greatly facilitated by the emergence of a relatively advanced number system, mathematics, and the calendar based on strides in astronomy.

The distinctive features of the new urban civilizations of Mesopatamia and Egypt can be set down at great length. Our interest, however, is in the commonalities of processes that we find and the play and interaction of forces and other elements discussed in previous chapters.

First observe *symmetry breaking*, the introduction of widespread adoption of agricultural practices. We can easily imagine that the first person (say a woman, collecting wild grains) had observed the falling to the ground of seeds and noticed the following year that new plants had grown where the seeds had come to rest. She, herself, then placed some seeds on the ground. Perhaps this first attempt at cultivation was unsuccessful; in the following year, no plants grew where the seeds had been placed, or only plants yielding inadequate or inedible grain. Eventually, however, some woman did successfully cultivate a fair amount of quality grain. Like any new technology, the practice of grain farming spread, no doubt being resisted, downplayed or ignored by many Upper Paleothic clans and groups, particular those in areas where prey, wild grains, nuts and berries were abundant and/or where land was inhospitable to cultivation. But even many of these clans and groups were won over as their population numbers mounted and led to pressures on available resources.

Where the soil was fertile and cultivation undertaken, we may speculate that a small village emerged. (The village of Jarmo is said to have had as many as 150 persons.) It is very likely that within that village there emerged at least one person skillful at crafting agricultural tools, and that he was engaged in such at least part-time. Likewise, there must have been at least one person skillful at and engaged (part-time at least) in shaping clay in pottery making. The archaeological digs at Abu Hureyra in what is now northern Syria suggest the existence some 7,500 years ago of role specialization by crafts, such as basket making, weaving of mats, and pottery making.

Hence, at a very early stage, there was *specialization* by occupation or craft supplementing previously developed specialization in activities between men and women. Moreover, when an individual specialized on a given task, typically he/she developed a superior skill at that task, and thus *economies of scale* resulted when the given task was performed for others in a community engaged in diverse activities. Thus, the impact of symmetry breaking brought on by the advent of agriculture (including the domestication of animals) was intensified by specialization and scale economies.

Moreover, in any given craft there developed different levels and extent of specialization, for example in the several types of pottery and tool making or of metal products manufacture. Thus there were different kinds of potters, tool makers, metal makers, and as a result further *hierarchical construction* representing still greater specialization, by occupation and by speciality within each occupation—construction that parallelled the emergence of different kinds of matter from the hypothesized original (Ur) matter, or that may be likened to the biologist's taxonomic classification of organisms.

There was also hierarchical construction in the buildup of political and social frameworks in the villages and later urban centers of the Neolithic era. When several individuals came to form a clan, a leader typically emerged to coordinate the activities of the individuals, to make decisions on migration, administer other affairs of the clan, or some combination of these. When later several clans came together to form a village, their leaders informally functioned as a body typically with some one identified to administer laws, enforce regulations, provide advice, organize and arrange defense against invasion, be involved in any of a number of other functions, or in some combination of these functions, each clan and each individual within it surrendering at least some sovereignty (independence in behavior). In effect, at least a three level power or authority hierarchy came to be constructed (individuals at the bottom, clan leaders at the nodes of an intermediate level, and the overall leader). As technology and know-how continued to mount, still more levels were added to the hierarchical power structure. Villages came to form a single body to: (1) protect themselves against invasion by enemies; (2) *coordinate* water use via irrigation projects; (3) plan a grain storage system in case of drought; or (4) achieve other objectives.

Eventually, as already remarked, a hierarchy may have been constructed with a king (or equivalent) at the top; the high priest, the military commander and aristocrats (major land owners) at a second level; scribes, master craftsmen, or their chosen leaders at a third; and so forth, with serfs and slaves at the bottom.

Within the hierarchical structures that did emerge, there were a myriad of interactions. In a typical one, the king interacted with the high priest, his military commander, and members of the aristocracy who in turn interacted with those below them and with each other. And those at the bottom interacted with each other and with their chosen leaders or representatives. That is, *interlevel* and *intralevel interaction* took place. Moreover, there was extensive interaction with the environment, the farmer and slave on a day-to-day basis adapting to the current weather conditions, and too often working the land with improper care and maintenance. In all likelihood the subsequent negative *feedback* effects were indiscernible daily, but not on a decade-by-decade basis. The king, aristocrats and others also interacted with the environment. They were responsible for irrigation construction and management and perhaps guided and regulated land and other resource use. The impact of their behavior on the environment in turn led to feedback effects, often cumulatively negative.

Much of the extensive interaction naturally was concerned with subsistence needs—the exchange at the village and city markets of food and to a lesser extent woven goods, pottery, tools and other items. Typically, the market tended to be at a central location, easily accessible to a relatively large population for those days and spatially organized so as to realize *spatial juxtaposition economies.* And within the capital city, those occupied in any given type of craft or other gainful occupation might have tended to locate next to each other to achieve *localization economies*—economies from cooperative action in acquiring inputs for their production, in seeking political favor and easing of regulation and taxation, and for other purposes. The localized areas, typically one for each kind of craft or other gainful occupation group, again tended to be spatially organized so as to realize *agglomeration economies*—economies, for example, from having close at hand the inputs (the products of other crafts and economic activity) that a given craft or economic activity required. The growth of urban civilizations and the survival of a dynasty—and there was a succession of dynasties in Mesopatamia and surrounding area from the Akkadian under Sargon (2,331 BC) and on—was thus dependent upon not only agricultural surplus but also the efficiency in management. In the wise use, direct and indirect, of that surplus, there was involved an elaborate maze of interactions to support the governmental machinery and in particular the military machine, which typically like Napoleon's army moved on its stomach.

The market served to mediate conflict—specifically, the *competition* among farmers and tradesmen in bartering or selling their goods and among buyers in purchasing goods. To avoid undercutting and to exercise monopolistic type control, suppliers may have achieved *cooperation* through formation of cartel-like groupings. In any case the market came to serve as a mechanism for achieving *coordination* among the many behaving units vying for gain.

As already remarked coordination also was required at all levels of administration—at the military level, in grain storage to avert famine when drought hit, and in *contingency planning* for other environmental catastrophes and for both military defense and offense.

Hence, when we examine an urban civilization like that of Babylon we find a situation that can be summarized as one of *coadapted meme complexes*. Ways, ideas and thinking embraced: specialization; scale, localization, spatial juxtaposition and agglomeration economies; hierarchical construction; and interaction (interlevel, intralevel, and interhierarchical, e.g. among the hierarchy of priests and that of government officials). Also, the urban civilization had reached such a level of complexity that the *principle of optimum loss of detail* was necessarily in operation, perhaps even *near-decomposability* when a representative or leader at one level had, for example, to report upon the conditions of his constituency to an upper level official.

In a very real sense, just as the village was a *survival machine* so was the urban civilization, though such civilization was to be either devoured by another survival machine or unable to fend off hostile competitors, invaders, and internal discordance—in effect, disintegration of the hierarchical power structure as a consequence of forces to be discussed below.

Other relevant aspects of the emergence of urban civilizations and kingdoms were also present. There was the process of *self-organization*, which characterized the development of most, if not all, villages, a new village often representing a *replication* of an existing village—in a real sense representing *self-reproduction* when inhabitants of the latter relocated because of the pressure of population on resources. On the other hand, in the *relocation process* random elements (noise) may have been present leading to a new idea for village structure, especially when expounded by a strong personality; or a new village may have had to assume a new structure because the fertility of its soil, water availability, climate, or some combination of these and other environmental elements differed greatly from that of a mother village. Thus *mutations* in village structure may have occurred. Whatever, within each village there was hierarchical construction. Moreover, many of the same forces operated on such construction in the villages as in urban centers, except that the scale of hierarchies was more limited. As a consequence, *self-similarity* may have been present in the power structures that emerged.

In the earliest urban civilizations, those of Mesopotamia and Egypt, the factor of *sensitivity to initial conditions* comes out clearly. One cannot claim that the early Neolithic humans that inhabited the Tigris-Euphrates region and the Nile river area were much different in mental capacity and know-how. But the strikingly different environments of these two regions led to sharply different civilizations. The Mesopotamian region was open and easily accessible to invaders. The Nile region was geographically isolated—grim desert barriers to the East and the West precluded invasions from these directions and the cataracts of the Nile made access difficult from the South. Moreover, unlike the Mesopotamian rivers which were unruly and could at times devastate entire villages and cities, the Nile was relatively calm. The Nile provided easy communication throughout its valley, fostering trade and in time political unification; and through its annual inundations it was a constant source of fertile mud for agricultural crops. Still more, unlike the Mesopotamian region, the Nile was blessed with abundant resources—stone for construction, clay for pottery, and gold for ornaments, and nearby were Lebanon copper and Sinai timber.

As a consequence of these differences in the physical environment, there were striking differences in political history—a succession of dynasties in Mesopotamia on the average short-lived, in contrast to the single political system in Egypt stable for thousands of years once Menes unified the many communities along the Nile. Both civilizations were polytheistic, one, the Mesopotamian, not associating its kings with divinity, the other attributing divine status to its pharaoh. Whether we can assign major influence to geographic factors for this particular outcome is problematic, but clearly, as we have already noted, the fact that the king in the case of Mesopotamia was not divine led Mediterranean scholars to study the stars in order to know the will of the gods, and thus to the development of astrology and astronomy, even perhaps with some influence on the development of mathematics and writing. Such innovative advances were absent in Egypt, the Egyptian civilization having imported writing and architectural and engineering know-how from the early Sumerian dynasty.

In these early times, weather conditions greatly affected the state of welfare and subsistence possibilities. These conditions are still unpredictable and beyond model replication by climatologists and other scientists, as certainly was the case then. Clearly they must have significantly affected agricultural yields, causing at times severe deficits at certain settlements. The Nile River inundations would have been no exception to fluctuations but since these inundations did provide sustenance for centuries, their fluctuations at least over units of time of no more than several years must have been bounded. Hence, given the unpredictable ups and downs of weather, it is tempting to speculate that the Egyptian experience over these centuries reflected *deterministic chaos*—the presence of a *strange attractor* in the Nile river behavior even though random elements were omnipresent. (However, recall that chaos theory is yet to be further developed to embrace the constant play of random elements.) Whether the Mesopotamian experience can be speculatively characterized as deterministic chaos or simply *turbulence* is an open question.

Perhaps in time with more archaeological digs, we will be in a position to speculate on the *fractal character* and *periodicity* or *aperiodicity* of the experiences of these early civilizations.

The above discussion smacks of geographic determinism. But this was not the case. What was the case was constant mutual interaction of an ever changing set of *coadapted meme complexes* and an environment—an environment generally slowly changing over the long run, though for Mesopotamia, one of violent fluctuations in the short run. In brief, the cultural *linguistic mode* was fully operating, exploiting and adapting to the environment. Within this mode, *measurement-control* processes were constantly being employed: in irrigation construction and maintenance, and allocation of the water resources; in several levels of the military establishment; in government administration; in the priesthood; and in other hierarchical subsystems. There was constant *flow of information* among these subsystems and their hierarchies, up and down, laterally and in diverse criss-cross fashion. *Participation* in decision making and policy formation was present, the exercise of which was highly skewed with little at the grass roots level.

Along with the above flow of information, there was constant creation of new ideas, tools and know-how, but not in any regular rate-determined manner that could be described by a differential equation. In essence, events and discoveries occurred in ordinal time, processes in general being *irreversible*, and thus *history* as an area of study became relevant. There were, of course, reverses and deterioration of constructed elements, such as the abandonment of villages when the fertility of soil was depleted, the loss of skills when crafts became obsolete from advance in technology (for example in the preservation of bodies at death), and all the civilizations did decline. That of Egypt did pass away even though for many centuries *stratified stability* characterized its development—for example, once the practices in tomb painting and in the coronation of a pharaoh were established they continued for thousands of years.

From another conceptual standpoint, the development of these early urban civilizations was, broadly speaking, an *unfoldment* of the *implicate order*. Emergent behavior, observed variables and factors which were formerly *unobservables*, and the development of technology, the arts, writing, and mathematics that came to be realized constituted additions to the *explicate order*.

Chapters 8 and 9 covered specific cultural subsystems of the early urban civilization—those concerned with painting and religion, respectively. With regard to the former we could suggest but not firmly establish a *Prigogine jump*; and with regard to the latter, there was gradual development of polytheism leading to hierarchical structures and specialization among the imagined gods and within the priesthood. When we examine other cultural subsystems, it is also difficult to find unquestionable evidence to justify a Prigogine jump hypothesis.

Take sculpture in which there were great works in the early urban civilizations, for example: *Kneeling Man-Bull from Jamdar Nasr*, c. 3000 BC; *Statuettes from the Abu Temple, Tell Asmar,* c.2700-2600 BC; *Gudea Worshipping from Telloh,* c.2100 BC; the items in the Temple of *Remeses* II, 1257 BC; down to the fine Assyrian relief productions in the early first millennium BC. But then we have the remarkable *Woman of Willendorf* and the *Man from Brno,* both dated around 30,000-25,000 BC; the *Mammoth from Vogelheid Cave, Wurttemberg,* c. 25,000-20,000 BC; the *Bison, Tue d'Audoubert,* c.13,000-8,000 BC; and *Seated Goddess*, Çatal Hüyük, c.5900 BC. Sudden transition cannot be firmly established.

Take architecture and associated engineering. Here again, an hypothesis of a Prigogine jump cannot be firmly stated despite the imposing zigurrat at Ur 2100 BC, and the monumental pyramids of Egypt, first constructed around 2600 BC. Building construction must have advanced gradually as villages appeared, at first mere hamlets, and grew in size. Evidence on this question is scanty, but two pieces of evidence suggesting gradualism are the great stone tower built into the settlement wall at Jericho, c.8000-7000 BC and the town plan at Çatal Hüyük (c.5650 BC and earlier) where houses were adjoined without streets to provide greater structural stability and defense against invasion.

At this point, we recognize that deeper understanding of the functioning and dynamics of an urban civilization requires many models, each considering a subsystem or part of the civilization, and as a consequence suggesting one or more 'causal' factors at work. For example, take the operation of the market for a basic commodity such as grain. At this market, there must have been a number of behaving units, each concerned with seeking a best deal (thus maximizing his/her satisfaction or utility). Each supplier was motivated to realize the highest possible price (in terms of money or other goods of exchange). Each buyer wanted to pay the lowest possible price. Together at each price, an aggregate supply existed, and an aggregate demand. Accepted economic theory, realistic for highly competitive markets wherein each behaving unit's action has a negligible influence on the outcome, states that the price will be determined at the point where aggregate supply equals aggregate demand. More sophisticated economic analysis then goes to explain the income, real or money, that each behaving unit realizes in a given unit of time, which then in one way or another provides the basis for taxation that is required to support the king and his staff, the military, the government establishment, and so forth. But note how this theory has little to say, if anything, about hierarchical structure and the participation of each behaving unit in that structure. Hence, one needs a model concerning the emergence, functioning and dynamics (change) of a hierarchical structure—and probably a different one for each kind of hierarchical structure, the military as against the governmental, since the objectives of each are different. Further, to construct zigurrats, pyramids and other physical structures, engineering-type models are required, and so are: (1) various specific socio-cultural models required for understanding the emergence of social values; and (2) behavioral models of the reactions of individuals to stimuli, not the rational economic man models, but models capturing physiological and psychological factors. In short, the understanding of the rise and decline of the early urban civilization requires an ensemble of a set of mutually constraining subsystem models—namely what Pattee designates a *heterarchical model.*

d. The Classical Greek Age

While there is much of interest in the record of the rise and fall of Middle East empires and kingdoms following the Babylonian and Egyptian: the Hittite (ca. 1600-1200 BC), Assyrian (ca. 900-612 BC), Persian (550-164 BC) and others, and the Cretan (ca. 2600-1200 BC) and Mycenaean (ca. 1600-1100 BC) civilizations, there is insubstantial evidence to establish any truly major advance, although we will have occasion to comment on certain significant steps within the Persian. The next truly major advance comprised the set of Classical Greek achievements. These achievements, it is to be recalled, were concomitant with the major advances in religion—the rise of Taoism and Confucianism in China, Hinduism and Buddhism in India, and Zoroastrianism and monotheism in the greater Mediterranean region. In Chapter 9 we have already considered monotheism as a major, path-breaking transition (Prigogine jump) in the Middle East, which also represents an instance of *stratified stability* as it has endured for over three millennia and spread to Europe, the Americas and other parts of the globe.

The gigantic advances stemming from the Classical age in Greece took place in a geographic region distinct from that in which monotheism developed, and in subsystems of human society other than religion. As already remarked, there was little development in Greece of spirituality and the real essence of religion; only a particularistic, sophisticated mythical polytheism emerged.

Classical Greece saw tremendous developments in architecture, sculpture, painting, drama together with poetry and music, philosophy, reasoning and rationalism, science and history, and man's level of reflection and consciousness.

In Chapter 8 we have already remarked on the major transition in painting that took place—the gracefulness and beauty of the works on vase and other ceramics and undoubtedly of the works of Polygnotus, Apolladorus and others that did not survive over time. Also noted were the great advances in architecture and sculpture. The buildings of the Acropolis and the Parthenon in particular were prime examples of the former—Istinas, Callicrates and Mnesicles being among the great architects. In sculpture there were: the wonderful Acropolis carvings by and under the direction of Phidias; the *Discus Thrower* c.450 BC, by Myron; *Aphrodite of Cnidos* c.340 BC by Praxiteles; and many others. All these embodied the daring, creative, freedom of artists, with outstanding beauty of form, expressiveness and humanism.

In the literary field, great advances were also scored. Greek invention of real drama reached an epitome in the classic tragedies and critical comedies, among others, of Aeschylus, Sophocles, Euripides and Aristophanes—for example, *The Oresteia, Oedipus the King, Bacchae,* and *The Knights,* respectively.

Intimately connected with drama was the further development of lyric poetry (subsequent to the path-breaking work of Sappho and others in the preceding decades in Greece) and music. In the latter, there were significant strides which went far beyond the practice of the Mesopotamians, Egyptians, Hebrews and other Middle East societies. In these societies music must have been an adjunct to: religious rituals and liturgy [for example, the blowing of the Shofar (ram's horn) and songs of praise by the Hebrews]; military attacks and celebrations; child nurturing; dramatic spectacles; and other cultural activities. As in other societies, music in Classical Greece, drawing upon practices elsewhere, did not become a fully mature and independent art as did Greek architecture, sculpture and painting. There were no abstract pieces, as a sonata, symphony or violin concerto, as exist today. Music as a pursuit was a development that was subsidiary to the development of the other arts and sciences. For example, Athenian dramatists were responsible for composing the music required for the singing and dancing in their plays, and they had to train the chorus. However, the first major steps in the formulation of music theory were taken by the mathematician and scientist, Pythagoras, who applied arithmetic ratios in the analysis of the structure of such intervals as octaves, fourths and fifths.

Beyond the arts and literary fields, there were other great innovations in the Greek Classical age. Pericles and other perceptive Athenian statesmen fashioned the world's first democracy, with the remarkable feature that each state decision had to meet with the approval of the assembly of citizens. Further, the Greeks gave birth to historical awareness. Herodotus, considered by some as the father of history, went beyond the use of previous chronicles and writings, and attempted to *explain* human behavior. While still accepting by and large the evidence of legends and articles, he asked: what are the lessons of history? Thucydides went further. Through careful questioning and sifting of other available evidence, he sought factual accuracy to undergird these lessons—the meaningful patterns of human behavior—and thus to enable man better to foresee the future.

The historians as well as artists were greatly influenced by what many consider the greatest innovation and achievement of all in the Greek Classical age, namely the development of philosophy—the intellectual scrutiny of the physical world and the role and actions of mankind within it. This development represented tremendous progress in the methodology of reflection. The philosophers probed the origin of the universe, not via myths but through logic and rational thought coupled with sifting of available and verifiable evidence—as was so evident in the works of the school of medicine led by Hippocrates and of Pythagoras and other mathematicians and scientists. They questioned the presence of supernatural forces. Moreover, they asked the question: was there a phenomenon of change and if so what were its causes? What was the basis of law? And from this questioning they moved into political and moral matters. The results were levels of reasoning, reflection, scientific scrutiny and humanism that had never been reached before. There was the teaching of Socrates, absorbed and greatly extended by Plato, his student, in a search for a grand design for the political state and society—a search that yielded a theory of ideal forms. In turn, Plato's student Aristotle sought to gather, analyze and systematize knowledge in all fields. The result of all this activity was doctrine and learning that has impacted the world for thousands of years.

Having sketched some of the key developments of the Classical age of Greece, we now examine them analytically in terms of the basic factors we find in physical and social evolution. First we observe *symmetry breaking*—the breaking in terms of the level of *reflection* and *consciousness* of man—the gigantic surge in philosophy with its search through reasoning, logic and science to understand the capacities, limits and nature of human beings and the prevalent forces in the physical universe. This surge was accompanied by: (1) inventions of analytical history to help forecast the future, and of drama, a *coadapted meme complex* embracing play-writing, poetry, music, acting know-how and choral singing, which was leaps beyond the rituals of the archaic civilizations; (2) new realms in medicine, science and mathematics; (3) beauty and harmony of architecture, sculpture and undoubtedly painting which in their expressiveness went far beyond that of the massive structures and sculpting of the Ancient civilizations; and (4) drastically new *hierarchical political structures* and *participation potential* within them. There was: *auto-catalysis*, *self reinforcement*, in each of these developments, involving *intralevel* and *interlevel interaction* in whatever hierarchical structures came to characterize each; and *crosscatalysis* among these developments. The latter phenomenon was more than the typical *agglomeration* forces which economists and regional scientists expound when discussing the rise of primary urban-metropolitan regions such as London, New York City or Tokyo. These forces were much more extensive, occurring within not only each cultural subsystem, but also the total system, embracing mutually reinforcing interactions among the subsystems. Take, for example, the effect of *participation potential* in the political reorganization upon philosophy and architecture, and the feedback effect of philosophy and architecture upon participation potential.

This is not to deny that there were a series of small perturbations in previous decades which anticipated developments in the Classical age—such as those from Thales' inquiring about the world's origin and Solon's reforms, both in the 6th century BC, and the new ideas and artistic know-how that the Greeks acquired from trading with other Mediterranean cultures. All these, however, were small perturbations. The amplification of these perturbations resulting in diverse crosscatalytic surges occurred at a place and time that were ripe for self-reinforcing and mutually stimulating outbursts, namely Greece and the last years of the 6th century and the early years of the 5th BC. Of course, one cannot claim that the amplification was predestined. There were *random events (noise)* that occurred which interacted with all these forces and other elements. For example, one may contend that the Greek victories over the Persians in 479 BC or the appearance of a particular *coadapted gene complex* in the being of Pericles was of great significance and indispensable to the amplification's realization. Moreover, one can maintain that the amplification reflected the right set of initial conditions, that is *a sensitivity to prevailing initial conditions*—the particular geographic setting, etc. Whatever, the amplification, or alternatively put, a *Prigogine jump*, or *punctuated equilibrium*, was realized, and a decidedly new *far-from-equilibrium situation*, a new set of changing *coadapted meme complexes* was attained.

Other commonalities were also present. The major shift of decision making and policy formation, in essence ruling power, to the Athenian assembly of citizens was equivalent to a Prigogine jump in the political subsystem. New *information flow* and authority flow patterns emerged as well as new *intralevel* and *interlevel interaction* within the new hierarchical structure that evolved. So did new inter-hierarchical interactions of this structure with other existing ones, and later with *new hierarchical constructions* such as the Delian League, and still later with the hegemonies of Sparta and Thebes. Within Athens and other Greek city-states, new *specializations* appeared—as represented by statesmen, philosophers, historians and those directly and indirectly engaged in athletic competition—as well as more highly specialized administrators, traders, architects, construction engineers, and so forth. Concomitantly, the Athenian economic subsystem saw increased *scale*, *localization* and *spatial juxtaposition economies.* The realization of such was dependent on expanded trade relations and prosperous commercial ventures (which added significantly to the size of the Athenian market) and on tribute from Athens' allies—all large enough to generate directly and indirectly the subsistence needs of those specialized in the military, arts and other nonfarming activities. Greater trade and tribute came about since in the preceding centuries the typical Greek city-state (polis) had, as a successful *survival machine,* encountered population pressure upon its *scarce resources.* In order to relieve such pressure, Greek colonies had been planted and sprouted all over the Mediterranean and Black seas and elsewhere—literally, a process of *replication (self-reproduction)* of the polis or city state, often on a smaller scale (*self-similarity*), within which the phenomenon of *self-organization,* particularly with respect to an independent political structure, could be observed. In the sense that each polis was not only an independent unit, but also a community of citizens to which the individual had strong allegiance, an allegiance that could dominate his own interests, it functioned as a unit (a *survival machine*) upon which the selective forces of nature and society combined impinged.

The significance of the mutual interaction of man (and his institutions) and the physical environment is clearly evidenced in the disparate development of Sparta and Athens. With mountains in the Northeast and West, and without good harbors for easy trade with other areas, Spartans faced severe obstacles in communicating with the outside world—to profit both from trade, cultural and technological advances elsewhere and to acquire directly and indirectly subsistence means to nurture artists, philosophers, scientists and the like. Population pressure pushed Sparta into conquering the adjacent fertile plain of Messenia and in concentrating its efforts on developing an impressive military might to control it. Little opportunity was afforded its citizens to devote time to the civil arts and philosophy. In contrast, Athens, with excellent ports and the fertile plain of Attica, was able to establish extensive communications and profitable trade with other Greek city states and the Mediterranean world, thereby to be exposed to and acquire the latter's knowledge and know-how. As a consequence, its arts and philosophy flourished and led the way in the Greek Classical age.

As ever, the forces of *competition* and *cooperation* and the tension they generated were present, whether we examine athletics and sports in this era, drama, or warfare—for example in the formation of the Delian League in a cooperative effort against the Persians, and in the intense competitive struggle between Athens and Sparta for Greek leadership in the Peloponnesian wars. The ever changing constraints of the physical environment and its *random* behavior were constantly at play, as were other chance factors (*noise*), like Socrates, a unique *coadapted gene complex* that nature stumbled upon.

The Classical age can also be viewed as a sudden unfoldment of a large chunk of the *implicate order* that constituted a major expansion of the *explicate order*, basic contributions to the latter being those of Plato, Aristotle and so many others. To depict the many processes (causal models) at play in the realization of the numerous contributions, a *heterarchical* model is once again required. Ideally, this heterarchical model should supplement whatever heterarchical model that might be constructed for pulling together the various causal strains behind the concomitant development of the Axial Age monotheism in the non-Greek part of the Middle East.

e. *The Hellenistic Era*

The Classical age gave way to the Hellenistic period taken by some to begin with the death of Alexander the Great 323 BC and continuing to that of Cleopatra 30 BC. During that period, the Greek culture was diffused, resulting in cultural cross-fertilization throughout the former Alexandrian empire which extended over the Mediterranean and far into Asia. This cross-fertilization did lead to a period of notable mathematical and scientific achievements that were not surpassed until the 17th century. There were: the works of Euclid (geometry), Archimedes (mathematics and physics), Aristarchus (heliocentric theory of planetary movements) and others; significant advance in the precision of measurement; and the systematic organization of the previous period's discoveries and insights into the nature of music to yield a coherent science of music—a theory so comprehensive as to become the foundation for Western music. There also were: (1) creation of more massive and grandiose architectural structures contracted for and subsidized by the diverse monarchies stemming from the dissolution of the Alexandrian empire; and (2) limited achievements in other fields—for example in poetry the works of Theocritus, and in philosophy and religion in the more practical and empirically-oriented approaches to living of *Stoicism* and *Epicurianism*. The former founded by Zeno proclaimed that to achieve happiness humans can and must live in harmony with themselves and with nature, and be patient, tolerant, compassionate and non-exploitative of others. And *Epicurianism* maintained that one could attain tranquility by pursuing a style of life based on reason and knowledge gained from sense perceptions, and avoidance of pain and trouble—withdrawal from the world being desirable. In total, however, these developments did not produce self- and mutual reinforcement that could be said to comprise a major outburst of creative activity. Yet the Hellenistic era, through the cross-fertilization of Greek and Asian cultures, constituted an important forward step in establishing a more international awareness and outlook among people.

The reestablishment of monarchy by Phillip of Macedonia in his rule of the Greek city-states and areas of Asia Minor which he conquered, and subsequently by Alexander and his generals who later carved up the Alexandrian empire, wiped out the basic spirit of democracy which had characterized the Greek Classical age. Nevertheless, major growth in industry and trade did occur and large cities sprang up as centers of government, economic activity and culture. Rulers and the upper classes amassed vast estates, and while the operation of these estates was prosperous, little of the gains flowed to the poor. The disparity between the rich and the poor mounted. This, coupled with the loss of the sharing of power in decision making within the Greek city-state region, and the inability of the new philosophies to reach the poor from lack of education led to very restive masses—masses in need of spiritual and emotional sustenance. These masses were ripe for embracing mystery religions such as the cult of Isis and Osiris and the one God religion of Judaism, in particular of the Judea-Christian sect which later became Christianity whose simple requirement for salvation and the blessings of heaven was faith.

f. The Roman era

i. *The rise of the Empire* Previous outbursts (Prigogine jumps) in man's development were based on shifts to (1) new technology and know-how in the cultivation of food for physical sustenance, (2) monotheism for religious sustenance and (3) reasoning, philosophy, science and the arts and literature for psychic-cultural sustenance. What was to take place to propel the outburst in the Roman era was to be of an entirely different character.

First, it was not based on truly great innovations in the technology of production physically to sustain population, although the Romans constructed a network of roads which greatly facilitated military operations, industry and trade, and engaged in creative architecture to be noted below. *Second*, it was not based on any great advance in religion. The Romans as a Republic had inherited a polytheistic set of beliefs from the Etruscans (e.g. the triad of gods, Juno, Minerva and Jupiter) and the Greeks, its methodology being adaptations of Greek legends. Consistent with their conservatism, willingness to compromise, desire to avoid disorder and general tolerance of different cultures, the Romans followed a policy of non-intervention in religious matters of others, while modifying their own religious inheritance in a number of ways in accord with changing social, political and economic conditions. Although there were: destruction of the Israeli nation, to wipe out forever revolutionary actions of the insurrectionists and Zealots; and sporadic persecutions of the Christians by Nero and others, the rule of the Romans while both a Republic and empire was one of permissiveness, except perhaps during the reign of Diocletian and his successors in the years immediately preceding Constantine's conversion to Christianity (312 AD). *Third*, it was not based on any great advances in literature, history, music, sculpture and painting, science, medicine and philosophy. There were the works of the early Romans (3rd and 2nd centuries BC)—the poetry of Ennius, the plays of Plautus and Terence, and the writings of the historian Pictor. Later, in the limited 'Golden Age' of Latin literature during the reign of Augustus, there were the contributions of Virgil, Horace and Ovid. By and large, these were a continuation and further refinement of the Greek advances—as were those of the historians, Livy and Tacitus. Greek influences were also very strong in Roman music, sculpture and painting; little new advance was realized in these fields. However, in architecture combined with engineering, significant advances were achieved. There was the embodiment of the rounded arch, vault and dome in architectural design which when combined with the Roman's engineering genius yielded solid constructions of massive proportions and beauty—such as the Colosseum and the Parthenon. Roman engineering built aqueducts to provide urban areas with efficient water supply; and, reflecting the urgent need for rapid and reliable transportation and communication for military control and trade of a rapidly growing and far-flung empire, the Romans constructed an outstanding network of long-lasting roads and bridges. Yet, despite their creative engineering associated with fine city planning, Romans achieved little in science and only a bit in medicine (the works of Galen). Finally, in philosophy there were the fine literary works of Cicero, and later the thinking of Seneca, Epictetus and Marcus Aurelius, a more religious-oriented form of Greek Stoicism. All in all, however, the above achievements, though notable in architecture and engineering, did not provide the seed and backbone for that surging, creative advance that nurtured and supported a far-flung empire—an empire covering England, all of continental Europe west of the Rhine and south of the Danube, northern Africa, the Mediterranean islands, and much of Asia west of the Euphrates—an empire that experienced two centuries of prosperity and peace, a phenomenon never previously or since achieved for this area. What were the critical factors that sparked and fed that surging creative advance?

The answer to this question lies in the creative law, government and administration that was evolved—a creative *coadapted set of meme complexes* that was concomitant with the construction of a dynamic, burgeoning *hierarchical order* backed up by military strength—an efficient and stable order over these two centuries.

There is much that is unknown about the early days of Rome, the works of Livy and other historians of these times bring replete with myths and legends untenable as fact. The founding of Rome is now generally taken as 753 BC, and the beginning of the republic as 509 BC when the kingship was replaced with its military powers being assumed by two magistrates named consuls. These two magistrates, primarily generals, were empowered equally, an arrangement that served to check abuses. They were elected annually by the Roman army as a voting body. Later, by law (367 BC), one of the consuls was to come from each of the two social classes, the patricians and the plebeians. There also existed a Senate, perhaps originally an advisory council to the king, made up of old and wise men, which later became composed of members from leading families. The Senate advised both the magistrates and the Roman people. In theory, the people were sovereign and had the power to legislate; in practice, because of the collective prestige of the Senators, the advice and deliberations of the Senate (when summoned by a magistrate to a meeting) were implemented by the magistrate or submitted by him to the people for enactment into law.

There emerged two popular assemblies. One was the *centurate assembly* where each person could cast one vote within a larger voting unit, and a majority of unit votes was required for a decision. This assembly voted on war and peace issues and elected the magistrate that exercised military functions, the voting power being allocated so as to maintain the political supremacy of the wealthy and the established nobility. The other assembly, the *tribal,* was a civilian one, more democratic than the centuriate; it did most of the civil legislating, and elected the magistrate not assigned military power.

Then there were two tribunates, the plebeian and the military, the former designed to protect citizens by means of granting their officials, the tribunes, the veto over unlawful acts of the magistrates. Soon thereafter the 'Law of the Twelve Tablets' was established, primarily to inform the people about where they stood in relation to the law, formerly unwritten. Other types of officials elected were: aedeles (assistants in managing the affairs of the city), quaestors (financial officials of the consuls), and censors (with the duty, among others, to draw up official lists of Roman citizens).

Ups and downs in the political power of different groups characterized the early period of Roman life and the days of the Middle Republic (264-133 BC). Throughout these times, there were elements of monarchy (magistrates), aristocracy (senators), and democracy (citizens voting in popular assemblies). When the Middle Republic era came to an end, the Senate strongly dominated the political hierarchy, having acquired great prestige from the successful conclusion of the Second Punic War.

By the beginning of the Middle Republic era, Rome had come to dominate Italy by conquest or alliances with city-states. The military ethos of the Romans, especially the aristocracy, was one factor in this outcome. So, too, were the effective checks and balances that the Roman political structure provided, although the significance of these should not be exaggerated. Equally, if not more important, was a factor that would persist throughout later times as well, namely the incorporation of conquered people into the Roman social and political system—in contrast to the Greek practice of excluding the conquered from citizenship and socio-political affairs. By the end of the Latin wars (340-338 BC), close to all the conquered people were given Roman citizenship—given at least a *potential* (perhaps in practice only a perceived potential) *for participation* in government. Moreover, the conquered towns were allowed to retain their local governments.

During the Middle Republic period, Rome's extension of military domination in the Mediterranean led to new administrative arrangements. No longer were incorporation via citizenship and alliances with city-states capable of working. Rome found it desirable to rule acquired territories directly as subject provinces. Each was governed by a senatorial magistrate with almost absolute power. Since the administrative apparatus was relatively lean, there was heavy reliance on local élite for managing disputes. Aside from the rules laid down by the conquering commander and magistrate, minimal intervention continued when additional provinces were added to the empire, and again the local cities were allowed to conduct their own affairs. However, with continued growth of the empire administrative arrangements had to become more complicated, the primary concerns of Rome still being the levying and collection of taxes and maintenance of order.

Starting with the Middle Republic, the crude tribal civil law of the Romans was transformed. As already remarked, Rome's leaders and philosophers were much influenced by Stoic thinking. Stoics maintained that all humans have at least some ability at reasoning and are accordingly able to perceive a universal natural law governing the whole world—a law with which each individual should harmonize his behavior. Such thinking implied a basic equality among all humans. When Roman domination came to extend beyond Italy, such thinking came to transform early crude tribal civil law (*jus civile*) into the natural law-based *jus gentium*—a set of principles common to all nations and thus applicable to all people, both Roman citizens and foreigners, ideally even slaves.

However, by the end of the Middle Republic, it was clearly recognized that the government was inadequate for the administration of a growing empire ruled by military-minded aristocrats competing for public office and glory. There was much discussion and debate over the future directions the Republic should take. Cato, the highly respected Consul, strongly urged a return to the old ways. His views failed to prevail. The forces for change, decline and later progress could not be stilled; nor could the intense rivalries within the Senate be controlled. The government went downhill. Turbulent times with assassinations and civil wars characterized the late period of the Republic (the years of the triumvirates and Caesar's dictatorship), continuing to the days when Octavian, later named Augustus by the Senate, amassed power. He may have been motivated by an intense covert desire for power, having appreciated the fact that concentration of power was an essential ingredient for maintenance of the empire; or he may have been driven by this factor in combination with others. In any case, he did come to recognize the essential need of people and officials (elected, appointed or otherwise) for participation in decision making and policy formulation, even if in fact such participation was only perceived by them.

Early on, after defeating Anthony and Cleopatra in a naval battle at Actium (BC 31) and the subsequent disintegration of Anthony's forces, Augustus became master of the Roman empire—a power position he was to hold for more than 40 years. He restored law and order to an unruly Roman state eaten away with corruption and murderous dealings. He set in motion a process aimed at establishing the best possible civilian government as a constitutional monarchy. To regularize his own position and to make it more acceptable to others, he proposed in 27 BC to lay down his powers as consul. This was rejected by the Roman Senate, and instead he was asked to administer large parts of the Roman empire. To make certain of conciliating the senatorial class, and to avoid any possible charge of monopolization, he vacated the consulship in 23 BC. In its place he was granted the power (rights and privileges) associated with the office of tribune, to which formally he, as a patrician, was ineligible, but which power was ratified by popular assembly. Through this tribunate power, annually bestowed upon him, he was able to participate in civilian affairs. He was also granted proconsular military powers, given the unofficial title of *princeps* (first citizen) and received numerous honors over the years. As a consequence, with his continuing control of the military, his position became unassailable—in effect his position and power became that of an emperor.

It is not necessary to present the details of the wise administration and leadership of Augustus. He encouraged local self-government and urbanism in the provinces under his immediate jurisdiction, and respected local customs. He followed in the footsteps of Caesar in establishing throughout the empire many Roman colonies where soldiers could settle to relieve the problem of what to do with military personnel no longer needed in times of peace. Citizenship was granted to these colonists and they in turn helped to spread Roman law and culture throughout the empire. He continued the practice instituted by Caesar of selecting an able successor—if necessary, by adopting an able young person as his son.

During the immediate years following Augustus' death, the provincial administration continued to function effectively despite the lack of high quality leadership from the Julian-Claudian and Flavian dynasties (27-96 AD). Claudius, however, did recognize that the efficient running of the empire was beyond the capabilities of an ordinary ruler. He created an imperial bureaucracy made up of professional administrators. Then followed the age of the 'Five Good Emperors' characterized by peace and one of unparalleled prosperity. By and large, these emperors were fair and humane, and concerned with the welfare of the empire; and during their realms, citizenship was extended to more and more subjects. However, it was not until 212 AD that the emperor Caracalla granted citizenship to all subjects.

At this point it is instructive to examine, in terms of the general factors discussed in Chapters 2-7, the rise of the Roman empire which reached its zenith in the last years of the 2nd century AD. There was *symmetry breaking*, the symmetry being that of a disjointed (in one sense, highly disorganized, chaotic-like) landscape of lightly interacting, largely independent entities: city-states, rural villages, farming communities and a few remnants of empires. The symmetry breaking involved a melding and consolidation of elements with low connectivity into a single unitary system—one involving cumulative aggrandizement and spatial spread. The amalgamation of the existing elements involved *hierarchic construction*. From conquest and alliances during the early years of the Republic that saw Rome's domination of Italy, the authority structure came to be composed of a top node (Rome) supplemented by a level of second-order nodes comprised of the leaders of the diverse Italian states. When the Republic and empire were subsequently extended, a new level was introduced into the hierarchical structure, namely a level of provincial administrative nodes. These nodes made it possible to avoid the high costs of administration and inefficiency from what would be in most instances too great an effective distance between the location of the controller and that of the controllee. In effect, the establishment of the new level of nodes constituted a reallocation of the top level's functions to achieve adequate performance and effective spatial control. Yet centralization at one location, namely Rome, was necessary for critical functions—for determining general military policy, wise road building, and level of taxation to support the central administrative functions.

In fact a number of hierarchical structures with different configurations and levels emerged in the required exercise of many functions—in the economic production of diverse commodities and provision of various services, and in the political administrative system. Eventually, there was created in the latter a set of distinct hierarchies, one each for justice, military affairs, finance and taxation, foreign affairs and internal affairs, each with its own state officials reporting back through their superiors to the emperor. The result was an elaborate administrative system, a huge complex hierarchical structure embracing a number of hierarchical subsystems, interconnected and interacting with each other and the environment (social and physical), which as already remarked was later to be imitated by the Roman Catholic church.

Naturally, Rome was the top node for a number of these functions, the provincial capitals for others and lesser cities and small towns for still others, particularly those of small importance. And these hierarchical structures interacted with each other in many different ways—as entities, but also with regard to their several levels. Rome, each provincial capital, each town and village typically experienced *scale, spatial juxtaposition, agglomeration* and even *localization economies,* reflecting *autocatalytic* and *crosscatalytic* processes and the play of *coadapted meme complexes.*

In many ways, but certainly not all, there was *self-similarity* but at different scales among the villages, towns, provincial capitals and Rome, both at the aggregate level and specific disaggregated levels. The many colonies set up by Caesar and Augustus paralleled biological processes of *replication, self-reproduction* and *self-organization,* each successful colony functioning as a *survival machine.*

As already noted, a distinguishing feature of the Roman empire considered by some to be a major advance in the civilization of empires was the extension of citizenship to the many conquered peoples and ultimately to all—a significant step forward in civil rights which was anticipated by the Persian empire under Cyrus. This constituted change in the *information flow pattern*, and in the *participation potential framework,* at least as perceived. Necessarily, in such a huge empire, much information did not flow because of limited capacity of *communication channels* and absorptive capabilities of the receiving units; and thus there were *density gradients* in the information flow field. In effect, the *principle of optimum loss of detail* (or the approach to an optimum) was operating everywhere and structures approaching or attaining *near-decomposability* were to be seen. This was clearly the case when, as just remarked, the emperor established provincial governments, delegated power to them and when in turn the provincial governors relied on local élite to handle local affairs that were directly inessential to the maintenance of the empire.

Associated with outbursts of development in the past have often been the appearance of truly unusual personalities—a Cyrus, a Socrates, a Pericles. In the Roman case there appeared the unique Augustus, another rare *coadapted gene complex.* One may be inclined to view his appearance as a *random (noisy) event* within a particular turmoil ridden socio-political setting (a particular *set of imperfectly or non-coordinated meme complexes)* that led to the formation of an administrative-legalistic scaffolding that generated two centuries of prosperity and peace and contained the numerous forces of disarray. It is of interest to ask several questions. Was the arena of internal turmoil and civil unrest into which Augustus entered as a player one of complete *turbulence*, or one of just *deterministic chaos* (that is, a bounded multi-dimensional region, a particular *strange attractor*), in which he was one of several highly *competitive* combatants, at times *cooperating* in coalitions to eliminate others? Or was the set of events he initiated and the outcome of 200 years of prosperity and peace, approached only once before or since by a Western European or Mediterranean empire, a concrete instance of *high sensitivity to initial conditions*? Or was this outcome simply a *joint play of chance and deterministic forces*? Or should this era of 200 years be treated as reflecting *stratified stability*—to be considered a stable *far-from-equilibrium situation* that was reached, later to fall into disarray, but nonetheless one that left a heritage of lasting achievements promising to survive so long as the phenotype man can master the hazards of his environment—hazards partly generated by his constant myopic interaction with the environment? These are questions to be addressed in the future by a team of interacting historians, scientists and noise and chaos theorists.

Part of the lasting achievements—for example, in engineering and architecture—represent elements of the *implicate* order which were *unfolded* and became part of the *explicate order.* These elements were previously *unobserved objects and variables* and unknown know-how in both the physical and social systems (for example the arch and its use in construction), which led to the *emergent behavior* and new social relationships found for instance in the development of bureaucracy.

There were other factors discussed in Chapters 2-7 that were involved in the rise of the Roman empire, which are obviously associated with those already mentioned, or which will be discussed below in connection with the decline of the empire. However, before we analyze the decline, we should briefly examine certain features of the Persian (Achaemenid) empire.

First, note that before the Roman the mighty empires that had previously been built up in the Western civilized world—the Assyrian, Egyptian, Hittite, Persian, and Alexandrian—did not, except for the Persian, exude the Roman administrative genius. The Persian reached its peak in a very short time, starting from 559 BC when Cyrus the Great inherited the throne. Early on, he organized his domain in an hierarchical manner. Twenty provinces were established with a satrap (protector of the realm) at the head of each, each satrap having a high degree of independence in administering justice and formulating financial and internal policy. Cyrus and the other early Achaemenid rulers were adept at conciliation and were willing to rule through local élites. They protected the institutions of their subjects, permitting them to maintain their own customs and religious ways. Also, they effected desirable reforms in law, greatly improved the coinage system and established a standard set of weights and measures. Equally important, Cyrus carefully planned and initiated the construction of a road system which Darius perfected and maintained in excellent condition, thereby permitting efficient military operations, providing a speedy courier system for communications, and facilitating trade within the extensive domain of the empire.

While on a number of occasions harsh measures were employed in governing conquered people, the early rulers were sensitive to the needs and well-being of those people, and overall the empire was prosperous, effectively run and relatively peaceful internally for well over a hundred years if not close to two hundred. It accomplished much in synthesizing the Near East cultures and helped pave the way for the Hellenistic advances that were to follow in science, medicine, religion, philosophy, finance, exploration and other areas.

As all top-heavy hierarchical structures that preceded and followed, it came to confront forces of decline—in this instance forces that started to emerge during the reign of Xerxes who succeeded the brilliant Darius. In Xerxes' later years his administration lacked the concern for continued progress characteristic of his predecessors—undoubtedly in part an effect of incentive-deteriorating affluence. The kings who followed Xerxes were relatively weak and less enlightened, and the empire came gradually to die. In the later years there were the typical occurrence of assassinations by power-hungry contenders for the throne, revolt of provincial governors, economic poverty and inflation, invasions and so many of the elements associated with the decline of other empires, earlier and later. The empire came to an end with the conquests of Alexander the Great in 331-330 BC.

In one sense, the Achaemenid empire was a perturbation resembling that of the Akheniton realm in Egypt—a perturbation that flourished for some time and then was squashed. However, the Achaemenid empire did demonstrate and still serves as a model of how it is possible for diverse populations, with different customs, languages, religions, social organizations and aims, each to prosper under the oversight and restrictions of a central government and the constraints that these populations in their unique ways impose on each other. It is an interesting question to ask what was the empire's direct and indirect influence, via the Hellenistic transitional period, on the formation and functioning of the Roman empire in the succeeding millennium—a question which we must leave unanswered.

ii. *The Empire's decline* We now examine the factors that led to the disintegration of the Roman empire. For understanding this we have many more records and data than we have on the decline of previous empires. Hence, we can probe more deeply into the process.

Historians and others have attributed the decline of the Roman empire to numerous causes, usually emphasizing one or a few. It is clear, however, that a number of basic factors were at play, often intensifying each other through *mutual interaction*. There were the barbarian invasions (considered by some to be simply noise), but the success of these invasions cannot be assigned to any lack of potential for countering them. The Roman military technology was certainly not inferior to that of the barbarians, the belief in and obedience to monarchy was often present, the population was clearly large enough to farm an adequate military force, and economic capability to support such a force existed. What was one of the direct causes of the Roman defeats was rottenness of government. Troops were inadequately trained and disciplined, too often poorly equipped and unpaid, and frequently without sufficient supplies. As a consequence, morale was low. The rottenness of government in turn was in part related to the political insecurity of previous decades when military leaders, especially those of the Praetorian Guard, intervened in political affairs, murdering emperors, replacing them with new ones in turn to be murdered. The Senate had for some time ceased to act with any political determination. Then, too, excessive and very costly bureaucracy had developed as emperors did away with the Senate's powers, and more and more delegated authority and functions to civil servants—especially after Diocletian, in his efforts to forestall the collapse of the empire, had expanded the body of bureaucrats to carry through the legislation and reforms he instituted.

Concomitantly, the administration of cities became increasingly inefficient. Urban élites became less and less motivated to provide the traditional unpaid civil service which had characterized and was a major source of strength in the days of the Republic. Perhaps even more significant for the deterioration of government were the breakdowns in the *flow of relevant information*. Emperors, whether or not competent, were too often screened from knowledge of happenings and prevailing conditions—knowledge which could have led to needed corrective measures.

Low quality of government and administration created undesirable social conditions, which in turn negatively impacted governmental operation. Many historians have noted the drag that stemmed from slavery. There were, of course, economic advantages to the slave system. It provided a cheap factor of production, the supply of which in the days of the Republic and early years of the empire was abundant—abundant because of the many conquests by Rome. Low cost of diverse services (provided by slave labor in the cities) and of agricultural production (from large inputs of slave labor) was possible, provided this labor was available and continued to be productive. But in the later years of the empire, the tempo of conquests declined and so did the supply of new slaves. At the same time, conditions under which slaves lived and worked in the rural areas were harsh, which partly accounted for their low rate of reproduction. The consequent diminished pool of labor, even after slave owners provided housing and other incentives to hold on to labor, caused agricultural production to be less profitable. The labor problem could have been counteracted by new and better agricultural technology, but landlords were not interested in such.

The decreased gains from agriculture were accompanied by the intensification of urban problems created by the resulting rural depopulation and migration to Rome and other cities where the empire had to provide free food for more mouths and more public entertainment. At the same time the empire had to resort to devaluation of the coinage to mint more money to meet the empire's mounting expenses, thus fueling a process of severe inflation. Taxation became heavier, intensifying still more the impoverishment of the poor and the ill effects of pestilences and disease upon them. Class tensions mounted, revolts transpired and civil disorder spread—and a number of other forces came to contribute to disintegration.

How embrace in one conceptual framework the many interrelated factors and causal elements residing in the several subsystems of the Empire—the military, the political decision-making, the governmental-administrative, the economic, the social and others? For this problem, the *heterarchical model* can be useful. It can help to specify more formally some of the interconnections among causal elements that can be hypothesized or identified—for example: (1) the deleterious effects of the affluence of political leaders and large landowners and in particular the deterioration over time of motivation for civic duty and drive for efficient administration partly as a result of such affluence; (2) the declining lot of the slave and the poor from such deterioration; (3) the consequent decrease in the supply of labor partly from the failure of slave population to reproduce because of its poverty; (4) the resulting decrease in agricultural productivity and thus the state's tax receipts and the consequent need for higher taxation of land-owners in the attempt to maintain the total revenue of the empire; (5) the debasing of the silver content of coinage from failure of higher taxation to meet the total revenue requirements; (6) the increasing rate of inflation from such debasement; (7) the further decline of the lot of the poor from such inflation; (8) the rising levels of social tensions and unrest from this further decline; (9) the resulting spread of revolts; (10) the need for the increased employment of the military to put down these revolts; (11) the need for increased revenues to pay for such further use and to provide the higher pay required by solders disgruntled by the inflation and inadequate provision of military equipment and supplies; (12) the decrease in soldiers' morale and willingness as well as ability to fight barbarians and maintain order because of low real income and lack of equipment and supplies; (13) the further tendency to increase the level of taxation to obtain more revenue for higher pay and to avoid such inadequacy; (14) mounting evasion of taxes and corruptive behavior stemming from the higher levels of taxation; (15) the increase in the supply of skewed information and misinformation to effect tax evasion; (16) the failure of top administrators to develop corrective measures and formulate good policy as a result of the distorted and/or inadequate information received; (17) the intensification of the empire's problems resulting from poor decisions to cope with the steadily growing problems within the empire; (18) the still greater levels of social tensions and disgruntlement that resulted; and so on, leading to a series of rounds of *crosscatalytic disintegration.*

In interrelating causal elements in the above manner, a heterarchical model can be extremely useful for understanding the decline—especially when a number of the elements are capable of being expressed in terms of a mathematical function relating a dependent variable to one or more independent ones. In some cases the function may be well-specified and able to yield a quantitative magnitude for the dependent variable. In other cases the functions may yield a quantitative outcome subject to a standard error of estimate, reflecting the joint play of *noise* as well as a process within the *deterministic mode*. And even if in most cases only a nonquantitative statement of relation (as a process within the *linguistic mode*) is possible, a heterarchical model (despite its restriction to *ordinal time*) can still contribute to understanding a changing situation and in particular to the need for *anticipatory planning*.

The collapse of the Roman empire has been considered by many to have been reached with the complete disintegration of the Western part by the end of the 5th century. The Eastern part, later designated the Byzantine empire, the outgrowth of Constantine's decision to transfer the capital of the Roman Empire from the West to the East in 324 AD, continued to exist for many centuries. In fact the period up to 632 AD, when the invincible Islamic forces attacked, was one of expansion and splendor. The Byzantine empire reached its height during Justinian's rule (527-565 AD), when the systematic codification of Roman law was prepared, a codification upon which the modern legal systems of most Western countries are now based. This codification in one sense was implied by the complexity of the social system that had evolved by the end of the Roman empire—an *unfolding* of a part of the *implicate order,* elements of which had been a continuing (*irreversible*) heritage of the Justinian reign. From 632 on, there was contraction and splintering of the Byzantine empire continuing until 1453 when it was conquered by the Ottoman Turks.

These Byzantine days saw significant achievements in architecture and art. Church construction, such as the Church of Saint Sophia, was designed, in sharp contrast to the Greek temple, to enrich the inward and spiritual experiences of the Christian worshipper; it employed the novel application of a dome centrally constructed upon a square building. Moreover brilliant mosaics were created, conveying intense piety or extreme majesty, often by distorting and elongating human figures. Also, the empire did preserve for the future much of the cultural advances of the classical Greek, Hellenistic and Roman periods, and did conduct effective missionary work among the Slavs.

In effect, the Byzantine empire was a *survival machine,* an effective truncated remnant of the Roman empire for some three centuries, but one whose internal weaknesses were not able effectively to combat the incursions of invaders (stronger sets of power-driven *coadapted meme complexes*) and was gradually eaten away over the eight centuries following its peak.

g. *The Middle Ages: The Early and the High*

From the collapse of the Roman empire in 476 up to 1050, the period of the Early Middle Ages often designated the Dark Ages, there is no record of any great surging advance, aside from the Byzantine mini-surge that reached its peak in the 6th and 7th centuries. The many years of chaos that ensued were occasionally interrupted by orderly times—for example the mini-Renaissance (Carolinian) when Charlemagne revived some aspects of classical antiquity, established Latin as a common language and thereby preserved the literature in this language. Also, there arose the well-operated communities of the monastic orders which served in their small way to maintain and promulgate high moral values and devout religious practices. And there was starting in the reign of Gregory the development of a new plainsong, the Gregorian chant, that was to become a central part of the Roman Catholic ritual.

In the European West, when there were many fierce invasions by ‘barbarians’ and Islamic dominance of the Mediterranean, when contact with Hellenistic culture was lost, and when indeed the years were economically bleak, real stagnation reigned. Yet the Early Middle Ages did provide the groundwork for two momentous developments to come. One, already mentioned, was the emergence of the papal empire, namely that of the Roman Catholic church in the West. The other was the beginning of a European culture distinct from those of the Byzantine and Muslim worlds—one that was built upon the isolation of the many different parts of Western Europe subsequent to the administrative, political and economic disintegration of the Roman empire. These isolated parts in the ensuing period of manoralism and feudalism were later to be unified to form a European system rooted in accord with new and different economic and social forces.

The formation of the European system may be said to have occurred during the period 1050 to 1300, frequently designated the High Middle Ages. Aside from the tremendous upsurge of power of the Roman Catholic church in the West which, as remarked in the previous chapter, reached its peak in the late 12th and early 13th centuries, there were no truly major advances in the social order. Yet there were a number of significant developments. First, there was the important complementary play of climate and agricultural technology during this period. For Northern Europe the weather conditions were quite favorable up to 1200 and at the same time there were in farming notable improvements—the use of a *heavy* plow, which was essential for *intensive* agriculture, the introduction of a three-field system of crop rotation, the use of watermills and windmills, and a number of other small advances cumulatively important. As a consequence, there was an enormous increase in agricultural output, decrease in malnutrition and poverty, and increase in the surplus from farming to support the population of cities. With such support for the growing cities, and with less incursions by foreign invaders during this period, national monarchies were able to develop in France, England and Germany, establishing institutions more representative and protective of the people—townspersons, clergy and nobles. Universities emerged and became centers of classical learning. Long-distance trade revived, commerce and economic enterprises flourished, and a new wealthy class of merchants arose. The Crusades, especially the first which was successful, provided a psychological uplift and confidence among the Western Europeans. It also reflected the growing political power of the Christian church of the West. Relative peace and prosperity reigned.

These High Middle Ages also saw a revival of intellectual development and an upsurge of creativity in literature, art and music. The period from the middle of the 12th to that of the 13th century was one of numerous translations into Latin of almost all ancient Greek and Arabic scientific knowledge, most of Aristotle's works and those of the major Islamic philosophers such as Avicenna and Averroës. Such allowed Grossteste (1168-1253) and Bacon (1214-1294) to make important advances in science with particular emphasis on empiricism. Simultaneously, there emerged scholasticism, the theory and practice of reconciling classical philosophy with Christian faith; the outstanding writings of St. Thomas Aquinas (1225-1274) on this subject were characterized by rigorous orderliness and deep intellectual probing. Concomitantly, there was growth in the use of vernacular languages, French, German, Spanish and Italian, in the poetic, lyrical and other literature that blossomed, the Divine Comedy of Dante (1265-1321) being a prime example. The mood of this literature was closely associated with the mood of the major architectural advance, the Gothic cathedral, that was attained. The Divine Comedy's mood, starting with despair in the poet's journey through hell and purgatory and ending up in the lofty realm of paradise with wisdom and salvation must have resembled the sensory experience of the lay person as he, beset with the troubles and anxieties of a harsh day-to-day existence, entered a Gothic cathedral and gradually imbibed the beauty and stimulating uplifting sensation of the soaring heights of a heaven-bound structure.

To understand better this interaction of literature and architecture, we should bear in mind the Romanesque-type constructions of the previous century or two. These constructions were designed to manifest the glory of God in a much more forceful manner than the basilicas, the large wooden, covered market-hall places of assembly of Roman times. The Romanesque monasteries and churches, however, were rather harsh and militant. Their essential features were rounded stone arches resting on massive piers, with firm unbroken walls and towers, few and small windows. The structures were very severe, relatively gloomy in appearance, with plain interiors occasionally relieved by brightly colored mosaics and frescoes. Often, inside and out, there were sculptured decorations and full length human figures in facades—figures which were elongated and grave with evocative power, a first manifestation of a revived interest in sculpting the human form.

The Gothic cathedral, with its uplifting sensation beginning with the gloom of existence and rising to the joyous heights of paradise, was so different in construction and impact. It was indeed a magnificent architectural structure, comparable to the wonders of a pyramid or zigurrat, but in addition it created a tremendously deep inward feeling far exceeding any that these other structures could evoke. With its innovative use of pointed arches, groined and ribbed vaulting, flying buttresses, lofty spires, and enormous stained-glass windows it must have produced for many worshippers an intense spiritual message that left way and far behind the world of logic and reason.

During these times, there were also great strides in music. Previously there had been little development, relatively speaking, since Hellenistic times. The Romans added little. Later, there was the Gregorian chant, a plainsong, the outgrowth of efforts to codify and organize traditional melodies—one of the achievements of the previous Dark Ages. But that was still homophony, the playing of only one melody without any harmonic background. The big advance, namely polyphony, the playing of two or more harmonious melodies together, experienced a major development during the 11th and 12th centuries. This advance culminated in the work of Léonin, *Magnus Liber Organi*, 1163, at the church of Notre-Dame, Paris that had become the musical center of Western Europe. It is to be noted, however, that polyphony had been anticipated in church music during the 9th and 10th centuries when there was exploration with two voices in parallel motion a fourth apart, and even earlier in folk music of which there is record of part singing, perhaps a result of participants having different natural voice ranges, an octave, third, fourth or fifth apart. The term, *Ars Antiqua*, has been used to designate the flowering of polyphony from approximately the middle of the 12th to the end of the 13th centuries.

Nevertheless, the bubbling creativity and relative peace and prosperity of the High Middle Ages—a prototype renaissance in itself, a period of many small-to-moderate sized mutually reinforcing perturbations about to burst into a truly major advance—had to pass through a period of abeyance. A crisis in the Middle Ages developed. There were: the bad harvests and economic difficulties in the first half of the 14th century; the horrible Black Death (bubonic plague) which decimated the population in the middle of the 14th century; the Hundred Years war (1337-1453) which devastated France and depleted the resources of England; the Schism in the Roman Catholic church (1378-1415); and then the feelings of insecurity arising from the Turks' capture of Constantinople in 1453 and conquests of neighboring Europe.

Yet while all this was transpiring and the economic/materialistic advance of Europe was slowed down if not decreased, the development of the arts continued—which was soon to result in the High Renaissance era noted in Chapter 8. Gothic cathedral building, which first began in the middle of the 12th century, flourished in the 13th, and continued to do so in the 14th and beyond. And by the 15th, new directions in architecture emerged. Brunelleschi, the initiator of Renaissance architecture, infected by the desire for a revival of Roman grandeur, designed new ways for the free use of classical architectural forms—for instance, columns, pilasters, pediments—to create new modes of harmony and beauty. Examples of the new architecture are: his *Cappela Pazzi, Florence,* 1430 and his dome of the *Florence Cathedral,* about 1420-36; Alberti's *S. Andrea in Mantua,* about 1460, and his *Palazzo Rucellai,* Florence about 1460; and Bramante's *Tempietto*, 1502-11. For another half millennium, Brunelleschi's innovations served as a major force in building design

Literary effort also continued to blossom, increasingly with a naturalistic bent, as witnessed by the great vernacular works of Boccaccio (1313-75) and Chaucer (1340-1400), the *Decameron* and *Canterbury Tales,* respectively. Compared to the Romanesque era, sculpture, too, took on a much more naturalistic flavor. Carvings of leaves, flowers, and other objects evidenced careful observation of actuality as did representations of humans more accurately proportioned and realistic in portrayal of facial expression. Examples are: G. Pisano's *Nativity and Annunication to the Shepherds,* 1302-10; *The Virgin and Child*, 1339 by a French goldsmith; Sluter's *The Prophets Daniel and Isaiah*, between 1393 and 1402;

Donatello's *St. George,* 1416, and *Herod's Feast,* 1427; Ghiberti's *Gates of Paradise,* 1425-52; and Michelangelo's *Pieta,* 1498-1500 and *David,* 1501-04.

As noted in Chapter 8, painting flourished beginning with Cimabue, Giotto, and Buoninsegna in the last years of the 13th century and first decades of the 14th, reaching its Renaissance zenith in the last years of the 15th century and early ones of the 16th.

In music, progress continued in the 14th century and later years, taking on a far more secular character. In France the *Ars Antiqua* gave way to the *Ars Nova* period. The latter was one in which there were many innovations of spirit and technique, development of more complicated rhythms, and a taste for harmonious thirds and sixths, with the *ballade, rondeau* and *virelai* as new secular forms. In Italy, polyphonic music became prominent for the first time, its three major secular forms being the *madrigal, caccia* and *ballata.* From these developments emerged music's Renaissance, viewed by some scholars as starting early in the 15th century with the shift of the center of European music to Burgundy. There Dufay, greatly influenced by the English composer Dunstable, freed the mass from its anchor to a plainsong tenor, extensively elaborated its settings and thus was instrumental in establishing the mass as a unified polyphonic frame. At the same time there occurred a significant dissemination of written music following the first publication in 1501 of polyphonic pieces made possible by the invention of movable type. This, together with the extensive travelling of Low Country musicians (members of the Franco-Flemish school) and their residence at princely courts throughout Europe during the last half of the 15th century, led to widespread interest in music, especially by the moderately wealthy classes. Other developments saw: standardization of four part vocal polyphony in the late 15th century, the fourth part being added below the tenor; and evolution of an independent instrumental idiom.

Against this background of 15th century happenings and works of leading composers—Okeghem (1410-97), Obrecht (1460-1505) and Desprez (1440-1521)—vocal polyphony soared to a pinnacle of expressiveness in the 16th century (as was witnessed in the performances of the Sistine Chapel choir). Also, instrumentalism burgeoned rapidly, national schools of music began to differentiate themselves from the rather international style that had resulted from the widespread activity and dominance of Low Country musicians, and Protestant church music made its appearance.

h. The growth and decline of the Papal Empire

Against the background of the above developments from the dissolution of the Roman empire through the Renaissance era we want to trace the forces leading to the construction and stagnation of the mighty hierarchic papal empire—an empire which reached its peak around the end of the 12th century and beginning of the 13th, and which began to fall apart after the death of Innocent III in 1216.

As the discussion in the above paragraphs has made clear, the Roman Catholic church of the West had tremendous impact upon literature, arts and science throughout the period (the High Middle Ages) of its gigantic growth as a power system; and in turn it was affected by the political, social and economic institutions which it conditioned. In what follows, we will not be interested in specific features, events and personalities *per se*, but only as they help us understand the broad set of forces operating through the period of growth and decline of this empire—particularly from the standpoint of acquiring insights for the general study of hierarchic construction and disintegration. We consider the experience of the papal empire to constitute the last complete case of hierarchic construction and disintegration in Western history. Subsequent to the Reformation and years of papal decline, a new set of perturbations in Western society have come to be amplified, resulting first in an Enlightenment stage and subsequently in the stages of Individualism, Restricted and later Unbounded. As indicated in Chapters 8 and 9 we currently have insufficient historical perspective to judge whether or not this outburst of Individualism has come to an end and is in a period of decline.

In Chapter 9 we traced the religious origins and the early development of Christianity from when it came into existence in the 1st century AD up to the heyday of the Roman empire. We alluded to its very strong appeal in its early years to slaves, the poor, the outcasts, and the underprivileged in general, as well as to both women and men, and even nobles, well-to-do, and the hated tax collectors. It had a simple requirement for redemption and salvation—faith and an inner attitude of surrender and openness to God. Later, after Paul's proselytizing, it came to be increasingly augmented with ceremonies and sacraments. We also alluded to its organizational structure, at first extremely simple with direct contact and communication between a bishop and a lay person, but later increasingly complicated with considerable indirect communication as other kinds of officials came to function in an elaborate hierarchical structure.

During its first two centuries the growth of Christianity was steady but slow. But after the death in 180 AD of the last of the Five Good Emperors, the decline in the Roman empire that set in nurtured rapid growth in Christianity. Among the religions competing for the allegiance of the distressed masses suffering from the times' political turbulence and economic hardships, and who greatly needed spiritual sustenance, Christianity had more to offer. Unlike Mithraism, a widespread mystery religion stemming from Zoroastrianism, it did not exclude women. Christianity was also newer than a number of others which had been around for centuries; it possessed a fresher and more forceful dynamism. Moreover, it had better and more far-reaching theories and views of salvation, formed local communities having greater concern for the life-sustaining needs of its members (e.g., care of the sick), and perhaps most important, a more effective administrative structure. Already, by the second century, the initial dichotomy of a religious leader (presbyter, priest, or bishop) and followers had become a hierarchy of more than two levels—bishops who presided over priests who ministered to the laity, with still lower rank deacons collecting donations and, in association with deaconesses, performing other community functions perceived to be of less religious significance.

The 3rd and early 4th centuries were not ones without problems for Christianity. There were intermittent periods of persecution—by Decimus in 250-51, Valerian in 257-60 and Diocletian and his successors in 303-11. These persecutions produced martyrs, which only came to inspire Christian leadership still more and impress the outside world. However, as already remarked, Constantine converted to Christianity in 312, and shortly thereafter extended complete freedom of worship to the Christians; and before the end of the 4th century, Theodosius had established Christianity as the official religion of the state, having outlawed the worship of all other religions.

These relatively sudden advances of Christianity also led to a still more involved hierarchical organization. Metropolitans (archbishops) appeared, having headquarters in the larger cities with authority over the clergy of an entire province, and in time patriarchs who presided over the oldest and largest of Christian communities (Rome, Jerusalem, Constantinople and Antioch and their surrounding areas). Further, in 445 Emperor Valentinian III decreed that all bishops were to submit to the jurisdiction of the bishop of Rome, the pope.

Note that the period covering the reigns of Diocletian (284-305) and Constantine (306-37) involved some notable reconstruction and reorganization of the Roman empire—a reorganization, however, that imposed still more burdens on the lower classes. These greater burdens were felt both directly and indirectly from the levy of higher taxes required to support the increased administrative functions and bureaucracy of the reconstructed order. Thus, the efforts at political restoration of the empire led to heightened impoverishment of these classes and reinforced the appeal of Christianity.

In the chaos and complete disintegration of the Roman empire during the 5th and following centuries, when the administrative structure was completely demolished by the barbarian invasions resulting in the breakdown of cities and appearance of isolated rural units, Christianity provided for the West the only element of unity, and thereby became more firmly rooted in the Western system. In many areas, its bishops, priests and other officials were the only administrative functionaries in place to confront day-to-day problems of people, however limited the experience and know-how of these officials.

While the rise in the power of, and system control by Christianity continued, there were at work forces of decline. Typical of a hierarchical structure that has won out in a competition due to its inherent efficiencies and then continues to grow in might, division sets in among the members of the élite, perhaps formerly highly cooperative, now competitive in the search for power and aspiration for top posts. Moreover, as a hierarchical structure becomes composed of more and more levels, impersonal relations more and more replace personal ones, and more and more possibilities become available for distorting the flow of information and for allowing corruption to take place and incompetents to hold office. And so it was with Christianity. There was early on the bitter clash between Arians and the Athanasians over the nature of the Trinity, and then a series of quarrels between the West and East over doctrine and power. At the same time, in their more worldly activity, officials were becoming increasingly separated from the spirit of the simple faith of Jesus and the Apostles. As a reaction, communal monasticism, aimed at preserving and practicing such faith and counteracting divisive forces, spread widely and became a dominant movement in the 4th century and thereafter. St. Basil (329-379) and St. Benedict (480-547) were two of its greatest leaders; the Benedictine order performed extensive missionary work, stressed the desirability and rewards of hard manual labor, and in its maintenance of literacy contributed much to preserving classical culture.

Until the reign of Charlemagne, there was little noteworthy advance or basic change in the Western church's situation. Gregorian the Great (r.590-604) did tighten the hierarchical structure, did formulate a more distinct theology and a strong Latin liturgy; and his reign did foster the development of the Gregorian chant. He and other bishops of Rome were able to collaborate gainfully with the Frankish rulers, Clovis and Pepin. And after 800, when Charlemagne was crowned by Pope Leo III as the official head of the Holy Roman Empire, the Christian Western church was able to achieve effective independence from the Greek-speaking Eastern church. As already noted, Charlemagne's reign, as a mini-Renaissance, did see some significant progress: encouragement of education in large part to upgrade the skills of the kingdom's administrative officials; the establishment of Latin as a common language; the nurturing of scholarship (a rebirth of antiquity) in a palace school that aimed at bringing more uniformity to Christian law and liturgy. Also, there was reform within the church's hierarchy and monasteries that resulted in the collecting and preserving of ancient manuscripts. But all this was another promising perturbation that was short-lived. The great empire that Charlemagne pieced together started to fragment upon his death; and disorder, lack of any effective central administration, papal corruption, isolated feudal communities and religious decentralization with local clergy appointed by the feudal lord and/or powerful laymen came to characterize the late 9th century. This was also a time when conditions were greatly aggravated by the successful invasions of the Vikings, Magyars and Muslims.

However, the seed for subsequent revival, creative advances and significant reform that led to an all-powerful papal monarchy was planted in these Dark Ages. In 909 in Burgundy the abbey of Cluny was founded, chartered to be completely independent of all feudal or secular lordship. Its first two abbots, Berno (910-27) and Odo (927-42), were able to establish extremely high standards for religious behavior, and thereby initiated a movement that was to have a vast influence. The movement spread throughout France and Spain; by 1049 there were 67 congregations of Cluniac houses and many other congregations and monasteries free of secular control. Thus, by the start of the High Middle Ages (usually set at 1050, but at times at 1000) there was an extremely strong lobby pushing reform of the clerical hierarchy. As a result, a number of popes who came into power did put in place significant changes—for example did effectively oppose concubinage and simony (the selling of spiritual things such as church offices). And in 1059 the college of cardinals (a special group of priests of the major churches in and around Rome that in the past had served to offer advice to the pope when called upon) issued a decree vesting in itself the right of naming a new pope—a right that hitherto had been exercised by secular authorities.

The reform of the papacy received its strongest support from Pope Gregory VII (r.1073-85). His efforts produced a crisis over the issue of lay investiture. He insisted that the authority to appoint abbots (the top officials of monasteries) and bishops must rest entirely within the Church itself, and that churchmen be free to obey canon law and not be subject to control and interference by lay people. The success of Gregory's efforts in obtaining the independence of the clergy represented a major advance of church power and upheld its claim to mediate over and control the whole of society. Moreover, later popes continued his reform efforts. The central government of the Roman church was reorganized. The college of cardinals was formally recognized as a consultative body, and together with other papal agencies came to constitute the well-organized papal *curia*—the papacy's administrative bureaucracy and court of law. Moreover, although the First Crusade is the only one that can be claimed to be successful, the Crusades did place the pope at the head of a large body of volunteers, as the explicit leader of Europe as a whole. This contributed much to the prestige of the papacy and the recognition of its power over many lands—power well beyond the limited territorial authority of any of the existing secular rulers in Europe.

With further development in church centralization involving the addition of new levels in its hierarchical structure, the Western church reached its zenith of power during the reign of Innocent III (1198-1216). He, like Gregory VII, was an extremely strong and capable administrator. Moreover, he asserted, still more vehemently than his predecessors, the plenitude of papal power—that all secular leaders were subordinate to the pope—and was successful in actions taken against what he considered to be improper behavior by rulers, for example Philip II of France and King John of England. His empire came to be an efficient ecclesio-commercial complex involving Lombard banking agents and having the financial resources able to support a bureaucracy that had not been equalled in competence and strength since Roman times.

The High Middle Ages was thus a period of not only great strides in papal power and lay religiosity but also, as remarked earlier, significant advances in technology, economic well-being of the people, development of national monarchies, growth of universities, and creative advances in literature, the arts and science. The advances were *crosscatalytic*. For instance, the religious ethic of hard manual labor together with advancing agricultural technology and favorable climatic conditions did much to realize the greatly improved economic conditions of the masses. And in turn the greater productivity of labor made possible the emergence of the multitude of monasteries that advanced the quality of religiosity and the vitality of Western Christianity, and strengthened the lay pressure for religious reform.

At this point, we could march through the set of concepts developed in Chapters 2-7, such as *symmetry breaking, hierarchical construction, stratified stability* and *irreversibility, bifurcation, . . . , autocatalysis, . . . , specialization, . . .*, and so on, and indicate how each can be viewed as involved in the building of the papal empire. But we have already done this in part and in a very general fashion in the summary of Chapter 9 for the set of religions, including Christianity, therein discussed. Moreover, to do so would be largely repetitious of what we have already done in this chapter in the discussion of the early urban civilizations, the Greek Classical age, and the Roman Republic and Empire. Instead, we wish here to focus on the forces of dissolution, disintegration and decline that set in. They may be attributed to, or associated with:

1. descendance and decay of leadership
2. demoralization
3. depersonalization
4. devitalization
5. dissension, disquietude and disbelief
6. despoliation of resources
7. deglomeration (economic)
8. deconcentration (functional)
9. development dynamics in general and
10. destruction and devastation from random natural events.

Let us examine the decline of the Western Roman Catholic empire from the early 13th century with regard to each of these mutually interacting phenomena—phenomena which are conceptually overlapping since generally acceptable, precise definitions of them do not exist.

Like his predecessors, Darius, Pericles and Augustus, Pope Innocent III may be classified as a unique *coadapted gene complex,* in essence a leader of tremendous capabilities and drive—characteristics which were not duplicated in those who followed. Thus, from biological considerations alone, the extent and power of the empire was doomed for at least some decline—a phenomena that characterizes any structure built up and elaborated by a unique leader whose administrative genius or near-genius is a random event not subsequently repeated or exceeded within a reasonable amount of time. While Innocent's successors perfected the papal administration in important ways, they lacked his administrative skills and cunning to preserve papal primacy under changing conditions—as clearly evidenced in 1303 when Pope Boniface VIII failed to bring King Philip IV of France to his knees and was defeated in the ensuing battle among the respective supporting armies. This is not to deny that rising nationalism (to be considered later) was increasingly winning people's allegiance, but the administrative ingenuity of an Innocent III was not present to counteract this phenomenon. It was not there to prevent the Babylonian Captivity (1305-78), namely the removal of the seat of the papacy to Avignon from Rome—the event considered by some to be the beginning of the toppling of the empire.

Frequently, such *descendance and decay* of leadership, as noted above, is accompanied by a *demoralization* present in those who inherit the top position of a mighty empire. This can transpire from the affluence that surrounds leaders' lives, their affliction with self-importance, their descent (often gradual and unawaringly) into political corruption and resort to unprincipled behavior to maintain the *status quo,* especially when they lack their predecessor's abilities. As with Xerxes who succeeded the administrative genius Darius of the Achaemenid (Persian) dynasty, and as with the Roman emperors who succeeded Augustus, and certainly those who came after the 'Five Good Emperors,' all of whom lacked the management capability of the genius or near-genius leader who preceded them, so too with the popes after Innocent III. Moreover, these popes (and their court) became preoccupied with secular affairs, and negligent in their primary obligation for spiritual nurturing. They eventually came to succumb to intrigue, corruption, unbridled luxury, concubinage, bitter internal fighting, and/or other behavior which came to undermine lay respect and allegiance—the Great Schism 1378-1417 being a classic example of jockeying for power.

Demoralization is frequently accompanied with *depersonalization*—the process whereby the initially close relations of a leader with his associates and constituency fade as the top officials who follow fail to: (1) concern themselves with the welfare of their subordinates and (2) recognize the essential functions the latter perform and their contribution to the good of society. This depersonalization characterized Xerxes' last years; it was also present in so many of the reigns of the emperors succeeding Augustus, for instance, that of the incompetent Nero, or later of the autocratic Diocletian.

It was present, too, on a widespread scale in the 14th and 15th centuries, times when the papacy was in general disinterested in piety, and when the local clergy were generally behaving in ways which failed to meet the personal spiritual needs of a laity emotionally and economically insecure consequent to the ravages of pestilences and hard times.

Closely associated (via crosscatalysis) with descendance and decay of leadership, demoralization and depersonalization is the *devitalization of a society*. The enthusiasm, the fervor, the ferment, the charisma, the driving force that characterizes a new cause and a brilliant figure at the head becomes dissipated as initial goals are attained. Once success accompanied by an all-powerful status is achieved, goals change. Preservation of the *status quo* becomes the order of the day. So it was in the lives of Xerxes, the heads of state including the 'Five Good Emperors' who followed Augustus, and Popes Honorius III, Gregory IX, Celestine IV. . . who followed Innocent III. Moreover, when lower calibre leaders follow a brilliant and extremely efficient one, and when their morality has declined and/or personal relations with staff and constituency have deteriorated, all within a devitalized society lacking commonly shared aims and ideals, *dissension* breaks out. Such dissension may occur among the nodes close to the top of the hierarchy or at lower-level nodes, including those at the bottom.

In the case of the Achaemenid empire, we see this during its last century in the scramble for ruling power—assassinations, insurrections, and constant plotting within the dynastic family and among the rulers of the satraps were common. After the reign of the Five Good Emperors of Rome we see the mania for power reaching an inordinate level, with extensive intervention and control by the military; in the forty nine years (235-84) over twenty different emperors were in power. In the case of the papal empire, such dissension and instability at the higher levels of the hierarchy reached a peak during the period of the Great Schism when for some years (1409-14) there was the spectacle of three contending popes.

Disquietude and disorder at the top levels of a hierarchy typically is associated, at least partially causally, with dissension and unrest at the lower levels—all questioning the validity of the *status quo*. Such dissension at the lower levels was present in the dying decades of both the Achaemenid and Roman empires. In the case of the decline of the papal empire, for which more extensive historical records are available (records much less preoccupied with invasions by external military machines since such invasions were few), lower level dissension is reported as extensive, especially at the lowest level.

To understand better this dissension, one must constantly bear in mind that preservation of the *status quo* in the power hierarchy requires exceedingly adept ability at literally continuous adaptation. In part this is because of constant, unpredictable environmental change. But also man as a biological organism is imbued with curiosity, for it is only through the curiosity of at least some members of *homo sapiens* that man has been able to survive. He always explores new ways of doing things. The ways may be (1) simply technological or (2) cultural when broadly viewed in terms of new social formations (groupings, organizations and institutions), and new religious orders, tastes, objectives, aspirations and the like—i.e. new negentropic arrangements and behavior. Put otherwise, there is an inherent *dynamic* in man and the state of his achievements which constantly questions any *status quo,* especially when a harsh environment prevails.

This questioning was pursued heatedly with respect to the advance in scholastic theology developed to support Roman Catholicism. Early contributors to the scholastic movement were Abelard and his student Lombard who attempted to treat theology as a science. But the main developments in reconciling faith and reason awaited, as already noted, the translations in the middle 12th century and later of Aristotle, Euclid, Ptomley and other classicists along with Arabic and Jewish commentaries and treatises. There followed the Golden Age of Scholasticism and the brilliant systematic work of Aquinas on the relation between reason's claims and the authority of divine revelation. However, by the end of the 13th century, the intellectual scaffolding which the scholastic movement provided the papal establishment was abruptly undermined by the condemnations of Aquinas' and later Bonaventure's analyses. These condemnations, coming from scholars pursuing a naturalist approach, chipped away at the so-called 'unassailable' papal theology. Moreover, the Renaissance, the Humanistic movement and the development of critical historical inquiry added fuel to the questioning of the existing order, thereby feeding and supporting dissension and other disquieting factors.

A more serious questioning of papal hierarchy and authority stemmed indirectly from another dynamic element, namely technology. In the 15th century there was the path-breaking invention by Guttenberg of printing by movable type. Together with the replacement of parchment by much less expensive paper which had earlier been achieved, this invention made possible a tremendous reduction in publication cost. It resulted in large-scale output of the Bible, first printed in 1455, and other literature. The laity who had received education could now read the scriptures of Jesus and his Apostles. [There had been during the High Middle Ages (1050-1300) a great outburst in educational activity, both at the primary and university levels, and subsequently a continuing proliferation of schools and writing in vernacular languages in contrast to Latin.] The educated laity could compare the original biblical texts with the statements and proclamations of the clergy. They could see the great discrepancies between the life styles of their local priests and papacy and that required by the early scriptures. All this greatly reinforced *dissension* among the Christian following. There had, of course, been discontent with the papacy and priests during the High Middle Ages. For example, during the good times of the 12th century Arnold of Brescia criticized clerical wealth and corruption, and there were the Cathari and Waldenses sects devoted to poverty. This dissension, however, was early absorbed by the outburst of religious vitality that characterized the good times of the High Middle Ages when there were active and spiritual bishops and a number of reforms, and when under the administrative genius of Innocent III the church had reached its zenith.

It was during the late Middle Ages, after 1300, when the good times disappeared, when the Black Death reared its ugly head, when the Hundred Years War took place, and when there was the Great Schism in the church that we see a rapidly mounting level of dissension. As remarked earlier (p. 274), Ockham's bitter criticisms of the papacy and condemnation of scholasticism had an impact on the later views of Eckhart who sparked a major mysticism movement in Europe, particularly in Germany. This movement questioned the relevance of the grandiose and majestic papal hierarchy and maintained that godliness can be attained by one's own inward-searching efforts.

Another anti-church movement stemmed from the critical writings of Wycliffe who in later life attacked the existing sacraments and urged Christians to shun the corrupt church, study the Bible and rely on their own resources and conscience to determine the path of the good life. His doctrine greatly influenced the preaching of Hus, leader of an uprising in Moravia. Hus distinguished between the institutional church headed by the pope and the true church of Jesus whose members were the true believers and saints, and not hypocrites. Even a pope could be an hypocrite.

Folk piety and discontent with church practice was further nurtured by the widely read work 'The Imitation of Christ' of Kempis (1441), a more orthodox statement, but one which urged followers to go beyond the authoritative aspects of the church and develop an inner believing heart and a discipleship of Christ in their lives.

Other *dynamic* forces were also at play: the exploration and overseas expansion of the Portuguese and Columbus' discovery of America; the growing tide of nationalism, partly reflecting the administrative failures of the papacy; and the rise of the middle class, urban centers, and city states and their increasing independence fostered by prosperous trade and new, organized commerce.

Dissension within a hierarchical socio-economic-political structure can often be related to inadequacy of resources, in particular *despoliation* (improper use of and failure to conserve resources), extreme pressure of a population on existing resources, or both. In the case of the papal empire, the good times preceding 1300 (i.e. the High Middle Ages) led to a large population increase such that in the 14th century a condition of overpopulation existed. The supply of food became inadequate, leading to considerable disquietude among the masses.

Disquietude and dissension within a hierarchical structure can also be generated by *destruction* and *devastation* caused by natural forces—*destruction* and *devastation* which can be completely or almost completely independent of a society's behavior, such as earthquakes, floods, climatic change, and pestilences, characterizable as noise or random events. In the post-zenith days of the papal empire, there occurred the bubonic plague, in part reflecting inadequate sanitation and related measures within the empire—but in large part unpreventable. The Black Death led the laity to harbor still greater resentment against and dissatisfaction with the existing order and church structure, especially where clergy fled the locale of their flocks when the Black Death hit.

All in all, the factors opposing *preservation of the status quo* of a highly-structured, elaborate hierarchical system are many, and they did operate on the papal empire, as well as the preceding ones. In effect, they bring about disintegration and dissolution, especially of an empire which proves to be *overly-concentrated,* and *overly-hierarchically-structured,* yet one which had been efficiently managed by the actions of a unique genius or near-genius administrator and leader.

Classical phenomena that crop up and induce, if not further stimulate, decline and decay are excessive taxation and inflation. Often the unique leader, because of his tremendous capabilities, has been able to put in place a far-reaching taxation system supportive of an elaborate bureaucratic structure and an extensive military. But once that leader passes on, his less competent followers are unable to establish revenue collections necessary for the support of the inherited bureaucracy and military. Typically, they are forced to raise the level of taxation, which generates greater resistance to tax collection and to efforts at tax evasion and diversion, which in turn may lead to still higher taxation. Sooner or later, manipulation of coinage or currency occurs, and the inevitable inflation—another incendiary phenomenon leading to empire decay on all fronts. Devaluation of currency and rampant inflation inflamed and magnified the decline of all three hierarchically-structured empires we have examined.

Once a top-heavy hierarchical-structured empire materializes from the energies and administrative deftness of a brilliant leader, theory suggests the need, early on, for (1) *deconcentration* of functions (e.g. taxation collection) and associated decision-making structure, and (2) *deglomeration* of production activities in an economic and regional science sense. In one way, this need seems to have been appreciated or perhaps only intuited by the highly competent Roman emperor Diocletian. For a time, he did realize a successful reorganization and reconstruction of the empire. He did effect deconcentration by establishing a tetrarchy, wherein the empire was divided into halves (East and West), and each half into two prefectures. (In this situation he formally ruled only the East, but actually dominated the entire empire.) However, while the system survived under his strong leadership, the deconcentration was too late to overcome the demoralization, depersonalization, devitalization and dissension that had preceded him. Moreover, the deconcentration at this late stage required a large increase in administrative staff and bureaucracy that had to be supported by still greater taxation and imposition upon the system of occupational and other inflexibilities. All this, however, just resulted in increased poverty among the poor, more internal dissatisfaction and dissension, and via feedback processes still more demoralization, depersonalization and devitalization. The reuniting of the empire by Constantine could not forestall the harsh force of decline and the ultimate entropic-type situation reached in the West in the tenth century of isolated manorial and feudal communities.

In the case of the papal empire, for over two centuries after Boniface's incompetent actions, no deconcentration was undertaken by the papacy to meet and adapt to the pressure of new forces, for example that of rising nationalism. Dissidence mounted and the stage was set for a series of critical writings and preaching—e.g. those of Luther, Zwingli, Calvin and others—that would lead in the European West to a spectrum of Protestant and Catholic national churches that irreversibly replaced the unitary Roman Catholic church. By the time the papacy and high clergy saw the handwriting on the wall and assembled the Council of Trent in 1545, it was too late to counter the forces of disintegration. While the Council did effect some notable reform within the Roman Catholic church, the outcome was the existence of the Roman Catholic church as one and only one institution among a number and one that was subject nation-by-nation to the different specific constraints each national government imposed. The persistent and persuasive criticism of papal rule and vigorous opposition to it thus came to underly the significant perturbation that, as discussed in Chapter 9, was amplified and came to generate the age of Enlightenment and the subsequent new far-from-equilibrium states of Individualism Restricted and then Unbounded.

So, as with Chapters 8 and 9, each of which concentrated on the examination of a significant subsystem, painting and religion, respectively, we find in this chapter's broad examination of major civilizations and empires the presence of many if not all the commonalities identified in Chapters 2-7. We now move on to the last chapter where we attempt further integration of materials and analyses already presented

References

Burns, E. M., Ralph, P.L., Lerner, R.E. and Meacham, S. (1986), *World Civilizations*, New York: Norton.

Chambers, M., Grew, R., Herlihy, D., Rabb, T.K. and Wollich,I. (1987), *The Western Experiences,* New York: Knopf.

Clark, K. (1969), *Civilization,* New York: Harper and Row.

Craig, A.M., Graham,W.A., Kagan, D., Ozment, S. and Turner, F.M. (1994), *The Heritage of World Civilizations,* New York: Macmillan College Publishing Co.

Crystal, D. (ed.) (1994), *Cambridge Encyclopedia*, New York: Cambridge University Press.

Eliade, M. (1978, 1982 and 1985), *A History of Religious Ideas*, Chicago: University of Chicago Press, Vols. I, II and III, respectively.

Frankforter, A.D. (1988), *Civilization and Survival*, Lanham, Md.: University Press of America.

Kagan, D., Ozment, S. and Turner, F.M. (1991), *The Western Heritage,* New York: Macmillan Publishing Co.

Kennedy, P.E. (1980), *Paleo-anthropology,* New York: McGraw Hill

Lewin, R. (1993), *The Origin of Modern Humans,* New York: W.H. Freeman and Co.

Lieberman, P. (1991), *Uniquely Human*, Cambridge, MA.: Harvard University Press.

McHenry, R. (ed.) (1993), *The New Encyclopaedia Britannica*, Chicago: Encylopaedia Britannica, Inc.

McKay, J.P., Hill, B.D., and Buckler, J. (1991), *A History of Western Society,* Boston: Houghton Mifflin.

Molleson, T. (1994), 'The Eloquent Bones of Abu Hureya', *Scientific American*, August.

Poirier, F.E. (1990), *Understanding Human Evolution*, Englewood Cliffs, N.J.: Prentice Hall, 2nd ed.

Smart, N. (1984), *The Religious Experience of Mankind*, New York: Charles Scribner.

11 Synthesis and implications for 2000 AD and beyond

11.1 Introduction

We bring this book to a close. What have we accomplished? What new understanding has resulted? What new insights have we gained? Can we better confront the problems of society—those of overpopulation and poverty, those of environmental deterioration, those from incessant physical strife and warfare? Can we move forward to meet the challenge of a Teilhard? Let us try to answer these questions.

In this book we have attempted three important advances in understanding—at least from the standpoint of the author, as he is able to gauge his own stock of knowledge and that of those in the arts, religion, sciences (social, physical and biological), and related areas and professions. These advances stem from: (1) recognition of the presence and pervasiveness of commonalities, which has received most attention in this book; (2) insights gained thereby for analyzing the ups and downs (swings) of systems (civilizations and empires); and (3) an appreciation of the necessity for the copresence in a complementary manner of the realms of both religion and science for an effective unitary system.

11.2 Presence and pervasiveness of commonalities

First, there are the commonalities, that is the pervasiveness of certain processes, originating in the physical world which have become omnipresent in the biological and social world as these have developed—processes that suggest, at least at some levels and to some extent, a unity of art, science and religion. To start, there is *symmetry breaking* and *hierarchical construction*. According to the widely accepted Big Bang and particle physicist's hypothesis, such was present in the physical world when: (1) the original Ur force was first structured into two types from the splitting off of gravity, and later was restructured to yield the four forces, gravity, strong, electromagnetic and weak, and (2) the original Ur matter was correspondingly first structured into two types from the splitting off of gravitons (yet to be verified experimentally), and later restructured to yield the many types of particles that have now been experimentally verified (electrons, protons, nucleons, etc.). This process was later supplemented with hierarchical construction in the formation of nuclei, then many types of atoms of successively increasing mass and complexity, as noted in the periodic table—an hierarchical structure which may be viewed as partially regular and partially irregular.

In the biological world, there was symmetry breaking with the random appearance of a particular molecule, the original replicator, which created a copy of itself—the breaking of the symmetry of nonreproducibility. From then on, there was hierarchical construction—for example, from noncellular organic matter to cellular, then to multicellular organisms, and in succession to organelles, tissues, organs, systems of organs, etc., until a complex system of organs coordinated by a brain capable of reflection and consciousness was reached. In the *social worlds* of insects, animals, humans and others, symmetry breaking irregularly but steadily occurred. In modern man, there was the transformation of society from one of hunting, foraging and related primitive subsistence-driven activities to one wherein agriculture (including the domestication of animals) to a large extent replaced these activities, even when advancing know-how led to the greater productivity of hunting and its related activities. Also, there was symmetry breaking in mystical and religious beliefs when the state in which polytheism was practiced was replaced by one wherein monotheism came to be prevalent, although in less advanced areas (particularly rural) polytheism persisted and still persists today to varying extent. Under polytheism, status hierarchies of gods developed, and in monotheistic faiths, hierarchies of officials.

Another set of commonalities in the physical, biological and social worlds has been associated with *specialization and scale, localization, urbanization and in general agglomeration and spatial juxtaposition economies (diseconomies)* in both hierarchical and nonhierarchical systems. Specialization and these economies have been discussed extensively in our examination of the rise and fall of empires in the social world. In the physical world also, these factors have been and are present everywhere. For example, there is specialization in the activities stemming from differences in properties among families of chemical elements and also among periods of them. The bonding the elements engage in parallel the different kinds of economies present in the social world. For example, the force leading to the bondings of like chemicals, such as hydrogen bonding, parallels the force of localization economies stemming from spatial juxtaposition of like enterprises. The forces leading to multiple bondings among elements—the formation of a double bond from a single bond, a triple bond from a double bond , etc. that leads to decreasing bond lengths and increasing bond strengths—parallel the forces lying behind scale economies (e.g. the strengthening of a firm's competitive position). The forces leading to the bonding of chemicals of different kinds and in different ways to form combinations, such as the hydrocarbons, parallel the forces that lie behind the social world's urbanization and general agglomeration economies. In the biological world: (1) specialization exists, for example in ant societies among the workers, soldiers and queens; (2) scale economies in the effectiveness of different size armies; and (3) urbanization and spatial juxtaposition economies in a society's macrostructure which in time may become constrained by diseconomies when the number of its units becomes too large. While we have viewed symmetry breaking as involving an orderly process, and at times considered a hierarchical structure as a subsystem of successively broken symmetries, existing orderly structures, whether hierarchical or not, can be subject to disintegration. Such disintegration can lead to *chaos*, as when smoothly flowing fluid breaks up into whorls and eddies, becoming turbulent and draining energy. We see this in the biological world when finely coordinated physiological systems break down, resulting in seemingly chaotic functioning as in fibrillation phenomena of the heart. The social world is no exception to disintegration phenomena. We saw it in the crumbling of the Roman empire, the reversal of the process of hierarchical construction, the smashing of stratified stabilities, and the high state of dysfunction reached in the late 9th century AD when the European world degenerated into one of isolated manorial and feudal estates.

It is interesting to speculate whether or not the situation of the late 9th century AD was not one of chaos (complete disintegration) but rather one of *deterministic chaos* (presence of a *strange attractor*), as turbulence and other physical phenomena can at times be viewed, and as heart fibrillation and other biological phenomena are often conjectured. Such speculation raises other very interesting questions, since in the physical world we have associated deterministic chaos and strange attractors with such factors as *sensitive dependence on initial conditions, bifurcation, fractal structure and dimension, self-reproduction, self-similarity at different scales, self-organization, relatively long-range correlations associated with density gradients, autocatalysis and crosscatalysis*. Are all these phenomena and processes also present in the social worlds? We have already noted the presence of some, for example, sensitive dependence on initial conditions when we observed the divergent time-paths and characteristics of the Mesopatamian and Egyptian civilizations. At the same time we have failed experimentally or empirically to verify others in a truly rigorous fashion, for example fractal structure and dimension in stock market behavior or in urban and other systems characterizable by a rank-size rule. It is not necessary to record where we have found and have failed or are yet to find these phenomena and processes in the social world. But clearly many if not all of them have or will be exposed.

In the biological world manifestations of commonalities with the physical world are by now abundant, particularly with regard to basic phenomena and processes. Yet we must emphasize the new development within the biological world which is not found in the physical world, namely *replication* and life itself. Associated with this development are, as we have noted, such elements as the *selfish gene, survival machines, mutation, competition, cooperation, coordination, allocation of scarce resources, altruism, extended phenotype, coadapted gene complex, species as an evolutionary unit, coevolution, punctuated equilibrium and peripheral speciation.* In the social world of man commonalities with these biological phenomena and processes have been well documented in the preceding chapters—where, for example, *meme* replaces gene, *coadapted meme complex* replaces coadapted gene complex, and cultural subsystem replaces species. However, the advent in the social world of man, and to a much smaller extent, that of animals, insects and others, introduced another basic phenomenon and process beyond the biological world, namely that of *consciousness and reflection* associated with brain development. It is in this regard that there emerges the challenge of a Teilhard—who we define to be a religious leader of any faith who like Teilhard, a Roman Catholic, is or was also a great scientist able to synthesize science and religion—who is able to question, if not prove the fallacy, of strict physical and biological determinism. To this challenge we turn later.

With the emergence of reflection and consciousness one may find it meaningful to associate Bohm's *unfoldment* process—the revelation to the human mind of what were once unknowns and unobservables (for example, how to make fire, or the force of gravity which until the experiments of Gallileo and later scientists was to most humans if not all an unknowable). Such revelation of unknowns and unobservables, formerly elements of the *implicate order,* builds up the *explicate order.* While it can be claimed that reflection and consciousness and the associated *anticipatory planning* and behavior is also present at a much simpler and more elementary level in insect, animal and other life, at a highly sophisticated level they exist only in humans. The concept of *heterarchical model*—a model that captures conceptually and in part quantitative connections of a multitude of causal chains such as embodied in the PSIG model (discussed in pp. 366-69 below)—is strictly one of human design only. For human society, such a model also implies a much more intricate pattern and control of *information flows.* While in the physical world information is transmitted via flows of photons, electrons, and other particles and density gradients, in the social world (and to a lesser extent in the biological) there are the controls of information flows from the exercise of the *principle of optimum loss of detail* and use of *near-decomposability* within hierarchical structures—and even more so within the intricate interconnections of complicated hierarchical structures that may be embodied in an heterarchical model conceived by a human mind.

11.3 Ups and downs (swings) of systems (civilizations and empires) and hierarchical structures

Beyond the identification and presentation of commonalities (and non-commonalities) among the physical, biological and social worlds, there is the second thrust of this book; it is concerned with the ups and downs of elaborate structures, hierarchical or not, and especially of empires and civilizations, as well as subsystems and systems within the several worlds we have examined—although this thrust has only been discussed in depth in Chapter 10.

We have already noted the factors that lead to the construction of hierarchical structures—structures that exist among sets of particles, forces, chemical elements, stellar bodies, organisms, species, humans, social groups, and within empires and civilizations. Beginning with replicating organic matter and going through organisms up to and including modern man, we have discussed: natural selection stemming from competition, alone and in conjunction with cooperation; use and misuse of environment; *Prigogine jumps* and *transitions to far-from-equilibrium states*; *stratified stability*; intricate and sophisticated survival machines; subcultures (for example, painting and music); *cultures as coadapted meme complexes*; and so on.

We have just discussed the parallel between: (1) scale, localization, urbanization, agglomeration and spatial juxtaposition economies and (2) the physical and other bond-strengthening forces, and how these economies and forces can lead to an elaborate hierarchical structure with its units inextricably linked in a most complicated fashion. But construction can also be followed by *dissolution* and *disintegration*. Specifically, in Chapter 10 we have associated with hierarchical dissolution and disintegration in the social world: *descendance and decay of leadership; demoralization*; *depersonalization*; *devitalization*; *dissension*, *disquietude* and *disbelief*; *despoliation of resources*; *deglomeration* (economic); *deconcentration* (functional); *development dynamics in general*; and *destruction* and *devastation* from random natural events.

Such phenomena—major downswings in the social world subsequent to major upswings associated with the development of hierarchical structures—raises the question of whether they parallel similar major oscillatory phenomena in the physical world. We are acquainted with the widespread acceptance of the Big Bang hypothesis—widespread because there are as yet no other hypotheses possessing better rationale and supporting evidence. The hypothesis is just a reasonable speculation given what we know about the universe—a speculation whose basis is currently being subject to serious questioning. Associated with the Big Bang statement is a set of highly speculative questions. One concerns the expansion and contraction of the universe. Is the density of the material in the universe great enough to exceed the 'critical closure' density—great enough to reverse the currently hypothesized expansion? If so, the universe would contract and converge entirely to approximately a point, or the equivalent under the force of gravity. Put otherwise, is the universe 'open', so that there exist no boundaries to limit expansion and the possibility of its attaining infinite volume, or is it 'closed' and with sufficient density to cause it to collapse?

Moreover, this question of expansion and collapse raises another one—namely, where and what is the missing mass (dark matter) that is generally conceived to be required to hold together the currently known configuration of galaxies? Then there is the inflationary theory of the universe designed to explain why, in the eyes of some, the current density of the material of the universe is dangerously near the closure density which if exceeded would lead to a big crunch of ever-increasing density and temperature. The inflation theory holds that for some finite period of time during the first instants of the universe's expansion the dominant form of matter in it exhibited repulsion rather than attraction which drove the universe very close to the critical divide, rather than away from it, as in the standard decelerating models (Barrows, 1992).

When piled one upon another, the above and other hypotheses, most being unable to be tested experimentally, suggest that an alternating expansion and contraction hypothesis in the company of still other hypotheses is not much inferior, if at all, to a non-alternating one. Given (1) the many commonalities among physical, biological and social systems that we have found, (2) the history of successive ups and downs, expansions and collapses of empires and civilizations, and (3) the birth, growth, and death of so many species, a social scientist finds an hypothesis of ups and downs, expansions and contractions of physical systems, including the universe, as good as any. He may consider it as reflecting just another commonality.

Yet, some modification of an 'up and down' hypothesis may be judged to be appropriate. We observe that in the collapse of most, if not all, empires and civilizations, a positive residue has remained. The art of writing remained after the passing of the early small-scale Sumerian empire, the predecessor of the present day calendar after the Egyptian, the application of participation potential after the Persian (Achaemenid), the use of the arch in construction after the Roman, and so forth.

In sum, the successive ups and downs (expansions and contractions) in the social world have often left in their wake increased understanding and know-how. Thus, over the very long-run one can point to cumulative extensions of the explicate order—an openness in the social world—although at times there can be loss of knowledge, and although *relative extension* of the explicate order may not be realized since the implicate order itself may be increasing in complexity and extension. Whether there exists in the physical world cumulative residues of successive expansions and contractions is a question that is very provocative but one that must be put on the back burner until the very hypothesis of universe expansion and contraction, let alone others, is examined at much greater depth.

11.4 The complementary realms of religion and science in a unitary system

Against the background of the many major hypotheses on the universe which remain to be tested and can be viewed only as speculative thought, we return to the basic question concerning the relation of religion and science. This question must have been raised by the early Mesopatamians, later the Greeks, the Romans, the Arabs, and certainly by the great spiritual and knowledgeable leaders of Islam and Judaism (such as Aveccina and Maimonides). It came to a head with the great scientific advances in Europe in the 17th and 18th centuries. As already remarked, Newton (1642-1727) viewed the amazingly complex system of celestial and terrestrial bodies as proof of the existence of God, of a tremendously creative divine mechanic; but he also maintained that all mystery results from ignorance and superstition. And before Newton, the brilliant mathematician Descartes (1596-1650) claimed that the truths of religion and morality could be demonstrated rationally—faith was not necessary. Moreover, he held that an analytic (mathematical-type) proof of the existence of God was possible—a God who had ordained an eternal set of laws. Later scientists, however, did away with or considerably reduced the relevance of God, either as a divine Being or concept. Darwin's evolutionary theory propounded in *The Origin of Species* (1859) contributed substantially to this decline, accounting in a significant way for the rise of atheism, as did the analyses of Freud (1856-1939) and the philosophers of the 19th century—Hegel (1770-1831), Feuerbach (1804-72), Marx (1818-83), and then Nietzsche (1844-1900) who pronounced that God is dead. In the 20th century, the reductionist view has come to the fore among evolutionary biologists. (Recall that reductionism is a theory that explains every complex phenomenon, especially in biology and psychology, in terms of the functioning of the simplest, most basic physical mechanisms that are in operation.) As already remarked (pp.20-23), E.O. Wilson, the founder of sociobiology and to many a reductionist, proclaims the underlying hereditary basis of social behavior, namely that the properties of complex societies can be explained in terms of the behavior of individuals (or units) comprising these societies, where each individual's behavior is explained by the biochemical properties of his/her cells, those of these cells by the properties of the individual genes—and so on down through the properties of molecules and ultimately those of protons and electrons. However, as we have noted (pp. 20-24), others, for example Lewontin et al., Gould, and Durham, do not accept biological determinism and more broadly reductionism; they insist that culture in many of its forms can and does play an important role, and hence that the social world is the result of the joint interaction of both cultural and biological factors.[1]

Similarly, Pattee, the system theorist, finds reductionism inconsistent with the complementary roles of his deterministic and linguistic modes.

The question of reductionism directly impinges on the relevance to social development and welfare of religion as a major component of culture. The reductionist, as a pure scientist, would relegate religion to a miniscule role. But certainly not a number of those who reject the reductionist approach, especially the Teilhards—the great religious leaders who at the same time have been outstanding scientists. As noted in Chapter 2, a Teilhard places great emphasis on consciousness and reflection as phenomena beyond the pure physical world—phenomena that give rise to philosophy, theology and spiritual thought and to creative arts in the broadest possible sense—a sense that incorporates literature, painting, architecture, sculpting, music and other artistic endeavors. A Teilhard may fully recognize that consciousness and reflection builds on experience and findings, not only findings from experience that leads to accepted knowledge and scientific conclusions but also findings from the experimental testing of new theories and hypotheses. But a Teilhard would certainly not admit that consciousness and reflection, as phenomena resulting from brain functioning, are wholly explainable in terms of scientific laws and relationships currently and yet to be unearthed. Nor would he agree that the outcome of the interaction of the billions of cells in the brain is wholly predictable by science—an outcome that would leave no room for mysticism and spiritual thought.

This last point can be well taken when one examines the state of science. Again, take the Big Bang theory to which currently all pure reductionists must go back to explain the existence of chemical elements and everything in the physical world. It is a theory whose days are limited, even though one may admit that it is the best around, and although it is highly likely that many of the scientific findings associated with and provoked by the Big Bang theory will carry over to the theory that replaces it. As already noted, there is the problem of missing mass in order to account for the configuration of planets, stars, star clusters, galaxies and galaxy clusters. More significant is the fact that the Big Bang theory revolves around a deterministic space-time path for the universe. Big Bang theorists are unable to specify the required initial conditions at the moment previous to the Big Bang, often suggested as one of infinite temperature. (For some discussion of the role of initial conditions see Chapter 7A.) They can only hypothesize conditions at 10^{-12} seconds after the Big Bang.

Even if with billions of dollars of investment in supercolliders and the like for particle research scientists will be able to trace the path of the universe for some time before 10^{-12} seconds, there will always remain the question of what were the initial conditions lying behind the existence of any earlier situation that may be speculatively identified. This points up the problem of *infinite regress* (discussed in Chapter 4) when a rate-dependent deterministic mode is employed—that is the problem of identifying the initial conditions z that yield in space-time the initial conditions *y* that are in turn required to yield the initial conditions *x* . . . etc. etc. and, finally required to yield the initial conditions *a* that are required say for the Big Bang theory. (In 1978 when particle physicists were still excited about the discovery of quarks, E.O. Wilson, in a reductionist-type manner, suggested that religion (mysticism) could possibly be given a role in evolution; it could be called upon to explain the existence of quarks.)

Note that the severe problem of infinite regress exists whatever the position one takes regarding the existence of randomness and noise—whether one claims like Einstein that 'God does not play with dice' where God can be interpreted here as Nature or an equivalent and thus rules out random mutation, or like Max Born (in Pagels, 1983, p.56) who states:

> If God has made the world a perfect mechanism, He has at least conceded so much to our imperfect intellect that in order to predict little parts of it, we need not solve innumerable differential equations, but can use dice with fair success.

More damaging to reductionism is the fact that while there has been tremendous progress in science, and while the explanation of many things and interactions has soared, it is the case, certainly true in the social world, that the number of *problems arising from and awareness of unknowable things and relationships has also soared*—and this situation is still likely to obtain if and when particle physicists reach their Holy Grail—a Unified Field Theory. We certainly cannot say that we have become relatively ahead in explaining the social world—the interconnected behavior and interaction of variables in the fields of economics, geography and regional science, sociology and social psychology, political science, anthropology, and other areas.

Further, there is the question: does, and, if so, to what extent, consciousness and reflection and their consequent effect on behavior in the social world come to affect the physical world and its evolution? That is, is there (or will there be) feedback to the physical system which affects and conditions the operation of the system's laws and their interactions—feedback say from present (or future) bioengineering and nuclear ventures? After all, humans represent matter and energy, and through consciousness and reflection exercise increasing control (however limited) over the behavior and interactions of the rest of the matter and energy in the physical world. And clearly, if humans are guided by mysticism and religion, then their mysticism and religion must affect as well as be affected by interactions within the physical world. Currently, some scholars find substantial evidence that an individual's psychological state (whether based on religion or not) can affect his physical state, his well-being as a functioning organism. Or consider the widespread, indeed almost omnipresence, of mysticism and religion during the ascending years of the papal empire; can it not be claimed that the mysticism of that empire's social world greatly conditioned individual and social behavior and thus came to have at least some impact on the physical world? While feedback from mysticism and religion may be admitted, would its magnitude be such as to impact the basic laws of physics?—an open question for the future.

In short, while *theoretically* advance in scientific knowledge, if sufficiently rapid and confined to an *abstract* world of little change, can reduce to insignificance the role of religion and mysticism in the social life of that world, in *practice* science has failed to address in the real world the problems of infinite regress, randomness, feedback and the constant enlargement of the implicate order (through discovery of hidden unidentified variables and relationships to be understood and explained). This failure means that science together with religion and mysticism must function in a complementary fashion in investigating behavior and interaction in that real world. It thus becomes incumbent upon us to evaluate accordingly the thinking of great religious leaders who were outstanding scientists as well—e.g. Maimonides, Avicenna, and Teilhard. Since we wish to do so with regard to the current advanced state of science, we shall concentrate on the writings of Teilhard, the most modern of these leaders. However, it is not necessary to accept one narrow view of Teilhard, a devout Catholic, namely that the Christianity of the West with its civilization is the only religion having ultimate survival capability. Our review of recent religious literature clearly indicates that Islam, perhaps one of the Far East religions, Judaism, or more likely a new, more integrative religion growing out of the ecumenical movement may well provide a core set of religious belief of the future.

In Chapter 2 we have dwelt at some length on Teilhard's thought. Recall: (1) the designation of the progressive psychosocial evolution of man as *hominization*; and (2) the increasingly elaborate organization over time of inorganic and organic matter as *complexification*. Such complexification, except for minor perturbations, has been according to Teilhard much more than random; with increase in knowledge and communication, and with migration, interbreeding, and cultural diffusion in general, it has and continues to lead to a *convergent integration*—to the emergence of *a single interthinking group* unifying a rich diversity of cultures and institutions into a single self-developing framework of thought—the *noosystem*. The development of the noosystem represents the extension of the biological into a *biopsychosocial sphere*. With knowledge increasing at an exponential rate and communications media developing apace, the bounds of the earth as a planet become more and more restrictive. They lead to the intensification of mental subjective activity, to higher biopsychosocial pressures and temperatures, to still more elaborate organization of differentiated structures, more finely and tightly integrated, and to still more selective evolution. In time, Teilhard contends, a state of *ultrahomanization* will be attained, reflecting a still more advanced stage of the process of cephalization based on the brain as the main organ of coordination. A *hyperpersonal mode of organization* will result in which human personality is of supreme importance. This personality will transcend the character of a highly individualized individual and consciously interact as a fully coordinated self with the outer world of men and nature. It will do so through more extensive interrelation and cooperation. Ultimately, the process of convergent integration within a global community of ultrahomanized personalities will approach closer and closer the ideal state of *Omega*.

As noted in Chapter 2, Teilhard's outstanding vision fails to come to grips with the economic and political forces at play in society—forces which are required to identify the step-by-step process needed for a closer and closer approach to Omega. Here is where economics, regional science, political and other social sciences can make a contribution. To do this let us first review the consistency of the Teilhard doctrine with the scientific and analytical materials discussed in the previous chapters.

As already remarked, there is nothing in Teilhard that basically differs from the views of Dawkins, for example those on the emergence of elaborate survival machines and coadapted meme complexes, and the writings of most other evolutionary biologists discussed in Chapter 3. He would share the views of a coevolutionist such as Durham and strongly disagree with the rather inconsequential role that a sociobiologist like E.O. Wilson assigns to religion. He would be in sympathy with the views of Pattee on the complementarity of processes within both the deterministic and linguistic modes, and on the principle of optimum loss of detail. As with Wolfram, he would indeed find cellular automaton operations extremely useful, as also the computer simulations of Artificial Intelligence experts, and would go along with the concurrent, distributed network and neural network and fuzzy system models of Hopfield, Smolensky, Kosko and others to shed light on the functioning of the brain. However, like Pattee he would conclude that the emergent behavior and creative spiritual thinking emanating from the brain is not a rate-determined process and is clearly not subject in its entirety to scientific analysis.

Also, Teilhard would not have any serious problem with Bohm's ideas, especially the unfoldment process whereby elements of the implicate order become revealed and part of the explicate order. He would view this process as an essential part of the growth of a noosystem, but might attach some qualifications to Bohm's notions of unending movement and undivided wholeness.

As a distinguished paleontologist and scientist, Teilhard, were he alive, would easily concur with most, if not all of the recent developments in particle physics, chaos and catastrophe theories, and noise research. He would easily incorporate in his thinking, if he had not already done so, the notions of symmetry breaking, field theory, and the build-up of hierarchical structures among particles, chemical elements and celestral bodies, the notion of strange attractors and most if not all the concepts associated with chaos and Prigogine-type theories listed on pp. 92-100, and in particular the joint play of chance and determinism in a number of processes.

He would, however, emphasize the limits to scientific analysis. He would point to the problems already noted of identifying: (1) initial conditions; (2) the many unknowns and interrelations in the physical system of which scientists are aware (e.g. the missing mass) and will become aware of with further advance in science; (3) the failure of science to recognize and identify the feedback of the human species on the functioning of the physical system; and in particular (4) the driving force of cumulative spiritual enlightenment and development from creative cerebral reflection—the force that can carry man step-by-step from where we are now toward his Omega. To him, the world is an open system and not immutable. Just as Dorner, the art critic, characterizes painting: as change-creating experience; as an evolutionary urge toward transformation of all traditional values; as growth continually involving a new 'rationality' characterized by a phase not of Being but of Becoming, so does a Teilhard when he speaks of spiritual enlightenment and development. His spirituality embraces the wonder of man (as did Newton when contemplating the mystery of the universe) as well as the deep-rooted anxiety that man experiences from the uncertainties of life that lead for example to prayer.

Along the space-time path that proceeds to Omega, Teilhard would most likely envisage increasing hierarchization and accept the notions (from economics and regional science) of scale, localization, urbanization, and spatial juxtaposition economies (diseconomies) as basic factors governing hierarchical construction. He would recognize Simon's near-decomposability as a element that is involved. He would have no objection to the principle of factor substitution in economics, nor with the interpretations of first-and second-order conditions (as discussed in Chapter 7A) with regard to space-time processes, and in general in the dynamical systems employed by economists and regional scientists. In fact, he would be most pleased to observe the functions of the Coordinator and Ombudsman in a number of the models covered in Chapter 7A. To him these functions place in bold relief the need for a spiritual leader (force) or equivalent body. The noted *Impossibility Theorem* of the brilliant economist Arrow, namely that a social welfare function cannot be scientifically established for a population having different preference patterns, and the implication that nations do not have any scientific basis for eliminating their conflicts (as discussed in Chapter 7A) merely reinforces the great need, he would claim, for a spiritual leader (body) to identify for a following (the world constituency) a proper combination of a system performance (objective) function *and an allocation configuration of subsistence goods and economic and social services that would satisfy the welfare criteria that society would set.* Such would serve to not only advance the aim of achieving equality in consumption and opportunity for individual development and participation in social affairs, but also foster morality, quality of life, security and other significant welfare criteria. He would probably view economics and regional science as among the most advanced social studies and sciences in that they employ most extensively mathematics and concepts and models from physics and other natural sciences, as well as the exceedingly important qualitative approaches of social sciences; and he might forcefully contend that if economics and regional science are far from specifying what is social welfare, then this must also be the case for other social studies and sciences. Consequently, he would maintain that these studies and sciences, too, must fall back upon spiritual leadership through use of ingenious cooperative procedures and other mechanisms to identify a much more desirable if not optimal welfare performance *and* allocation combination.

In his concept of hyperpersonalization, Teilhard envisages each individual effectively interacting directly, and most often, indirectly with all others—an interaction configuration similar to the one Bohm (1990) has in mind when he expresses the need for a *common coherent culture*. Teilhard thus would find of great appeal the concept of participation potential as developed in Chapter 7 that can be stated, at least primitively, in mathematical form and be partially operational, quantitatively speaking.

In proceeding from the present noosystem to a future state of noogenesis, a modern Teilhard would be concerned with the steps that are to be taken, in particular with economic and social policy that is required to move the system along desirable directions. He then might find extremely useful computer simulations of outcomes of different policies. In particular he might be favorably disposed to the use of comprehensive and integrated models from economics, regional science and other social science areas. He might find them of great value in order to avoid policies and steps that might be detrimental to society and to identify a set of possible policies that might be desirable. Any such model would need to be capable of examining impacts of policies concerned both with the overall performance of a system and its parts *and* the allocation to individuals (and groups) of goods and economic and social services. In this connection a comprehensive framework such as the PSIG (the peace science integrated global) model might have appeal. See Figure 11.1 (and discussion in Isard, 1990). At its bottom we have depicted *subnational linkages* and models—models primarily economic (such as the Leontief input-output) that can yield quantitative estimates of relevant magnitudes at the local, community, subregional and regional (within nation) levels. Relevant magnitudes would cover: number of jobs, output by industry, family income, food consumption and so forth. As indicated by a vertical arrow, these models can supply as well as receive information from relevant national models (such as national econometric models). The models for each nation can also be interconnected to yield an overall global model (such as LINK, or that of the World Bank) which then can indirectly receive from and provide feedback information to all local and subnational levels. The national models can generate information on Gross Domestic Product, employment, investment, consumption, government expenditures, exports and imports. At all the above levels there are also the *social accounting models (SAM)* which can yield distributional impacts—impacts upon different socio-economic groups whether categorized by family income (for example, below poverty level, above poverty level but poor, lower middle income, and so forth), ethnicity, race or other variable. Information from SAM models can then be usefully transmitted to *demographic models* which are in turn of value for formulating measures and policy to fight overpopulation, poverty, and other key aspects of the population problem.

Returning to the *global/national economic models* circle of the figure we note that combining significant information flows from it with equally important information flows from the *global/national political models* circle can yield significant *socio-politico-economic impacts.* These impacts stem from the generated levels of: (1) educational and health expenditures to enhance the productivity and well-being of a population; (2) drug and crime control expenditures; (3) infrastructure investment expenditures; (4) police (military) expenditures for security purposes; and (5) other welfare program expenditures.

Control and enhancement of environment quality would be another major concern of a Teilhard. At the bottom of Figure 11.1 we note models that project emissions of air pollutants (for example, CO_2 that currently is viewed as contributing to earth warming and H_2S that contributes to acid rain), water pollutants and others which then impact climatic and the ecological systems which in turn affect health, food supply, labor productivity, and so forth.

It is not necessary to go into any further detail to demonstrate to a modern Teilhard or his associates and staff the usefulness of these models to help identify the steps and policies that are needed to proceed along the path toward an ideal state. The one critical place where morality, ethics, spirituality and religious ideals provide basic inputs into the system which are not only indispensable but unavoidable is depicted in the central box designated *Conflict Management: Integrated World Policy Development.* Here is where key tradeoffs are made. Here is where the decision is made on which economically, politically and socially feasible combination of performance *and* allocation configuration to pursue—that is to set up as the objectives for the next period at a given decision point of time. Here is where constraints, as guidelines, are set on ethical, moral and other behavior of individuals and groups within the social system—in particular, constraints on tendencies for unbounded or excessive individualism and group independence. These guidelines would directly and indirectly reflect expressions of love of humans for one another and for nature and the surrounding environment. Here is where spiritual and religious forces combined with accumulated knowledge on workable conflict management procedures and mediation and negotiation processes would have preeminence—clearly to help eliminate unnecessary poverty, environmental degradation, and physical violence, strife and warfare.

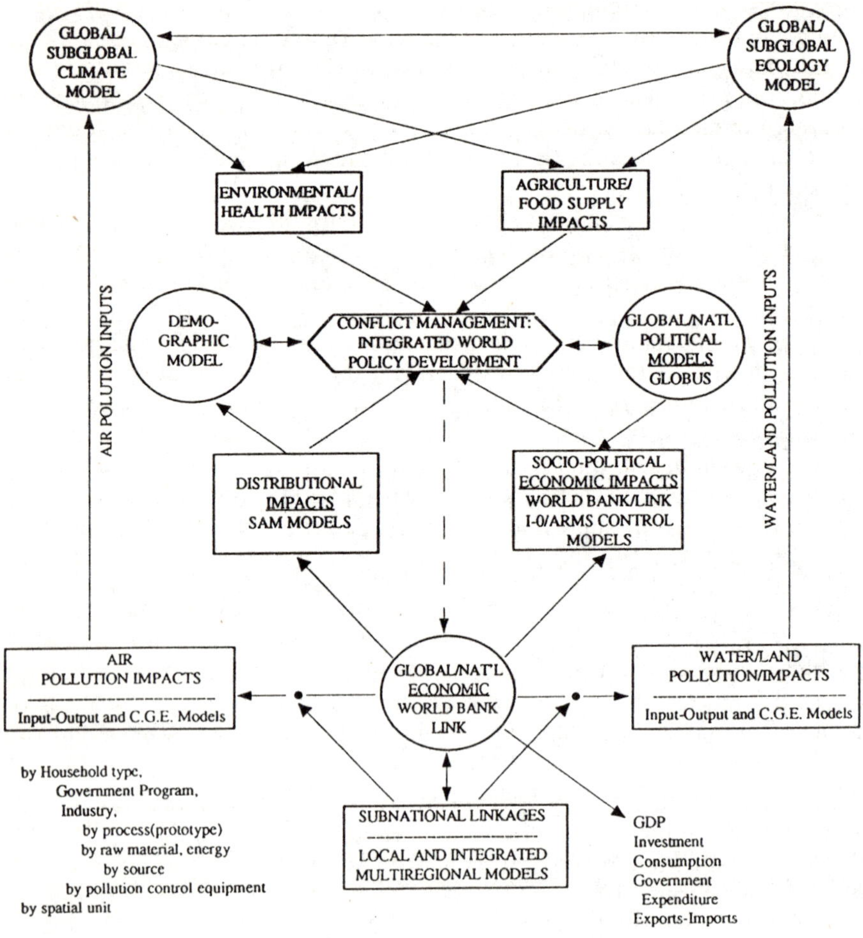

Figure 11.1 The PSIG (Peace Science Integrated Global) model
Source: Isard (1990)

It should be noted that the PSIG model is in essence an heterarchical model. It embodies many causal chains in the diverse submodels that it contains and helps to fill in the details of a space-time path. These submodels, however, are not fully, or even nearly fully connected. The cognitive-linguistic-spiritual mode, the involvement of many subjective factors, is everywhere present. Yet the model represents a comprehensive framework, and an integrated one in the sense that the key parts of a system can be set down as well as the diverse interconnections among them, though many of the interconnections can only be captured qualitatively. Hence, only ranges of possible outcomes from the operation of an integrated model can be obtained, and many shortcomings can be found, a result of our inadequate knowledge of much of what we assign to randomness.

It is extremely important to note that the PSIG model is only one of many comprehensive and integrated models that can be designed. It clearly is biassed in the direction of the knowledge base and research training of the author.

Nor can we expect any comprehensive and integrated model to project the ups and downs of the path toward Omega or other ideal state. We can imagine at certain periods of time great strides along this path, later to be succeeded by retrogression—perhaps some orderly hierarchical deconstruction based on societal transitions found to be efficient and otherwise desirable. Perhaps, also, there will be encountered hierarchical disintegration from such factors as inadequate planning, excessive technological advance, and incompetent leadership (after all, man is not infallible), as was realized in the decline of empires and civilizations in the past. Can mankind expect another period of 200 years of peace and prosperity or even more than that, before a decline sets in? Additionally, devastating effects upon humanity can occur from uncontrollable random elements, natural catastrophes, major climatic change, ecological disruptions, plagues and the like as man moves along the path toward a more ideal state.

Once more we must state that when we speak of Teilhard, we do so generally. As we have remarked, a number of effective religions or sets of mystical or spiritual beliefs may qualify for providing the core set of beliefs in the future. However, it may also be that two or more religions may come cooperatively to provide core sets of beliefs, each playing a central role in a culture. Within each there may emerge an outstanding figure—outstanding both in scientific analysis and spiritual message—in essence the equivalent of a Teilhard or Avecinna or Maimonides. Each may have his/her own Omega and space-time path to it. Even within each religion (or set of mystical or spiritual) beliefs, there may be differences leading to sects, each with its own conception of a heavenly (or subheavenly) state. However, among these Omegas, heavenly or subheavenly states, there would be many commonalities. And in the current situation of our world, there is no need at present to assign preference to any when we speak of a Teilhard and his associated spiritual, mystical and/or religious message.

11.5 Implications for 2000 AD and beyond

Thus far this book has by and large dealt generally with developments and processes since the Big Bang. It is natural for the readers to ask: what are the implications of this study of the long past for the near future, say the next century? Some readers may be content with a discussion of broad forces that might be at play. Others, however, would like the specification of a set of well-defined steps that might be taken to combat the urgent problems confronting a macro-regional or global society.

No one can predict the future. At one extreme there may be extensive nuclear devastation or like destruction that annihilates the human race, or leaves the planet with one or more tiny pockets of *homo sapiens*. At the other, the global society may come to enjoy another 200 years of peace and prosperity.

a. *The thinking of those who forecast the future: the problems the global society faces and measures to be taken*

To begin, let us hark to the warnings and proposals of current respected figures in world society. There are a number of such figures, but space limitations of this book compel us to concentrate on the thinking of two. We choose to record briefly the thinking of Oscar Arias, a former president of Costa Rica, a Nobel laureate, whose suggestions and recommendations are heeded much throughout the world community. We also refer to the views of another world leader with a different personality and background, namely Robert S. McNamara, a very successful business executive when the head of the Ford motor company, a forceful participant in U.S. policy deliberations while Secretary of the U.S. Department of Defense, and an effective administrator of a world institution when the Director of the World Bank—a leader who admits to serious mistakes in actions he proposed and helped implement and has learned from experience.

In a recent address Arias (1994) speaks of the trend toward increasing globalization, toward a gradual transformation of man's world into a global neighborhood or village—a phenomenon reflecting the shortening of distance resulting from technological advance and the growing permeability of geographic and cultural borders—borders that are gradually losing relevance. McNamara views the world as becoming increasingly interdependent in 'which nations are entwined with one another economically, environmentally and with regard to security' (McNamara, 1995, p. 4). Until the time when appropriate control, educational and other measures are taken, both Arias and McNamara also foresee: (1) the continuation of the population explosion; (2) accelerating pressures on the environment and as a consequence its certain degradation; (3) mounting growth of inequalities and a constant struggle for human security against global poverty; (4) growing military threats among developing nations, fostered considerably by the export to them of nuclear and other weaponry primarily from developed nations; and (5) persistent inadequate financial and other assistance from the developed nations to the less developed ones required to overcome the internal problems of the latter. These leaders set forth a vision of *collective security*. To achieve this security they propose among other actions: (1) strengthening and reorganizing multilateral institutions such as the United Nations; (2) the development of effective conflict-resolving mechanisms; (3) world-wide demilitarization to curb violence and more importantly to free large quantities of resources for educational, medical and other social programs for human development, particularly in developing countries; and (4) major effort at the preservation of the global environment to allow sustainable development for all.

Let us probe more deeply into the phenomena which Arias, McNamara and many others speak of and examine the consistency of their thought with the commonalities discussed in this book. (Of course, other commonalities may be set forth by other analysts as having major force or may come to light in the future.)

Take globalization. It is indisputable that at least in certain respects such is occurring. Communication technology is growing at a rapid, and some would state an alarming rate—such technology being *coadapted meme complexes* of increasing complexity and often hierarchical structure. For example, this advance is revealed in the tremendous increase in the interconnectedness that has taken place in recent years in the world's financial community, the potential for the enlargement and diffusion of its supporting knowledge base mounting unabated. At one time, namely in year 1990, many informed analysts claimed that Tokyo had come to surpass New York as the leading world financial node, and that in fact Tokyo's emergence as the top node represented the addition of a new level in the world's financial hierarchical structure—a structure with New York, London, Frankfurt and two or three others constituting the second level. Whatever the view of this development, some futurists would characterize the rapid rise of financial and other global interconnections of increasing complexity within society as evidence of the world's entrance into an era of *cyberspace*.

As yet cyberspace has not been rigorously defined. One portrayal considers it an environment 'in flux, where shared values collide with discussions of diversity, where freedom of expression, freedom of association, and equality are paramount and attempts at censorship are vehemently opposed and/or routed around' (Ogden, 1994, p. 714). Its skeleton is the *Internet*, a network of networks than can in effect cement together the information society—all via its electronic superhighways, its communication facilities, its computer complexes, its interactive digital libraries and so forth. A geographer or regional scientist confronting this cyberspace, or whatever the designation of the perceived world of which the Internet is a skeleton, must admit to a sudden revolution in the concept of space. From one point of view, 'space' has been enlarged and from another partly supplanted by 'a vast web of telecommunications whose fixed points are computers of many sorts,' wherein 'traffic is information, and the vehicles that navigate its highways are intelligent software animals called knowbots' (Batty and Barr, 1994, p. 699).

Clearly, the Internet, as a major technological advance, has and will continue to foster a tremendous unfoldment of the *Implicate Order* that has and will constitute a major development of the *Explicate*, including a recognition of the soaring stock of relationships which are at each point of time unknowables. The potential for both good and bad of the Internet and other aspects of cyberspace is widely recognized.

Associated with the globalization trend are the many commonalities found in the history of empire building: increasing *specialization*, for example in the cornucopia of current investment channels (equities, bonds, mutual funds, derivatives, etc) that brokers and other financial operators advocate and offer; steadily rising *scale economies* (as reflected in the growth of giant financial institutions and multinationals); mounting *urbanization* associated with *spatial juxtaposition economies* (as recorded in the growing percentage of the world's population in megalopolises, metropolitan regions, and cities); yet a sizeable amount of efficient *deconcentration* of industrial activities, other functions and habitat patterns based on *deglomeration economies* (lower rents and housing prices, less congestion, closer proximity to nature, etc.) which counter the *agglomerative*, land-use-intensifying forces just mentioned.

Underlying these phenomena are communication channels exponentially increasing in capacity and the concomitant growth of information flow. Given the slow rate (if any) of increase in the storage and analytical potential of the brain's neural system, a smaller and smaller chunk of the total information flow accessible by any single, efficient decision maker at any decision point of time necessitates more extensive filtering of this flow, greater reliance on the use of the *principle of optimum loss of detail* and on the organization in a *near-decomposable* manner of the hierarchical structures which that decision maker is on top of.

How far the globalization trend and its associated phenomena will go and how long it will endure is an unanswered question. Without doubt, a successful confrontation of the problems that will come to plague society will require the emergence of a truly outstanding leader—clearly a unique, *coadapted gene complex* of tremendous ability to confront the conglomerate of issues that increasing hierarchization generates. For as already noted, the globalization trend has been accompanied by a population explosion and dangerously mounting pressures on the environment—basic problems that the globalization process as yet does not seem to be able to combat. Population numbers, estimated at 5.61 billion in mid-1994, is projected to be between 8-9 billion in 2025 and some 12.6 billion in 2100 (*Future Survey Annual*, 1995, pp.5-6)—growth that may also generate severe international migration problems. Even more serious is the problem of environmental degradation, a problem that the population explosion intensifies. More people means more consumption, in turn more extraction and use of scarce resources, and more potential for generation of pollution and disturbance to the environment—disturbance that portends the extinction of species in the wild, destruction of marine biological diversity, non-sustainable exploitation of the forests, and so forth. The *selfish gene,* as embedded in competitive survival machines—the individuals, groups and organizations of human society even when groups and organizations include altruists—is at work. The propensity to 'free ride,' to exploit, to pollute in situations where one behaving unit contributes only infinitesimally to environmental poisoning, which when pursued by all behaving units in an unchecked manner, does lead to serious degradation. Recognition of this effect by all is essential for environmental sustainability and involves the development of a stewardship ethic in society. It means that, even if only by appealing to the selfishness of behaving units, international agreements, laws, regulations, conservation and biodiversity strategies, formation of world-wide alliances and benign coalitions and other diverse means of cooperation will be required for the survival of the human race. At the same time, *anticipatory planning* and foresight more systematic than ever before—planning that dispenses with fragmented decision making and involves greater *consciousness* and *reflection*—will be essential for efficient management at the global level down to the many local ones—all with thorough appreciation of the complexity of the ecosystem. Yet, despite the extensive and mounting discussion of this problem and optimistic talk—about how revolutionary advances in computer facilities and their interconnection with software engineering, control systems, integrated circuits and telecommunications will be able to monitor and model environmental conditions and control the level of pollution and cope with unhealthy resource exploitation and so forth—the quality of the environment continues its steady decline and today sustainability remains far from being attained.

Closely allied with the population explosion and environmental degradation problems is the persistence and again steady growth of a set of inequalities among the world's population—inequalities in income, the means for subsistence, opportunities for education, access to health facilities, protection against persecution, etc. and in general in *participation potential* with regard to the numerous decisions that affect people's welfare, and in the exercise of that potential. The absolute differences, between those relatively few at the top node of the hierarchy(ies) and the billions at the bottom level of the hierarchy(ies) who are scratching out a subsistence, seem ever to grow. Again, while talk about striving for universal standards and human rights, about conserving and preserving native people and cultures, and about narrowing and desirably eradicating inequalities continues unabated, these goals are little realized and the world seems to move further and further from their attainment. Accordingly, *dissension*, *disquietude* and *disbelief* appear on the rise. Talk about *demilitarization*—especially of the developed countries as a means to provide funds for assistance to developing nations—has hardly led if at all to that occurrence. The reality in 1995 is one of reduced military expenditures in at least some of the developed nations, but at the same time increased exports of arms from them to less developed nations. Moreover, some proliferation of technology and know-how for producing nuclear missiles (if not actual production) has occurred in recent years; and certainly the potential for military threats and actions inclusive of conventional, nuclear, biological and chemical warfare has magnified throughout the world—an outcome clearly disadvantageous to collective security and indeed to the security of the very developed arms exporting nations.

The persistence of the above problems and others raises the question of whether the current globalizing world society is soon to be or already is at a phase of *dissolution* and *disintegration*—this despite the clear globalization trend evidenced in financial markets, the mounting number, size and power of multinational corporations, the rapidly spreading worldwide telecommunication facilities, etc. Is the world society at a point of bifurcation, one path leading to progress and peace (perhaps one or more centuries of such as was realized by the Roman and the Achaemenid empires), the other path leading to disintegration and dissolution of its multinational institutions and other structure, and even of the structure of nations and elements of them?

Let us examine possible disintegration and dissolution factors. Already, one has clearly appeared in the globalized financial community. In 1991 and thereafter the node that had been at the top of the intricately complex financial hierarchy—namely Tokyo—was found to have been extensively infected by malpractice and corruption which has resulted in considerable *demoralization*. There has been inadequate and improper policing of this critical segment of the economic system.

Previous to this demoralization had been the dissolution and disintegration of the powerful Soviet Union and the Warsaw Pact bloc which has generated considerable instability in Eastern Europe with spread effects to the rest of the world. The concomitant demise of the Cold War lifted the strong constraints that the competing Two Big Powers had upon a number of developing nations under their hegemony with the consequence that the latter have increased their militarization spurred on, as already mentioned, by short-sighted political leaders in developed nations failing to curb, and often facilitating undesirable arms exports.

In the last years of the twentieth century, there has occurred increasing terrorism, its operation more diverse and of greater magnitude—a serious dysfunctional aspect of the globalizing world, one that is encouraged by the emergence of cyberspace, Internet and all its advanced and increasingly sophisticated technology. Drug production, trade and consumption is still on the march, better heeled and more strongly entrenched in society than ever before; and the spread of crime, particularly in the former Soviet Union, has not been mitigated. The resulting dysfunction and demoralization has led to significant *depersonalization*. Less and less prevalent has become the sense of neighborliness among local populations—as so boldly and harshly revealed: (1) in central cities where people on the scene of a violent crime fail to intervene, participating merely as bystanders; and (2) where nations tolerate, again as bystanders, genocidal ethnic cleansing of neighboring countries (as was the case in the former republics of Yugoslavia)—critical pieces of evidence for the hypothesis of waning humaneness in world society.

All these downswing factors in addition to excessive (and inefficient) *deconcentration* to outlying areas of non-poverty population and viable economic activity (in essence flight from the problems of the central cities of megalopolises and metropolises that intensifies their problems) contribute to the general *devitalization* of society. Together they can be said to paint a bleak picture of the future global society.

Yet, the outlook cannot be stated definitely as bleak. The one element typical of major declines of societies, empires and civilization that does not seem to be present in the 20th century and certainly its last years is *descendance* and *decay* of uniquely capable integrative and inspired leadership. As one scans the world in 1996, one cannot discern such leadership at the global level. Nor has there been any for many decades. One can point to a Franklin D. Roosevelt in the United States, a Charles de Gaulle in France, a Winston Churchill or Queen Victoria in England, a Mahatma Ghandi in India, a Nelson Mandela in South Africa and others—but none have had the stature of the Cyrus/Darius dyad of the Persian empire, Augustus of the Roman empire, and Innocent III of the papal empire. Hopefully, such a leader may emerge in the years ahead—just as Augustus rose out of the throes of the turbulent Roman state in the first century BC—capable of coping with the exceedingly complicated problems of a globalizing society and staying on top of them for a number of years. Perhaps the leader may sprout from a major 21st century ecumenical movement whose strength may lie in effecting a rebirth of the sacred out of a morally debilitating sway of a highly materialistic and over-technologized society—a leader who effects a delicately balanced, harmonized coplay of religion and science, of the linguistic and deterministic modes.

Hopefully, too, whatever leader emerges, he/she may be able to pass on his/her power (formally and/or informally bestowed) to a well-trained and highly capable cadre of successors—another 'five good emperors'.

If the above leadership arises and a period of high prosperity and peace transpires, inevitably, but unpredictable as to time of occurrence, there would come to the fore new leadership with capabilities inadequate for the administration of the inherited realm. For this new leadership, the inherited realm would be overorganized, overconcentrated or both and with functions too finely articulated and complexly interconnected. Further, this leadership would likely be unable to: (1) exterminate corruption; (2) manage with its staff the fisc; (3) combat escalating inflation or its equivalent during its incumbency; (4) plough through the gridlock of opposing claims and pressures of innumerable lobby groups; (5) adapt flexibly to the changing socio-political situation; (6) confront devastating environmental and associated destructive phenomena (earthquakes, floods, dramatic weather conditions, nuclear mishaps, etc.) with humaneness and wisdom; (7) temper dissension and police the potentially disruptive actions of dissidents and terrorists; (8) forestall political fragmentation that splits the global realm into a set of highly competitive blocs of nations, each tightly coalesced (wherein macro-regionalism supplants the nationalism of the past); (9) curb politically haphazardous divisiveness that results in a chaotic jigsaw pattern of local, tribalistic-type, or ethnocentric authorities with overlapping jurisdictions, conflicting allegiances and a highly confusing multiplicity of interconnections; (10) still the formation of a still more disaggregated society resulting in part from noise of significant magnitude in which supraindividualism is rampant; (11) constrain the rise of an excessively tight and orderly decentralized or Fundamentalist-type society, one operating at a very low economic, material-oriented level yet spiritually more uplifting; or more likely, (12) control a combination of several or more of these control elements.

Whatever the specific space-time path that is realized, it would necessarily be one of global dissolution and disintegration to a significant degree—one, to reiterate, that is inevitable but hopefully deferred until after some 200 years of peace and prosperity.

One may ask, out of curiosity: what might be the residue of that 200 years of peace and prosperity—a residue like that of writing left by the Sumerians, and the calendar concept by the Egyptians? Would there be an outcome that would facilitate analysis at a much deeper level of causality, perhaps within a more elaborate and linked heterarchical framework? And with regard to the subsequent global dissolution, would there have developed the means to uncover potential points of stratified stability that crop up that could be garnered to modify the downswing and preclude further disintegration?

All the above discussion assumes of course that no nuclear holocaust or similarly destructive phenomenon has already made extinct the human species.

b. *One set of possible immediate and specific steps for peace and prosperity*

As already remarked, in asking for the implications of this study for the future, some readers might be impatient or dissatisfied with the above discussion of broad forces at play, and insist on the specification of a set of well-defined steps to combat the urgent problems of our globalizing society. It is of course impossible to predict with certainty any space-time path that could occur, let alone the fan of innumerable possible space-time paths that can take place. Yet let us indicate how one might *structure* a response, where the specific steps cannot be assigned any significance whatsoever, but where the methodology behind the structure can.

Recall Arias's and McNamara's goal of collective security, attainable through the effective functioning of a set of international or multinational institutions, or perhaps a world government with appropriately constrained powers. This new world governing structure—perhaps a reorganized and strengthened United Nations coupled with a more effective International Monetary Fund, World Bank, International Court and/or other institutions—would operate to keep and maintain peace, protect the environment, achieve sustainable development, enhance and secure civil rights, eliminate terrorism, adjudicate international conflicts, and so forth. It would be: a global norm setter; a conflict manager, mediator and arbitrator; an adjudicator; a broker; a network operator; a coordinator; a convener; an international peace-keeper; and a seat of power, influence and knowledge-distribution in an information-hungry electronics world. It would establish and administer environmental, financial, communication and other regulations pertaining to the globe. It would bond nations and regions together while preserving diversity of religion and other cultural dimensions. It would guard against excessive urbanization and excesses in other processes. It would encourage integrative scholarship and have a research arm designed on the one hand to conduct research of immediate and direct use for formulating its policies, and on the other to stimulate basic research at universities and relevant institutions—research both of a comprehensive nature within a heterarchical framework such as depicted by the PSIG model and of more limited scope to attack in depth a particular problem, such as the drug problem.

All this would be a huge assignment. The reader can rightfully ask: how can all this be attained? To illustrate, recognizing fully the unpredictability of the specific space-time path of the globalizing world at this point of time, and that what follows is pure speculation, let us start by examining at some detail the possible attainment of international peace-keeping, one of the critical subgoals foremost in 1995 in the minds of Arias, McNamara and so much of the world's population. Posit that this task—a task which none of the major developed nations is willing to undertake alone but which most, if not all, have at times considered urgent—is to be undertaken by a reorganized and strengthened United Nations serving as a future world governing authority.

There are several critical aspects to the assumption and the efficient operation of this activity by the current United Nations. One critical aspect is the size and cost of this activity. In a paper on peace-keeping operations and related aspects, Klein and Marwah (1995) first present the detailed Table 11.1 on the costs to UN of peace-keeping operations from 1948 to 1993. They then carefully examine background materials for projecting the future numbers for personnel and total costs. Taking into account the rough and partial estimates made by Urquhart (1993), Krasno (1994) and Steinbrunner (1995) and other relevant information, they reach the tentative conclusion that a standing UN force of one million would cost between \$50 and \$100 billion at 1994 prices—\$50 billion being approximately 0.2 percent of Gross World Product (GWP). To analyze the impact of such an operation on individual nations, they first consider several ways in which these personnel requirements and costs might be assigned to nations. Excluding most poor countries (each of whom accounts for 0.2 percent or less of world output but in total, 17.49 percent), they suggest a distribution of total cost of \$50 billion among the remaining 42 countries in several ways, three of which are noted below for 10 selected countries:

(1) in proportion to their shares of GWP as listed in the third column of Table 11.2 (shares of GWP are recorded in the first column);
(2) in proportion to the size of a country's per capita Gross Domestic Product (GDP), as listed in the fourth column of Table 11.2 (per capita GDP are recorded in the second column); and
(3) as an average of the corresponding entries in columns 3 and 4.

As can be expected, when the base is share of GWP alone, U.S.A. and Japan have the largest costs, namely \$15,756 and \$7,878 billion, respectively (column 3, of Table 11.2) and poor countries the least (for example, \$0.103 billion for Bulgaria). When the basis is per capita GDP alone (column 4 of Table 11.2), Switzerland, even though a very small country with a low level of population, has the highest cost; its per capita GDP is higher than that of U.S.A. Clearly, 'ability to pay' by country must be based on both share of GWP and per capita GDP. The weights to be attached to these two criteria, or others that might be introduced, would undoubtedly be a source of conflict and subject to negotiations. The simple 50-50 rule (equal weight to each) yields the country costs of column 5.

Table 11.1 Peace-keeping operations (as of October 21, 1993)

Symbol	Since	Military/ Police Strength	Fatalities	HQ	Estimated Annual Cost UN (millions)
UNTSO	1948	220	28	Jerusalem	$ 31
UNMOGIP	1949	39	6	Srinagar Rawalpindi (India-Pakistan)	7
UNFICYP	1964	1,255	165	Nicosia (Cypress)	45
UNDOF	1974	1,107	31	Damascus (Golan Heights)	36
UNIFIL	1978	5,266	193	Naqoura (South Lebanon)	146
UNIKOM	1991	361	---	Umm Qasr (Iraq-Kuwait border)	75
UNAVEM II	1991	79	3	Luanda (Angola)	40
ONUSAL	1991	1000	2	San Salvador (El Salvador)	35
MINURSO	1991	350	4	Laayoune (Western Sahara)	37
UNPROFOR	1992	26,160	59	Zagreb (Croatia)	1,200
UNTAC*	1992	22,000	55	Phnom Penh (Cambodia)	750
ONUMOZ	1992	7,000	6	Maputo (Mozambique)	290
UNOSOM II **	1992	28,000	81	Mogadishu (Somalia)	977
UNOMUR***	1993	81	---	Kabale (Uganda/Kwanda)	4
UNOMIG****	1993	88	---	Sukhumi (Georgia)	23
UNOMIL†	1993	426	---	Monrovia (Liberia)	40
UNMIH*****	1993	1,267	---	Port-au Prince (Haiti)	50
UNAMIR§	1993	2,608	---	Kigali (Rwanda)	98

* Withdrawal 15 November, 1993; the cost is one-half of a two-year estimate.
** 17,700 US troops in support of UNOSOM II were not under UN command or authority.
*** Six-month period
**** Scaled back after cease-fire broke down
† Seven-month period
***** Six-month period
§ Peak strength

Source: United Nations Peace-keeping Oeprations, *Information Notes, 1993: Update No. 2 (UN Department of Public Information).*

Table 11.2. Suggested country cost of a $50 billion expenditure for a United Nations peace–keeping operation (for selected countries) (in U.S. dollars)

Country	Percent Share of 1993 GWP (1)	1993 Per Capita GDP (2)	Country Cost Based on Percent Share of 1993 GWP (billions of $) (3)	Country Cost Based on Relative Level of 1993 Per Capita GDP (billions of $) (4)	Average of columns (3) and (4) (billions of $) (5)
Australia	1.40	$14,002	$0.848	$1.843	$1.346
Austria	0.67	$13,952	$0.406	$1.837	$1.121
Brazil	2.25	$4,592	$1.363	$0.604	$0.984
Bulgaria	0.17	$3,200	$0.103	$0.421	$0.262
China	3.00	$404	$1.818	$0.053	$0.936
Egypt	0.40	$1,609	$0.242	$0.212	$0.227
France	5.00	$16,122	$3.030	$2.122	$2.576
Germany	6.50	$16,097	$3.939	$2.119	$3.029
Japan	13.00	$14,955	$7.878	$1.969	$4.923
Poland	0.55	$1,999	$0.333	$0.263	$0.298
Saudi Arabia	0.50	$7,836	$0.303	$1.031	$0.667
Switzerland	0.80	$20,996	$0.485	$2.764	$1.624
Turkey	0.45	$3,973	$0.273	$0.523	$0.398
U.K.	4.00	$14,538	$2.424	$1.914	$2.169
U.S.A.	26.00	$20,749	$15.756	$2.731	$9.243

Source: Klein and Marwah (1995) Numbers rounded

The country burdens of a UN peace-keeping operation may go beyond the direct costs suggested in Table 11.2. First, the suggested costs might be met by a country from the military budget that might be expected to have been approved were the UN peace-keeping operations not in effect. That is, a country's total military budget might remain the same while its assigned costs for the UN operation might be matched by a corresponding decrease in expenditure on military arsenal. In this case, indirect effects might be minimal.

Second, if the suggested nation's costs were not partially or fully offset by reductions in its military program, the operation of the UN peace-keeping activities could well lead to expansion of the nation's export trade and an increase in its GNP, especially if a nation were to have unemployed labor, capital and other resources as was the case for most nations in 1995. To examine such effects, Klein and Marwah employ the powerful LINK model in ingenious ways. One way which is conceptually simple but which clearly indicates that the question of impacts and indirect effects is researchable and can lead to useful estimates involves the following assumptions and procedures. Assume that UN peace-keeping activities in year 1994 involving a force of one million and $50 billion is undertaken, with contributions allocated according to shares of GWP, from the 23 countries having the largest shares. Assume also that domestic defense spending in each country remains the same—that is, is not reduced by the required contribution. Still more, assume that the increment of $50 billion in 1994 is increased in subsequent years to grow at the same rate as real gross world product. With these assumptions Klein and Marwah compare the effects on exports and Gross National Product, as generated by the LINK model. Such effects are recorded by selected years in Table 11.3. When compared with levels of exports that are projected without the UN peace-keeping operations, we see that by 1999 exports are estimated to have increased by $4.25 billion in North America (Canada and U.S.A.), by $3.32 billion in the Developed East (Japan, Australia and New Zealand), by $1.10 billion in the Latin American and Caribbean area, . . . , and by $13.68 billion in the EC countries of Europe.

These amounts correspond to increases in exports of 0.4, 0.6, 0.4, . . . , and 0.7 percent, respectively. The respective estimates of GNP increases are $17.13, $16.93, $9.36, . . . , and $42.88 billions, corresponding to percent increases of 0.24, 0.30, 0.74, . . . , and 0.48.

However, the above increases may be associated with inflation and other undesirable effects. Klein and Marwah examine these in the several scenarios they consider. They find for the scenario just described that there is no serious inflation; for example, the inflation would be up by 0.1 percentage points in some OECD countries, but unchanged in others. For this scenario, substantial increases in trade flows and real gross world product are the major effects.

As fully recognized by Klein and Marwah, the above projections would be different for other scenarios. For example, if full employment of labor and other resources obtained in each of the 23 countries, there could be significant inflation. However, the contribution of their paper is not in the particular estimates obtained but in the demonstration that at least one powerful methodology does exist (despite criticisms that may be levied at it, or substitute methodologies that may be proposed) which can effectively attack and project cost and related impacts of a UN peace-keeping operation, as well as other operations to be discussed below.

Table 11.3 Increases in exports and GNP from a $50 billion UN peace-keeping operation scenario, by selected areas, 1995-99 (in US dollars, Dec.1994)

Area	1995		1997		1999	
	Amount	Percent Increase	Amount	Percent Increase	Amount	Percent Increase
North America						
Exports	$1.37	0.2	$3.31	0.4	$4.25	0.4
GNP	$10.26	0.16	$16.07	0.24	$17.13	0.24
Developed East						
Exports	$1.27	0.3	$2.26	0.4	$3.32	0.6
GNP	$7.92	0.15	$13.79	0.25	$16.93	0.30
Latin America and Carribbean						
Exports	$0.36	0.2	$0.78	0.4	$1.10	0.4
GNP	$6.47	0.61	$7.91	0.68	$9.36	0.74
Africa						
Exports	$0.18	0.2	$0.43	0.3	$0.71	0.5
GNP	$2.66	0.52	$3.91	0.73	$4.56	0.81
West Asia						
Exports	$0.36	0.3	$0.53	0.4	$0.83	0.5
GNP	$1.71	0.50	$3.04	0.77	$3.45	0.80
Eastern Europe and States of former USSR						
Exports	$0.33	0.3	$0.75	0.6	$1.23	0.9
GNP	$0.16	0.01	$0.23	0.02	$0.33	0.03
EC (EU)						
Exports (EC)	$3.42	0.2	$8.09	0.5	$13.68	0.7
GNP (EU)	$15.26	0.19	$28.93	0.34	$42.88	0.48

The areas are comprised as follows: *North America* (Canada, U.S.A.), *Developed East* (Japan, Australia, New Zealand), *EC* (Belgium /Luxemburg, Denmark, France, Germany, Greece, Ireland, Italy, Netherlands, Portugal, Spain, U.K.), *Latin America and Caribbean* (Argentina, Bolivia, Brazil, Chile, Columbia, Ecuador, Mexico, Paraguay, Peru, Uruguay, Venezuela, Caribbean), *Africa* (Algeria, Egypt, Ethiopia, Gabon, China, Kenya, Libya, Morocco, Nigeria, Sudan, Tunisia, and others), *West Asia* (Iran, Iraq, Kuwait, Saudi Arabia, and West Asia Oil Importers), *Eastern Europe and Former USSR States* (including Bulgaria, Czechoslovakia, Hungary, Poland and Romania). Excluded from this table are *South and East Asia* and the *Mediterranean* areas and their countries.
Source: Klein and Marwah (1995) Numbers rounded

Having taken a first essential step in achieving the vision of Arias, McNamara and others, can we take another? Can we take a second one in going from here to there? Can there be created an independent financial base for the UN, a base adequate to meet its peace-keeping operations and at least some others—one that is both economically feasible and politically reasonable, and one that is capable of enhancement as required by world events?

First, we must recognize that reliance on voluntary contributions by nations has been a serious obstacle to the achievement of peace-keeping and other objectives of the current UN. For example, the suggested country contributions of Table 11.2 are unlikely to be realized, and therefore cannot be viewed in any way as establishing a firm financial base for the United Nations. Many proposals have been put forth to meet this need. A number involve payment for the use of the global commons and resources. For example, with regard to the use of air space, there are proposals of taxes on

(1) international air travel
(2) international air cargo
(3) international postal services
(4) international telecommunications

and associated taxes on

(5) use of airports (departure fees), and
(6) passports.

With regard to the use of international waters and seabed (ocean resources) there are proposals of taxes on

(7) maritime cargo shipments on international waters
(8) maritime travel on international waters
(9) ocean fishing activity
(10) mining activity and other exploitation of ocean and seabed resources.

Then there are other proposals involving the taxing of undesirable activities such as

(11) dumping toxic waste and other pollutants in the oceans
(12) international arms transfers
(13) consumption of nonrenewable energy resources, for example, a carbon tax per barrel of oil or its equivalent in coal, or a levy per ton of carbon emissions.

Still other revenue raising suggestions are the holding of a monthly (or annual) United Nation's lottery, special drawing rights (SDRs) on the International Monetary Fund, fees on Antartica activities, and so forth.

While a number of these and other proposals should be studied carefully, one proposal, namely, the Tobin tax, does have a number of desirable attributes and considerable appeal for establishing an independent financial base for the UN. The Tobin tax, originally proposed in 1978, was designed to dampen and hopefully eliminate speculative runs on the key world currencies that had adversely impacted the real world economy and caused disruptive effects on national economies. To reduce the volatility of exchange rates Tobin proposed a small uniform tax on foreign exchange transactions, at least in all key currencies, thus to make pure speculative activity more costly and decrease its volume. At the same time, the tax would represent a one-time very small cost to long-run investments involving foreign exchange transactions and thus have negligible effect on such. Additionally, the tax would allow national monetary-fiscal authorities to pursue long-run socially desirable policies without confronting the possibility of being subject to upsetting and impeding reactions from disorderly financial markets. Both these positive effects of a Tobin tax would remain to complement the provision of an adequate and independent UN revenue-raising capability.

What revenue would a Tobin tax generate? According to several analysts,[2] the current (1994) average daily amount of foreign exchange transactions per working day might be crudely set at $1 trillion U.S. For 240 working days per year, this adds up to $240 trillion U.S. of foreign exchange transactions annually. At the very, very low tax rate of one tenth of one tenth of five percent (a fraction corresponding to the decimal value 0.0005), this would yield $120 billion U.S. If half of this sum were to be lost through (or used up by) exemptions, evasions and collection costs, there would be left $60 billion U.S. annually for UN operations—expressly to meet the $50 billion for the UN peace-keeping operations scenario examined by Klein and Marwah, with a sizable amount left over for social development and other UN programs. Increasing the rate to one tenth of one percent (a fraction corresponding to the decimal value of 0.001) would yield $120 billion to meet the budget for peace keeping and a number of other UN programs, and would permit transfers and allocations to other international institutions for their programs.

Of course, opposition to a Tobin tax would arise from the short-term, highly speculative operators and institutions, but the many gains to most if not all countries that would be garnered would be so large as to overcome resistance by the small number of disadvantaged individuals, businesses and groups.

Also, some analysts have surmised that resistance and opposition would arise from existing international bodies, for example the International Monetary Fund and the World Bank.[3] With careful planning, such reaction should be avoidable. The functions of both these institutions are critical to the healthy development of the Less Developed Countries (LDCs) and others. The provision of these functions should be expanded, and a share of the Tobin tax revenues should be allocated for just this purpose. Both these bodies have tremendous expertise in financial and related economic matters. This expertise should be fully exploited by the United Nations (or other world government). Furthermore, this expertise would allow these two institution to apply desirable checks and balances on UN financial operations.

Other scholars might project resistance by some of the LDCs. A number of these countries may be strongly wedded to different objectives, for example economic growth, creation of jobs and elimination of poverty, all or largely without due regard for effective environmental management. However, given the tremendous potential of the Tobin tax for generating revenue—not only today but also tomorrow when foreign exchange transactions are very likely to have grown significantly—it should be possible to meet at least in part several of the LDCs objectives—indirectly via corresponding UN programs and directly via agreed-upon shares of the total revenue to be distributed to the LDCs. Since these countries would generate a very small number of foreign exchange transactions—most of the revenue would come from transactions at the financial centers of the developed countries—the result would be that each LDC could be receiving a much larger share of the pie than would be indicated by the fraction of total transactions taking place within its borders.[4]

For the tax to be effective—and for its tremendous revenue potential to be realized—political figures in the major developed countries along with representatives of other countries would need to agree on a uniform tax level, or on a narrow range of permissible country tax levels. In essence, these figures and representatives would constitute a unit having authority, one to be housed in and be part of the United Nations. The existence of the unit, as a United Nations unit, together with an independently-run collection and amassing of transactions tax revenue would in a sense represent a yielding to a small degree of sovereignty by the participating countries of the United Nations, a very desirable movement in the minds of many.

Along with the two steps concerned with costs and revenue aspects, there needs to be introduced conflict management methods or procedures to reach agreement on: (1) a uniform tax level or a narrow range of permissible tax levels; (2) weights to be attached to the different objectives of all the participating nations and international institutions; and (3) many other conflictual issues such as the required revision of existing treaties, the need to avoid double taxation, regulations to prevent tax evasion, the manner and agents of collection and the distribution of the proceeds to programs and cooperating institutions, and so on.

Fortunately, many if not all of these conflictual issues can be directly and effectively attacked and at least partially resolved. Over the last two–to–three decades there has been great progress in evolving workable negotiation procedures and conflict management techniques—for example the construction of 'yessable' propositions[5], the use of a single negotiation text, and the development of the priorities–determining method.[6] No one of these procedures and techniques is generally applicable. However, once a particular problem is identified and well-defined, at least one of the procedures and/or techniques can be found suitable. (See Tables 11.4, 11.5 and 11.6, respectively, for *Selected Principles of Negotiation, Selected Principles of Mediation, and A Partial Listing of Quantitative Conflict Management Procedures).* Such would then constitute a third effective step toward an Arias/McNamara-type vision.

While it is possible to expand at length on how each of the set of steps could function, it is sufficient for this chapter to indicate the way in which a set of meaningful steps can be structured in an integrative manner that may be required of a key political-spiritual figure engrossed in the process of leadership and specific decision making and policy formulation. To reiterate, the particular set of steps described must be regarded as just one of innumerable space-time paths that could be currently viewed as possibly realizable.

11.6 Concluding remarks

We now bring both this chapter and book to a close. In this chapter we have once again gone over the commonalities in art, science and religion. We indicated how these commonalities became more and more explicit as the universe moved from the symmetry breaking and hierarchic structure originating in the early matter/energy state, and came in time to undergird the system of chemical elements. With the later appearance of the biological system, self-reproduction, self-organization and other commonalities became more explicit, at least to the human eye. Finally, as a consequence of our consciousness and ability at reflection we as humans are able to and do perceive the presence of the set of commonalities in, and governing the complicated, intricate and highly interconnected social, biological and physical processes of reality (but only reality as we are able to experience and contemplate it).

Given the reflective capacity of our brain with its billions of cells, we inquire in this chapter whether the ups and downs of systems, a cycle phenomenon, is a commonality and in particular is it present in the physical universe when viewed as a hierarchical structure—as well as in civilizations, empires, animal societies and other hierarchical structures in the social and biological worlds. A social scientist may be inclined to answer in the affirmative, but solid evidence concerning the missing mass, the hypothetical critical closure density and the rate of expansion of the universe are yet to be obtained, and the validity of the inflationary and other

Table 11.4 Selected principles of negotiation

1. Preparation and analysis phase
   a. Understanding enhancement
   b. Third-party intervention
   c. Misinformation spread
   d. Monolithicity attainment

2. Proposal-making phase
   a. Interdependence appreciation
   b. Opponent sensitivity
   c. Pathology awareness
   d. Long run–short run trade-off
   e. Deterrent threat
   f. Unilateral-reward action
   g. Time exploitation
   h. Approximate cost-benefit balance sheet
   i. Monolithicity attainment

3. Interaction phase
   a. Continuous evaluation
   b. Stance–issues separation
   c. Strategic element highlighting and questioning
   d. Trust/respect/knowledge building
   e. Time exploitation
   f. Third-party intervention
   g. Realism in the face of failure

4. Final packaging phase
   a. Enhancement of implementability

Source: Isard and Smith, 1982.

Table 11.5 Selected principles of mediation

1. Preparation and analysis phase
   a. Own understanding enhancement
   b. Participant understanding enhancement
   c. Monolithicy attainment
   d. Ground rules (agenda) settings

2. Interaction phase
   a. Flexibility in approach
   b. Tread lightly in early rounds
   c. Tension releasing
   d. Impartiality establishment and maintenance
   e. Minimal aim
   f. Single negotiating text
   g. Compromise constraint
   h. Exploitation of fractionation/logrolling/side payment potentialities
   i. Encouragement of inventiveness
   j. Encouragement of analysis by participants
   k. Test for fairness and stability
   l. Time exploitation

3. Interaction phase
   a. Enhancement of implementability

Source: Isard and Smith, 1982.

Table 11.6 A partial listing of quantitative conflict management procedures

1. Compromise over proposed actions (outcomes)
   • in one step *or* a sequence of steps
2. *Min total of:* ranks (highest rank = 1), rank concessions, percentage concessions, percentage goal shortfalls, absolute concessions, absolute goal shortfalls.
3. *Max total of:* rank improvements, percentage improvements, percentage goal achievements, absolute improvements, absolute goal achievements, *or* utility.
4. *Min the difference in:* ranks, rank improvements (concessions), percentage improvements (concessions), percentage goal achievements, absolute improvements (concessions), *or* absolute goal achievements.
5. *Max the min in:* rank improvements, percentage improvements, *or* absolute improvements.
6. *Min the max in:* rank concessions, percentage concessions, *or* absolute concessions.
7. *Max equal:* rank improvements, percentage improvements, absolute improvements, *or* goal achievements.
8. *Min equal:* rank concessions, percentage concessions, absolute concessions, *or* goal shortfalls.
9. Changing actions to "if. . . then. . ." policies.
10. Achievement of minimum requirements (satisficing)
11. Median efficient joint action
12. Concession along efficiency frontier
13. Split the difference in action space or outcome space
    • one step *or* a sequence of steps
14. Weighted average in action space or outcome space
    • one step *or* a sequence of steps
15. Alternating leader-follower
    • in action space or outcome space
16. Leadership principle
    • in action space or outcome space
17. Aggressive follower principle
    • in action space or outcome space
18. GRIT (reciprocated tension-reducing actions, a sequence of)
    • in action space or outcome space
19. Incremax (maximizing in each of a series of small improvement steps) in action space
    • with split the difference, weight averaging, alternating leader-follower, *or* GRIT
20. Incremax in outcome space
    • as in 19, above
21. Decremax (maximizing in each of a series of small concession steps) in action space
    • as in 19, above
22. Decremax in outcome space
    • as in 19, above
23. Equidistant movement in action space
    • regarding improvement or concession
24. Last offer arbitration (with incentive to think of others)
25. Hierarchical programming (relaxed or not)
26. Zeuthen concession (least to lose goes first)
27. Method of determining group priorities

•Procedures 2-8 may or may not involve weights to be assigned to the relevant item of each participant.
Source: *Isard and Smith, 1982*

theories still to be tested.

Another controversial area, regarding which we carry further the discussion in previous chapters, concerns the validity of reductionism, and the relative roles of science and religion in understanding past and current development of the world society and its numerous subsystems. If one were to imagine a great social figure at hand, steeped in both scientific analysis and theology and religious practice—a modern Maimonides, Avicenna, Teilhard or equivalent—he would be able to incorporate into his own theology most, if not all, of the commonalities discussed in this book. He would not confront any major inconsistencies and conflict between doctrine and science. However, when he confronted today's perplexing problems of the pure physical scientists, for example their inability to establish initial conditions in their deterministic models (the problem of infinite regress) and others already noted, and man's inability to develop scientifically a pure measure of social welfare (the Impossibility Theorem), he would need in the foreseeable future to fall back upon religious and ethically oriented actions as guiding principles—principles that were essential in social development in the past and are currently [as explicit in the central (conflict management) box of the PSIG model of Figure 11.1].

In brief, both religious and scientific principles have been and will continue to be complementary elements in any unitary society that we can foresee.

We would be remiss if we did not ask what the implications of this study are for the future. We would certainly expect most, if not all, of the commonalities discussed in this book to continue to be in force and possibly some new ones that our limited perceptive abilities would come to identify. However, given the incessant play of the mutation of genes, coadapted gene complexes, memes and coadapted meme complexes, and in general noise that can disrupt delicate balances in our systems, prediction becomes impossible. What we can do, however, is project the future, that is predict it on the basis of assumptions regarding the continuance of, or specified changes in key structural aspects of development. Thus, as discussed in this chapter, if an analyst assumes that: (1) technological advance in cyberspace and the forces for globalization continue; (2) that no devastating natural phenomena occurs; (3) that there appears a great political-spiritual leader who is able to (a) attack effectively the problems of population growth, environmental degradation, inequality and poverty, (b) significantly advance civil rights and demilitarization, (c) effectively combat nuclear terrorism and accidents, (d) play a decisive role in a world government (a strengthened United Nations or the equivalent) attained by a set of steps such as the highly speculative one sketched in the previous section; and so forth, he/she may project many years of peace and prosperity. However, that analyst's assumptions cannot be taken as realistic, and his/her projected outcome would be rejected outrightly by many as pure speculation and by others as subject to a very low probability. Other sets of assumptions and projected outcomes, some very pessimistic and some not, have been advanced by other analysts. Each, however, must be assigned a low, if not a very low probability, even assuming the absence of the ever present and unpredictable noise, although from future experience and scientific advance some phenomena currently classified as random may come to be understandable.

In effect we cannot predict the future. At best we must fall back upon historical experience and science to identify forces, relationships and structure which will be present in the future with higher rather than lower probability, and accordingly propose a set of procedures to project outcomes of higher rather than lower probability. These would constitute steps for 'improvement' in society, fully recognizing that such improvement and social welfare are subjective concepts and best given concrete representation (that is defined) by an outstanding political-spiritual leader and his/her associates.

Putting aside specific speculations about the future, we trust that our discussion of commonalities in phenomena and processes among many fields of study makes possible the transfer of much knowledge among these fields—transfer that contributes to further understanding within each. We hope that this allows us to perceive better the structure and function of the world in which we live, the ups and downs of it and its parts, thereby to provide a better basis for decision making, policy formulation and planning with regard to some of society's critical problems. Lastly, our recognition of the many unknowns in our world, our failure to identify many relationships among variables (known and unknown) governing our complexly evolving world leaves room for much cultural development, in particular for science and the sense of wonder, mysticism and religion to be necessary complements in designing our future space-time path.

Notes

1 It should be borne in mind that the differences in statements made by scholars from diverse disciplines tend to reflect the perspectives, terminology and stock of knowledge in those disciplines. A scholar from a given discipline has greater ability to write in detail, more effectively and at greater length on matters dealt with by his discipline. He tends to write of contributions from other disciplines less effectively and in a manner that may often understate his actual appreciation and evaluation of these contributions to explanation of phenomena.
2 See *Futures*, 1995, Vol. 27, No.2, March and items cited therein.
3 See Shome and Stosky (1995) and Felix and Sau (1995).
4 One additional point. As of 1995, none of the highly developed nations—U.S.A., Great Britain, France, Germany, Japan or other—has imposed a Tobin or similar type of tax on foreign exchange

transactions. This is partly the result of the globalization of the financial community. If a speculator in one of the above nations wishes to exchange his currency for another (say dollars for yen), he need not do so in or through Wall Street or Tokyo; he can do so indirectly at other financial markets that trade in dollars and yen via a third currency. Thus if a tax were levied on developed country Z's transactions, a speculator or investor residing in Z could carry through a desired transaction at the financial center of another country, thereby evading the tax.

5 See Fisher and Ury (1981).

6 See Saaty (1988, 1991), Isard and Smith (1982), and Isard (1992).Also, for coverage of many useful negotiation procedures and practices, see issues of the *Negotiation Journal*; and for listings and discussions of possible conflict management methods and techniques see Raiffa (1982), Isard and Smith (1982), and Isard (1988).

References

Arias, O. (1994), 'Human Security: Our Common Responsibility,' ECAAR Japan Symposium, Tokyo.

Arrow, K. J. (1987), 'Arrow's Theorem,' in Eatwell, J., Milgate, M. and Newman, P. (eds.), *The New Palgrave: A Dictionary of Economics*, London: Macmillan, pp. 124-26.

Barrows, J.D. (1992), 'Cosmology, Theories' in Maran, S.P. (ed.), *The Astronomy and Astrophysics Encyclopedia*, New York: Van Nostrand Reinhold.

Batty, M. and R. Barr (1994), 'The Electronic Frontier: Exploring and Mapping Cyberspace,' *Futures*, Vol. 26.

Bohm, D. (1990), Interview in Louwrien Wijers and John Pijnappel (eds.), *Art Meets Science and Spirituality*, London: Academy Group Ltd., pp. 26-33.

Felix, D. and Sau, R. (1995), 'On the Revenue Potential and Phasing in of the Tobin Tax' UNDP Conference on the Tobin Tax, October.

Fisher, R. and Ury, W. (1981), *Getting to Yes*, Boston: Houghton Mifflin

Future Survey Annual, 1995, Washington, D.C.: World Future Society.

Futures, 1995, Guilford, England: IPC Science and Technology Press, vol. 27, No. 2, March.

Isard, W. and Smith, C. (1982), *Conflict Analysis and Practical Conflict Management Procedures: An Introduction to Peace Science*, Cambridge, MA: Ballinger, reprinted by the Department of City and Regional Planning, Cornell University, Ithaca, New York, 14853.

Isard, W. (1988), *Arms Races, Arms Control and Conflict Analysis*, New York: Cambridge University Press.

Isard, W. (1990), 'Progress in Global Modelling for World Policy on Arms Control and Environmental Management,' *Conflict Management and Peace Science*, Vol. 11, No. 1, Spring.

Isard, W. (1992), *Understanding Conflict and the Science of Peace*, Oxford, U.K.: Blackwell.

Klein, L.R. and Marwah, K. (1995), 'Peace-Keeping Operations and Some Related Aspects,' ECAAR Japan Symposium, Tokyo.

Krasno, J. (1994), *A United Nations Rapid-Deployment Permanent Force, Cost Analysis*, a Report for the Yale University, United Nation Studies Program, Parliamentarians for Global Action, 1994.

McNamara, R.S. (1995), 'A Vision of U.S. Security in the Twenty-First Century', ECAAR Japan Symposium, Tokyo.

Negotiation Journal, New York: Plenum.

Ogden, M.R. (1994), 'Politics in a Parallel Universe: Is there a Future for Cyberdemocracy,' *Futures*, Vol. 26.

Pagels, H.R. (1983), *The Cosmic Code*, New York: Bantam.

Raiffa, H. (1982), *The Art and Science of Negotiations*, Cambridge, MA: Harvard University Press.

Saaty, T. L. (1988), *Decision Making for Leaders*, Pittsburgh: RWS Publications.

Saaty, T. L. (1991), *Prediction, Projection, and Forecasting: Applications of the Analytic Hierarchy Process in Economics, Finance, Politics, Games and Sport*, Boston: Kluwer Academic Publishers.

Shome, P. and Stosky, J.G. (1995), *Financial Transaction Taxes*, Washington, D.C.: IMF Working Paper No. 95/77.

Steinbrunner, J.D. (1995), 'Reluctant Strategic Realignment,' *The Brookings Review*. Washington, D.C.,Winter.

Urquhart, B. (1993), 'For a UN Volunteer Military Force,' *The New York Review of Books*, vol. XL, June 10.

Index